Transboundary River Governance
in the Face of Uncertainty

MAJOR RIVERS AND DAMS OF THE COLUMBIA BASIN

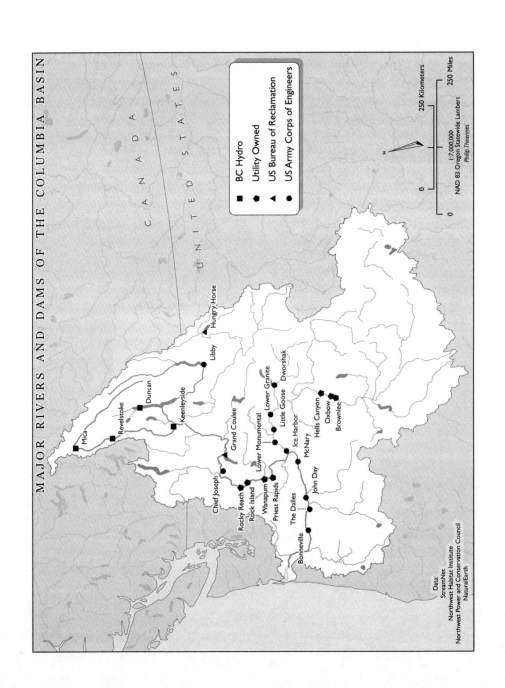

CANADA

UNITED STATES

Mica

Revelstoke

Duncan

Keenleyside

Hungry Horse

Libby

Chief Joseph

Grand Coulee

Rocky Reach

Rock Island

Wanapum

Priest Rapids

Lower Monumental

Lower Granite

Little Goose

Dworshak

Ice Harbor

Hells Canyon

Oxbow

Brownlee

McNary

The Dalles

John Day

Bonneville

Legend

■ BC Hydro
⬠ Utility Owned
▲ US Bureau of Reclamation
● US Army Corps of Engineers

0   250 Kilometers
0   250 Miles

1:7,000,000
NAD 83 Oregon Statewide Lambert
Philip Thoennes

Data:
StreamNet
Northwest Habitat Institute
Northwest Power and Conservation Council
NaturalEarth

Transboundary River Governance in the Face of Uncertainty: The Columbia River Treaty

A PROJECT OF THE UNIVERSITIES CONSORTIUM ON COLUMBIA RIVER GOVERNANCE

edited by Barbara Cosens

Oregon State University Press
Corvallis

The paper in this book meets the guidelines for permanence and durability of the Committee on Production Guidelines for Book Longevity of the Council on Library Resources and the minimum requirements of the American National Standard for Permanence of Paper for Printed Library Materials Z39.48-1984.

Library of Congress Cataloging-in-Publication Data
Transboundary river governance in the face of uncertainty : the Columbia River treaty : a project of the Universities Consortium on Columbia River Governance / edited by Barbara Cosens.
    p. cm.
  Includes index.
  ISBN 978-0-87071-691-1 (alk. paper) -- ISBN 978-0-87071-692-8 (e-book)
  1. Canada. Treaties, etc. United States, 1961 Jan. 17. 2. Water resources development--Law and legislation--Columbia River Watershed. 3. Water resources development--Social aspects--Columbia River Watershed. 4. Water resources development--Environmental aspects--Columbia River Watershed. 5. Water resources development--Columbia River Watershed--Citizen participation. 6. Watershed management--Law and legislation--Columbia River Watershed. 7. Watershed management--Social aspects--Columbia River Watershed. 8. Watershed management--Environmental aspects--Columbia River Watershed. 9. Watershed management--Columbia River Watershed--Citizen participation. 10. River engineering--Environmental aspects--Columbia River Watershed.
I. Cosens, Barbara, 1955-
  KDZ642.C65A3 2012
  346.7304'69162--dc23

                                        2012015122

 Oregon State University Press
121 The Valley Library
Corvallis OR 97331-4501
541-737-3166 • fax 541-737-3170
http://osupress.oregonstate.edu

CONTENTS

Acknowledgments

This book would not have been possible without the energy and enthusiasm of the members of the Universities Consortium on Columbia River Governance who planned and carried out the 2009 Symposium on Transboundary Governance in the Face of Uncertainty: The Columbia River Treaty. Funding for the symposium was provided by the University of Idaho College of Law and the Thomas Foley Institute at Washington State University. In 2009 the Consortium members were: University of Idaho, Barbara Cosens; University of Montana, Matthew McKinney; Washington State University, Edward Weber; Oregon State University, Aaron Wolf and Lynette deSilva; and University of British Columbia, Richard Paisley. The Consortium has since grown to include Craig Thomas, Alan Hamlet, and Lara Whitely Binder at the University of Washington, Nigel Bankes at the University of Calgary, and Sarah Bates at the University of Montana, who have all contributed to continuing efforts to provide a neutral forum for a cross-border dialogue on the Columbia River Treaty. Nicholas Sackman, Water Resources graduate student and law student at the University of Idaho, provided invaluable help both with the logistics of the symposium and editing the contributions to this book. These acknowledgments would not be complete without expressing my deep gratitude to my husband, Eugene Allwine, for his constant support throughout this project. Finally, a heartfelt thank you to the peoples of the Columbia River Basin whose vision and hope for the future serve as a constant reminder of the power of a collaborative dialogue.

Barbara Cosens

Introduction to Parts I, II, III

Barbara Cosens, Lynette de Silva, Adam M. Sowards

This book is an outgrowth of the first University of Idaho College of Law Natural Resources and Environment Symposium held in 2009, which focused on the issue of transboundary water governance in the face of uncertainty. The symposium used the natural laboratory of the Columbia Basin, shared by the United States and Canada, as a focal point for discussion and the following question as the point of integration for contributions: *How do we design and implement governance of international watercourses in the face of uncertainty?*

The symposium was developed in collaboration with researchers from Oregon State University, University of Montana, University of British Columbia, and Washington State University. Representatives of the first four of these universities and the Universities of Washington and Calgary have joined to form the Universities Consortium on Columbia River Governance and continue to work with stakeholders in the basin to inform issues of governance. The Consortium has evolved to focus on three efforts within the basin: (1) to provide a yearly forum for an informal cross-border dialogue on the future of the Columbia River Basin; (2) to connect Consortium university research to stakeholders in the basin; and (3) to engage students at Consortium universities in relevant research and curriculum both to serve the basin and to provide a natural laboratory for understanding issues of water basin governance.

This collaborative effort among academics and the people of the Columbia River Basin was inspired by a moment in time. Joint operation of the river for the purposes of hydropower production and flood control is governed by a 1964 treaty between the United States and Canada (reproduced in the Appendix). Certain of the flood-control provisions expire in 2024, and either country must provide ten years notice should it seek to terminate the treaty. Thus efforts are underway to understand and predict changes within the basin and to determine whether those changes warrant modification of the treaty. The expiration of provisions that have protected basin residents and businesses from flooding, and the need to review the treaty provide a window of opportunity and a potential for open dialogue not otherwise present in the daily operation of the river under a treaty that, for its intended purposes, works. At the same time, the absence of any immediate crisis provides both the basin and the researchers with time to explore and undertake a measured dialogue.

This volume is a contribution to the effort to explore the question of water governance in the face of uncertainty. The specific application to the Columbia River allows the reader to explore that question in the realities and constraints of a real international basin. In the Columbia River Basin, as elsewhere, political boundaries were drawn without consideration of river basin boundaries. In fact, 276 surface water resources cross international boundaries (Oregon State University Program in Water Conflict Management and Transformation, 2011). Over the next decade, several contributing factors could trigger rapid change and social and economic instability in these international watersheds, placing greater demands on competing water interests and a greater need to cooperate across jurisdictional boundaries. The contributing factors explored in this volume as a subset of drivers of change are: changing values; empowerment of local communities; a threatened and deteriorating ecosystem; and climate change. Uncertainty in these social and ecological factors challenges traditional approaches to governance of transboundary water resources—approaches that rely on the certainty that historic data concerning water supply, demand, values, and ecosystem health can be used to predict the future; approaches that protect sovereignty through clear upfront rules for dividing benefits across political boundaries rather than flexibility to adapt to change and foster system resilience.

Part I of the volume begins with an exploration of the treaty itself, how it is implemented, and the social changes reflected in both increased local empowerment and changing values since 1964. Contributions from historians Jeremy Mouat, Paul W. Hirt, and Adam M. Sowards provide bookends to Part I by placing first the treaty and then the changing values of the people of the Columbia River Basin within the larger context of environmental history. The intervening articles describe how the treaty is implemented by those charged with the task, and changes in local empowerment, capacity, and values viewed through both legal changes and the voices of those in the basin.

Part II addresses two of the prominent ecological changes underway in the river: the health of the anadromous fish (only present in the U.S. portion of the river) and the current stalemate in efforts to address their decline, and climate change. Each of these topics is dealt with thoroughly elsewhere in the literature and the contributions here provide a mere sampling of the issues faced in these areas. They provide a window on the complexity of the uncertainties surrounding the ecological system and the difficulty of addressing that uncertainty without coordination at the scale of the basin. These issues should also caution us that an inflexible international agreement based on predicted change may be inadequate to address future surprises, even those comparable to the changes since 1964.

Part III begins the exploration of the future for international cooperation in the Columbia River Basin from the viewpoint of two experts with considerable experience in the basin. Their views provide a real-world grounding before the academic explorations of Part IV. At first glance, the reader will see redundancy in the description of the geographic, hydrographic, biological, and social setting of the river, as well as the description of the treaty and its impact on the basin, in Parts I, II, and III. But read closely, with a view to each author's background. The differences in these descriptions tell as much about the different values and viewpoints in the basin as each analysis of the treaty. Understanding this is essential to the future of this and any international basin.

To understand the flow of the contributions to this volume, it is necessary to begin the story with the river itself. As described by James D. Barton and Kelvin Ketchum, both with considerable experience in implementing the treaty, the Columbia River Basin covers 259,500 square miles, with 15 percent in Canada and the remainder in the United States. Portions of seven states (Washington, Oregon, Idaho, Montana, Nevada, Wyoming and Utah); the province of British Columbia; the aboriginal lands of First Nations in Canada; and fifteen Native American reservations in the United States lie within the Columbia River Basin. To express runoff from the Columbia River Basin in terms of its average annual flow of 200 million acre-feet at the mouth would be misleading. The year-to-year variability in unregulated peak flow is 1:34, compared to a mere 1:2 on the Saint Lawrence River or 1:25 on the Mississippi River (Hamlet 2003). Hirt and Sowards note that this variability translated to substantial storage potential in the eyes of early twentieth-century boosters and engineers.

In 1805, when Lewis and Clark made their way down the Columbia River to Astoria, there were, of course, no dams. Salmon fisheries sustained the native population. Falls slowed upriver migration of salmon and provided excellent fishing locations. Each year thousands of Native Americans from numerous tribes gathered at locations such as Celilo Falls (now drowned by The Dalles Dam) to fish and trade (Landeen and Pinkham 1999; Hirt 2008). Mary L. Pearson describes the effect on native culture. Competition from commercial fishing and an influx of canneries began in 1866. The U.S. Army Corps of Engineers began transforming the river for navigation with locks at the Cascades as early as 1896, with numerous dams to follow (White 1995). As noted by John Shurts, most dams in the U.S. portion of the river main stem served to generate hydropower and aid navigation, but did not store substantial water. Barton and Ketchum describe that in 1948, even though the total flow was close to average, runoff occurred rapidly and peaked with a flood in May that destroyed the town of Vanport, Oregon, with an estimated flow of over 1 million cfs (average peak flows are less than half that rate). At the time of the 1948 flood, total storage capacity on the Columbia was about 6 percent of the average annual flow, as noted by Anthony G. White. Barton and Ketchum compare this to the Colorado, which has a storage capacity of over four times its average annual flow or the Missouri, with storage capacity over two times its average annual flow. The approach at the time, implemented by the U.S. Army Corps of Engineers, was to address flood control through storage. The problem: the best remaining storage sites were in Canada.

This brings us to the starting point of the contributions to Part I: the history of negotiation of the 1964 Columbia River Treaty between the United States and Canada. Jeremy Mouat's contribution tells us that, even before the 1948 flood, the International Joint Commission formed by the Boundary Waters Treaty between the United States and Canada was directed to study the possibility of storage within Canada to provide flood control or power benefits to both countries. The Columbia River Treaty that would form the framework to accomplish this task was not adopted until 1964. Mouat describes what may have been the largest obstacle to its completion: the fact that the three new dams contemplated would all be in British Columbia and the majority of the flood control and hydropower benefits would be in the United States. Between 1961 and 1964 negotiations between the federal government of Canada and the province of BC led to a solution that would turn the operation and benefits under the treaty over to the province and divide benefits between the U.S. and the province allowing sale of Canada's share of power not needed by the province. The treaty has been hailed throughout the world as a pinnacle of international cooperation on freshwater sources, as described by Barton and Ketchum. White's contribution gives us a view of the provisions of the treaty itself.

One further complication would need to be addressed before the treaty could be completed. In 1964 the Pacific Northwest did not require the amount of power the new projects would generate. As Chris W. Sanderson notes in his contribution in Part III, this was solved when Congress authorized construction of the Pacific Northwest–Pacific Southwest Intertie allowing sale of power to utilities in the southwestern United States, with a preference for sale to northwest utilities (Pub. L. 88–552, August 31, 1964; Vogel, Part IV Chapter 1).

The treaty contains no automatic termination date or renegotiation clause and 2024 is the earliest date either party may terminate. At least ten years notice must be provided; hence the importance of a thorough review of the treaty before the year 2014. The operating entities are undertaking studies to inform options to be explored by 2014 and have begun a process of stakeholder input (U.S. Army Corps of Engineers and Bonneville Power Administration 2009). Certain of the flood-control provisions, paid for upfront by the United States to cover sixty years, expire in 2024, leading to consideration of whether the time is ripe for modification of the treaty. The contributions in this volume look at this question by exploring what has changed since 1964 and where the uncertainties lie in contemplating basin needs after 2024. Generally, the sources of change are in five areas: (1) change in values concerning the river, assessed by Matthew McKinney and his students through a series of interviews, and analyzed by Hirt and Sowards by, in part, looking at changes in the law; (2) change in empowerment of local communities and, in particular, of Native American and First Nation governments, described by Barbara Cosens and given personal meaning by the narratives of Mary L. Pearson (Native American) and Garry Merkel (First Nation); (3) change in the viability of populations of anadromous fish that spawn within the Columbia River system and gridlock in the existing forum for dispute resolution concerning fish versus dams, described by fish biologist Chris Peery and analyzed through the lens of the resulting litigation by Carmen Thomas Morse; (4) climate change, analyzed by climate scientists Anne Nolin, Eric Sproles, and Aimee Brown; and (5) change in population and energy demand, described by John Shurts. Some of the important points raised by contributions are summarized in the following paragraphs, supplementing where appropriate from the literature, in each of these categories.

Change in values concerning the river: Two approaches are used: historians Hirt and Sowards examine the adoption of new laws reflecting a change in societal values concerning the environment; and Matthew McKinney describes the results of a reconnaissance-level survey of stakeholders in the basin done by students at the University of Montana.

Like all historical moments, events, and documents, the CRT was situated in time and place. Hirt and Sowards emphasize how the CRT fits within a century-long pursuit of economic efficiency by purportedly controlling nature to enhance individual, corporate, and national wealth. However, as they also note, the treaty came at a moment in time when cultural and political values were in the midst of a paradigm change to bring issues of equity—for nature, for Native peoples, and others—into the calculus of natural resource management. They conclude that the CRT was entirely a product of its time, reflecting a limited set of values and goals predominant in water-development programs of that era. They argue that moving forward will require balancing efficiency *and* equity. In addition, Hirt and Sowards look at a more subtle yet pervasive change in laws that reflect a trend surely to impact any effort to update the Columbia River Treaty. Beginning with, or resulting in, the passage of the Freedom of Information Act in 1966 and the National Environmental Policy Act in 1970, the expectation of the public for access to and participation in governmental decision making began to increase dramatically in the United States.

A reconnaissance-level situation assessment of stakeholders in the Columbia River basin done by students at the University of Montana under the direction of consortium member Matthew McKinney confirmed this expectation of public input within the basin. The initial assessment identified several key perceptions. First, if measured by the 1964 goal of flood protection and increased power production, the Columbia River Treaty has been an outstanding success. Second, among the key issues identified by stakeholders that were not addressed in 1964 but should be in the future were the health of the salmon fishery and participation by affected communities, Native American tribes, and First Nations. This perception is paralleled by the dramatic change in empowerment among basin communities.

Change in empowerment of local communities and in particular, of Native American and First Nation governments: Enhanced empowerment and capacity of basin communities, suggesting that they have the capacity to participate and are likely to demand participation in any decision on whether and how to modify the treaty, is reflected in the following changes since 1964: (1) legal recognition of the treaty rights of certain Native American tribes to participate in the harvest and management of Columbia basin fisheries within the United States, set forth in contributions by Cosens, Merkel, and Pearson; (2) constitutional recognition of the rights of First Nations in Canada in 1982, set forth in contributions by Merkel and Cosens; (3) legislative recognition of the Columbia Basin Trust in Canada in 1995, set forth in contributions by Cosens and Merkel; and (4) establishment of the Northwest Power and Conservation Council in the United States in 1980, briefly discussed by Cosens and set forth in more detail in a contribution by Shurts in Part III.

Displayed in the contributions by Pearson and Merkel is raw candor. Their perspectives as Native American and First Nation members respectively carry with them the history of the Pacific Northwest and all that has brought us to this juncture and moment in time. This gives us the space to listen deeply to the voices of the "River People," to learn the plight of salmon, and to come to terms with the human component of the Columbia River Basin ecosystem. Through Pearson's account of the River People, we begin to know what fish represent to the indigenous people, what their life was like before the building of dams and the signing of agreements and treaties. We begin to get a sense of what spiritual, cultural, and emotional connections their society has with salmon. And we hear what the River People need in this pre-negotiation period and the potential renegotiation of the Columbia River Treaty. The intent of these chapters is not to burden the reader with some sense of guilt nor righteousness, but they do request that the reader listen wholeheartedly and without resistance.

This listening approach brings us to terms with "now," since the process in moving forward may actually have little to do with projections into the future, but in fact have more to do with coming to terms with the present. With the thoughts presented by these authors on issues related to the Columbia River Basin, listen without judgment. This describes where we are now. Our collective experiences have brought us here. In fact, listening is an art, championed in practices of negotiation, meditation, conflict management, and interpersonal skill development.

Deep listening requires us to listen with every pore of our being, beyond thoughts and emotions. Over time, through this kind of intense listening, there is the potential for a shift that can be transformative, that can result in stakeholders being more inclusive, and that can lead to thinking beyond ones' own individual interests. From these social changes, we turn to the ecological changes in the basin.

Change in the viability of populations of anadromous fish that spawn within the Columbia River system: The decline of anadromous fish in the Columbia River system has been extensively documented elsewhere and is addressed here in contributions by Peery and by Thomas Morse. The blockage of migration from Canada and the reservations of certain upper Columbia River Native American tribes was a *fait accompli* by the time of the 1964 Columbia River Treaty as a result of the completion of Grand Coulee Dam. As noted by Peery, in the remaining portion of the basin, harvest of chinook salmon declined from a high of two million in the early 1880s to less than one hundred thousand when they were first listed in the early 1990s. The salmon fishery in the Columbia River basin is now supported by over two hundred hatcheries. Thomas Morse details the ongoing litigation concerning operation of the federal dams and salmon recovery, which points to relative gridlock between the two competing values. It is difficult to argue that these changes were not foreseen in the decision to dam the river (Bottom et al. 2009), but it is clear that there is a rising desire to revisit that decision.

Peery writes of the system's physical and biological mechanisms. We learn how the networks of constructed dams and impoundments and manipulated flows have affected the dynamics of the Columbia River system. Through this approach, Perry also reaches out and asks us to come to terms with the human dimension of the Columbia River ecosystem and the impact we have on the biological processes across the basin. He asks us to be "mindful," to have awareness, to be conscious, especially as the potential renegotiation of the Columbia River Treaty approaches. Peery suggests using this timeframe to make others aware through the dissemination of information, and through identifying data gaps, so these data needs can be filled.

One caution raised at the symposium by participant Thomas Leschine is important to note: "It is uncertain whether degraded salmon ecosystems remain sufficiently resilient to respond positively to ongoing restoration programs, or have shifted to a stable, low-productivity state that may persist regardless of the climatic regime." Under the definition of resilience—"[t]he amount of disturbance an *ecosystem* can accommodate without shifting to a fundamentally different structure, function and feedback mechanisms" (Leschine 2009)—it is possible that we have so altered the ecological system of the Columbia River that salmon restoration in any way resembling a natural system is impossible, creative governance notwithstanding.

The post-1964 law with the largest impact on operation of Columbia River dams on the United States side of the border is the Endangered Species Act adopted in 1973. NOAA Fisheries (then National Marine Fisheries Service) began listing anadromous fish in the Columbia River system in 1991, and today eight salmon and four steelhead species that rely on habitat within the basin are listed. Although numerous factors impact these species, operation of dams for hydropower has been identified as a major factor, and operation of the Federal Columbia River Power System (that part of the hydropower system at federal dams in the U.S. portion of the basin) has been the subject of numerous Biological Opinions and subsequent challenges, resulting recently in what some refer to as operation of the river by the federal district court. This is detailed in the contribution by Thomas Morse. The ESA and subsequent listings reflect a change in values and provide an indication of strong interest in giving voice to issues concerning anadromous fish in any negotiation concerning operation of dams on the river. The current gridlock in the judicial system may be a further indication that the solution will come through a form of governance able to adapt to changing values rather than a lawsuit. Further, as detailed in Shurts' chapter in Part III, meeting the concerns of those interested in anadromous fish requires consideration of the run of the entire river, thus raising the possibility of inclusion in treaty issues. Yet no one has yet articulated what might induce Canada, whose anadromous fish runs were cut off in 1942 by completion of Grand Coulee dam, to collaborate on their return to health in the U.S. portion of the river.

Climate change: Water planners have long relied on data from a historic period of record to project water supply into the future. It is the seasonal variation, and the year-to-year variation that can be forecast within the degrees of historical variability, that the type of agency (or "entity") level operational planning envisioned by the 1964 Columbia River Treaty handles well, as detailed in the article by Barton and Ketchum, which describes the current adaptive capacity of the river operation planning.

Climate change takes us out of the range of variation that can be predicted based on historic behavior (Hamlet 2003). Most current discussion on climate change focuses on reducing emissions of greenhouse gases. This is an important goal. However, due to the lag in impact, even the most aggressive efforts at reduction in emissions will not prevent continued impact for the foreseeable future. Climate experts recommend planning for adaptation through use of scenarios that represent a range of possible futures, rather than projections based on historic behavior of a system (Solomon et al. 2009). Thus, given the range of potential temperature and precipitation changes, governance that is adaptive to climate change must include authorization that allows managers to respond to actual outcomes ranging from the best- to the worst-case scenario (Hamlet 2003).

Modeling by the Climate Impacts Group presented by Alan Hamlet at the symposium and published elsewhere suggests that precipitation may not change dramatically within the Columbia River Basin, albeit substantial uncertainty is associated with this statement. However, changes in annual snowpack can be predicted with greater certainty and are already underway in the basin as documented in this volume by Nolin, Sproles, and Brown. The basin relies on snowpack as natural storage that, similar to reservoirs, moderates summer flows. With climate change, reduction in snow-water equivalent may be as much as 35 percent in the U.S. portion of the basin and 12 percent in the Canadian portion by 2060 (Hamlet 2003). This reduction in natural storage means that the artificial storage configuration in the basin will be insufficient to reap the power benefits available in the past. In particular, summer production that serves utilities in the southwestern U.S. will decrease if the current configuration is maintained (Hamlet 2003).

Moving out of the historic water supply regime has impacts beyond power production. The Columbia River Treaty provides an excellent framework to address high flow. However, it does not address low flow under a climate change scenario. Adaptation to climate change for other uses such as irrigation and fisheries requires response by multiple agencies in the U.S. with no framework for coordination. Irrigation occurs during the summer season when the flows are lowest if storage is insufficient. As Shurts notes, the result of failure to address low flows: fish and farmers will bear the brunt of climate change if no effort is made to adapt.

Change in population and energy demand: Energy demand and development has not proceeded as contemplated by the treaty drafters in 1964. As described by Shurts, at that time, planners expected the rapid growth in power demand that followed World War II to continue. This would mean that new thermal generation would have to rapidly replace hydropower as the dominant source of energy in the Pacific Northwest. Conservation nationwide in the wake of the 1970s energy crisis altered this picture, but not before the commitment of major expenditures on development of nuclear power had been made in the Pacific Northwest. The major overestimate of demand and underestimate of the cost of nuclear power plants led to a financial debacle the region is not anxious to repeat and the plants were not completed (White 1995). As a result, hydropower remains the dominant energy source in the region and the value of the system has grown dramatically. With the current push to develop non-carbon sources of energy, it is likely to become even more valuable. The draft power plan released in September 2009 by the Northwest Power and Conservation Council indicates that "the most cost-effective and least risky resource for the region" to meet electricity demand over the next twenty years "is improved efficiency of electricity use" (Northwest Power and Conservation Council 2009). If this projection proves true, it is likely hydropower will remain at the core of Northwest energy production through any near-term scenarios.

The articles in Part III form the bridge between the story of the Columbia River Treaty and the changes since its 1964 ratification and the academic analysis

(Part IV). Part III looks at the future of the treaty through the eyes of experts with a long history in the basin. One author, Shurts, raises the prospect of change; the other, Sanderson, of stability. Both represent views widely held in the basin. Both must be understood and given attention in any effort to revisit the 1964 Treaty.

As both a scholar and a practitioner in the U.S. portion of the Columbia River Basin, Shurts views the treaty through the lens of possibility. He asks: What changes in the basin since 1964 appear inconsistent with both the future envisioned then and the future predicted today? What avenues for accommodating change are available without revisiting an international treaty? Do the changes in the basin call for international or merely domestic action?

As an expert on both the treaty and energy law and policy in Canada, Sanderson calls attention to the enormous benefits joint development of the Columbia River have brought to the region, and asks: Is the risk of destabilizing an effort that works to the benefit of many too high?

An introduction by McKinney and Edward P. Weber precedes Part IV, but a brief discussion of the volume as a whole is warranted here. The primary themes running through this volume are a description and acknowledgement of the major changes in the social-ecological system of the Columbia River Basin since the treaty was ratified in 1964; the struggle to confront the question of whether those changes are so fundamental as to warrant disrupting a cooperative effort at transboundary operation and benefit sharing of a river that, for its purposes, has worked; and the difficulty of addressing transboundary river governance when faced with high levels of uncertainty. Readers will find a split between those close to the daily operations of the original negotiation of the 1964 treaty and those who have suffered harm from the things it does not address. But read carefully to avoid that being the limit of your evaluation. Those close to the 1964 treaty are reminded on a daily basis how difficult is it to achieve and maintain the current level of cooperation across an international border and how easy it might be to lose all benefits. Their concerns must be addressed at the same time the legitimate issues raised by those harmed are heard. Understanding these views is aided by the voices of academics in Part IV. The ability to analyze more objectively how the changes and impacts fit within global changes in governance of transboundary waters provided by McCaffrey et al. and Craig W. Thomas, and specific analyses of almost fifty years of implementation of the 1964 treaty provided by Eve Vogel provide some perspective to use in evaluating the contributions in the first three parts. Vogel's insights into the question of when it is appropriate to raise issues to the international level and when it is better to leave the issue to a more open collaborative process provide an excellent framework for considering the contributions of Gregory Hill et al. and Tanya Heikkila and Andrea K. Gerlak, who analyze smaller-scale collaborative efforts within the U.S. portion of the Columbia River Basin. Balancing the level of response, the degree of uncertainty versus flexibility, and calls for greater democratization in resource decision making with stability reflect the challenges the people of the Columbia River Basin face in moving beyond 2024. It is also clear that the basin benefits from the fact that it does not start with a clean slate. The lessons of fifty years of treaty implementation combined with the substantial increase in local capacity and experience in collaborative efforts will serve the basin well in entering a dialogue on what the next century of river governance should be.

Works Cited

Bottom, Daniel Kim Jones, Charles Simenstad, and Courtland Smith. "Reconnecting social and ecological resilience in salmon ecosystems." *Ecology and Society* 14(1): 5 (2009). http://www.ecologyandsociety.org/vol14/iss1/art5/

Hamlet, Alan. "The Role of Transboundary Agreements in the Columbia River Basin: An Integrated Assessment in the Context of Historic Development, Climate, and Evolving Water Policy." In *Climate and Water: Transboundary Challenges in the Americas,* edited by H. Diaz and B. Morehouse (Dordrecht: Kluwer Academic Press 2003).

Hirt, Paul. "Developing a Plentiful Resource: Transboundary Rivers in the Pacific Northwest." Chapter 6 in *Water, Place, & Equity* edited by John M. Whiteley, Helen Ingram, and Richard Perry (Cambridge, MA: MIT Press 2008).

Landeen, Dan, and Allen Pinkham. *Salmon and his People: Fish and Fishing in Nez Perce Culture* (Lewiston, ID: Confluence Press 1999).

Leschine, Thomas. "Salmon Fisheries on the Columbia from a Resilience Perspective: Past, Present and Future." Presentation at 2009 University of Idaho Natural Resources and Environmental Law Symposium, Transboundary Governance in the Fact of Uncertainty: The Columbia River Treaty; powerpoint available at http://www.uidaho.edu/law/newsandevents/signature/nrel-symposium/2009-nrel-symposium/Presentations

Northwest Power and Conservation Council. *Sixth Northwest Power Plan, Plan Overview* (February 2010). http://www.nwcouncil.org/energy/powerplan/6/default.htm

Oregon State University Program in Water Conflict Management and Transformation. Transboundary Freshwater Dispute Database (2011). http://www.transboundarywaters. orst.edu/database/

Solomon, Susan, Gian-Casper Plattner, Reto Knutti, and Pierre Friedlingstein. "Irreversible Climate Change due to Carbon Dioxide Emissions." *Proceedings of the National Academy of Sciences* (PNAS) 106: 1704 (February 10, 2009).

U.S. Army Corps of Engineers and Bonneville Power. *Columbia River Treaty: 2012/2024 Review: Phase 1 Technical Studies* (2009). http://www.bpa.gov/corporate/pubs/ Columbia_River_Treaty_Review__2_-_April_2009.pdf.

White, Richard. *The Organic Machine: The Remaking of the Columbia River* (New York: Hill and Wang 1995).

Part I: The 1964 Columbia River Treaty and the Changing Voices Since 1964

The Columbia Exchange:
A Canadian Perspective on the Negotiation of the Columbia River Treaty, 1944–1964

Jeremy Mouat

Several years ago a scientist claimed that "[t]he Columbia River Treaty, with its focus on an 'engineered river' for flood control and winter hydropower, marks a clear transition in the Columbia's history from a natural river to a managed water resources system" (Hamlet 2003, 271). Arguably the treaty was simply the last stage in a much longer process that had the same goal. But the scientist was right: the point of the treaty was to complete the process of turning the river into an organic machine, to use Richard White's memorable phrase. The treaty was made necessary by the fact that the river's flow took it from Canadian territory into American. The question of who had the authority to engineer the river's flow was key to the course of the treaty negotiations. It is worth stressing that the treaty created not just a "managed water resources system," but one that was continental, or at least one that no longer existed in two distinct national jurisdictions. Although those in favor of the treaty frequently used the phrase "cooperative development,"[1] the treaty plainly reflected longstanding American plans for the Columbia River.

Through the mid-twentieth century, a series of dams was constructed on the main stem of the Columbia River within the United States to generate hydroelectric power and to assist in flood control and irrigation. The river was being harnessed, as various advocates had insisted it ought to be. As the scale of these projects increased and as scrutiny of the potential of the river grew more intense, the "inefficiency" of the natural system became more apparent to the planners and the boosters (McKinley 1952; Pitzer 1994; Ficken and Woods 1995; White 1995; Lang 1999). The source of this "inefficiency" was the natural environment, whose regulation of the flow of the Columbia River ran counter to efficient power production. The river's flow peaked sharply during the late spring and early summer, as warmer temperatures melted the snow that had fallen during the winter. Much of this seasonal flow came from that section of the Columbia north of the 49th parallel, within Canada. The uneven flow of the river meant that the generation of power was not constant; surplus water spilled over the floodways during the periods of heavy flow, and periods of low flow could prevent turbines from generating power at full load. The purpose of the treaty was to deal with this problem: it dictated the construction of storage dams in Canada, dams that would conserve some of the river's peak flow, holding it back so that it could be used to generate power through the winter, when the flow tended to be much reduced.[2] The treaty also provided money to Canada in exchange for the construction of these dams.[3] The negotiations that culminated in the treaty lasted twenty years, from 1944 to 1964, and have attracted a good deal of scholarly attention. This chapter re-examines these negotiations from a Canadian perspective, to argue that their course and outcome can only be understood if we situate them in a broad context. Earlier studies have tended to assume that it is possible to account for the course of the negotiations by looking closely only at the personalities and the politics of those directly involved. In addition, most accounts tend to assume (from various perspectives) that the treaty was unavoidable and that alternative arrangements were either implausible or unattainable. In fact, the treaty negotiations were never a foregone conclusion; there was nothing inevitable about the outcome (Krutilla 1967; Swainson 1979).

★ ★ ★

The background for the treaty lay in the early discussions and study in the United States for the development of the Columbia River Basin. This process was underway by the 1920s and culminated in the "308 report" that appeared in 1934, *Columbia River and Minor Tributaries* (House Document 103, 1934).[4] In its nearly two thousand pages, the two-volume study outlined, as its subtitle indicated, "A General Plan for the Improvement of the Columbia River and Minor Tributaries for the Purposes of Navigation and Efficient Development of Its Water Power, the Control of Floods, and the Needs of Irrigation." This was followed shortly after by another investigation commissioned by the National Resources Committee and undertaken by the Pacific Northwest Regional Planning Commission. This second study dealt "with immediate and urgent problems in the Columbia Basin and particularly with the policies and organization which should be provided for planning, construction, and operation of certain public works in that area."[5] This immediacy and urgency was due to the fact that major dams on the main stem of the Columbia—the Bonneville and the Grand Coulee—were being built by federal authorities, and the former project was nearing completion. After considerable debate, the plans for the Columbia River articulated in the 308 report and specified by the Pacific Northwest Regional Planning Commission led to the Bonneville Project Act of 1937, federal legislation that created the Bonneville Power Administration (BPA) (McKinley 1952; Voeltz 1962; Norwood 1980; Pitzer 1994). (By contrast, a quarter of a century would elapse before a Canadian entity with similar jurisdiction to the BPA emerged to deal with power development on the Canadian section of the Columbia River, and by then President Eisenhower and Prime Minister Diefenbaker had already signed the Columbia River Treaty.)[6] Then in 1943, Congress directed the Army Corps of Engineers to review "the 308 Report submitted in 1932 and assess it in the light of the completed dams and the newly formed Bonneville Power Administration" (Billington and Jackson 2006, 192).[7] Six months later, in March 1944, the Canadian government agreed with an American proposal to request that the International Joint Commission (IJC)—the body that deals with "boundary waters" issues between Canada and the United States—"to determine whether a greater use than is now being made of the waters of the Columbia River System would be feasible and advantageous."[8] After receiving this request, the IJC established the International Columbia River Engineering Board. It would be the body that would study the question of the river's development, a study that turned into a fifteen-year project (International Columbia River Engineering Board 1959).[9]

The reference to the IJC in 1944 is typically cited as the point at which formal scrutiny of "Upper Columbia River Development" began, and it is thus seen as the first step in the process that led to the treaty. However the earlier American studies of the Columbia had plainly set the stage. Those studies had made it clear that, as one authority noted in 1952, "The regulation of the Columbia system within the United States for power and flood control . . . cannot be fully achieved . . . without the development of Canadian storage . . . [and] the integrated operation of structures" (McKinley 1952, 111).[10]

Consideration of a joint plan for the Columbia's development came during the Second World War, and at a time when relations between the two countries were changing. The fortunes of war help to explain why that change was underway. Canada had declared war on Germany in September 1939. The subsequent collapse of France and other European nations under the Nazi onslaught, and the beleaguered status of Britain, encouraged Canadian authorities to reflect on the fragile nature of their own country's defenses and its ability to withstand attack. In 1940 the Canadian prime minister and the American president signed a formal defense pact (the Ogdensburg Agreement), establishing a common continental defense system. The parallel agreement on defense procurement the following year encouraged a far closer alignment of the economies of the two countries (Stacey 1955; Granatstein and Cuff 1974; Granatstein 1992; Perras 1998). In the context of the war then underway, these agreements attracted little criticism. But they would encourage some soul searching north of the border once the war was over.

If the war had brought its share of hardship and death to Canada, its economic impact was largely positive. The country had not suffered any damage to its infrastructure as a result of bombing or shelling, and overall the war had encouraged dynamic growth as well as a far greater degree of industrialization. In many respects Canada appeared to be in excellent shape as the war came to an end. There was a mood of confidence, a sense of pride in the country's war effort, as well as a belief that in the future the country would play a greater role in world affairs. Canada's participation in the discussions that led to the foundation of the U.N. seemed to justify this optimistic mood. But this newly confident state confronted some difficult questions.

The United States' dominant position within the emerging postwar world order was plain. This new reality rendered irrelevant the way in which Canada had approached its relations with the United States ever since the Treaty of Washington in 1871.[11] In essence the Canadian policy had been to invoke the British connection as a counter-balance to the looming American presence. This strategy was no longer possible; as a political scientist pointed out a decade later, in 1956: "[F]or the first time in Canada's history, we are faced with the full, naked force of American influence without the modifying influence of any counter-pressure" (Ferguson 1956). This came just as Canadians were becoming far more interested in finding ways to withstand that influence.

During the 1950s and 1960s Canadians began to re-imagine the nature and meaning of their country. This was not a straightforward process, as events in Quebec through the 1950s and the 1960s suggest. Grappling with the nature of national identity was fraught with tension and controversy in English Canada as well (Igartua 2006). Increasingly, these debates in English Canada revolved around the nature of the country's relationship with the United States. The dominance of American popular culture, especially in the new media of radio and film, was a source of growing concern. Canadians had had substantial misgivings about the American presence prior to the war, but these re-emerged with far greater force after 1945. This led to the first of several royal commissions that would attempt to work out a blueprint for Canada in the postwar era, the Royal Commission on National Development in the Arts, Letters, and Sciences, appointed in 1949 (Litt 1991; 1992). A subsequent study, the 1957 Royal Commission on Canada's Economic Prospects, catalogued the growing dominance of the Canadian economy by the United States.[12] And a series of public controversies underscored a growing unease in Canada over the extent of the American presence in Canada, notably the parliamentary debate in May–June 1956 on the construction of a natural gas pipeline across the country, as well as the suicide of a Canadian diplomat in Egypt, a death blamed largely on McCarthyism.[13]

The victory of the Conservatives under John Diefenbaker in the 1957 federal election, after more than twenty years of Liberal rule, was evidence of the growing nationalist mood in Canada.[14] Nor did this escape the notice of the American government. On the eve of a presidential visit to Canada in 1958, for example, Secretary of State Dulles wrote a memo to President Eisenhower, pointing out that the aim of the upcoming trip was "the improvement of our relations with Canada under its Conservative Government." Dulles noted that:

We seek to establish the same mutual confidence and close working relationship with the new government that we enjoyed with the Liberal Government for 22 years. The attainment of that relationship is, however, somewhat impeded by the existence of vocal, widespread criticism of the United States and its policies. In large part this criticism owes its origin to Canadian nationalism. It has been further nourished by the election campaign as well as by the current recession in Canada. A major manifestation of this has been a tendency to assert Canada's independence of the United States. Some members of the government have been prone to play upon the emotional response that such assertions evoke and to try to make the United States the whipping boy for many of Canada's ills (Dulles 1958–60, 686–87).[15]

The six years in which Diefenbaker and the Conservatives remained in power in Ottawa were critical ones for the Columbia River Treaty negotiations. The nationalist mood—with its accompanying suspicion of American motives and behavior—was a significant factor influencing the Canadian side during those negotiations.

Regardless of the change in government at the federal level, however, water resources fell under provincial jurisdiction.[16] It follows that any account of the Canadian position in the negotiations that led to the Columbia River Treaty must acknowledge the role of the provincial government. From 1952, W. A. C. Bennett and his Social Credit Party ruled British Columbia, although such was the force of Bennett's personality that his views tended to be those of the provincial government.[17]

Prior to entering politics, Bennett had been a merchant in the province's interior, and this background influenced his politics in several important ways. He was dedicated to the province's growth and development (nouns which he believed could not possess any negative connotations), and he especially wanted to encourage the expansion of the province's infrastructure, including roads, railways, and electrical supply (Tomblin 1990). Bennett was also suspicious of the motives and actions of the province's urban business elite and opposed too heavy a reliance on debt financing.[18] He was also a flamboyant showman. For example, when he declared the province debt-free in the summer of 1959, he ostentatiously staged a "bond-fire" of the cancelled bonds, which were piled on a barge on Okanagan Lake near his Kelowna home, where they were to be ignited by a flaming arrow shot by Bennett (he missed). In fact, the debt's disappearance was more a matter of a change in financing techniques, as Bennett—his own Minister of Finance—turned from direct to indirect debt financing (Sherman 1966, 179–208; Mitchell 1983, 277–80). He would approach the Columbia River Treaty negotiations in much the same spirit.

★ ★ ★

When Bennett's Social Credit government assumed power in 1952, British Columbia embarked on a twenty-year trajectory predicated on the relentless pursuit of growth. In Ottawa, by contrast, very little changed, at least until the election of Diefenbaker in 1957. Up until the mid-1950s, the federal government did not appear to be especially concerned with pushing the Columbia River project, choosing instead to wait for the IJC-commissioned report on the river.

Fifteen years passed before the International Columbia River Engineering Board presented its report to the IJC, in 1959. In the meantime, proposals to develop the Columbia continued to be referred to the IJC. After the Columbia River flood in 1948, the American government brought forward a proposal for an upstream dam on the Kootenay River, in Libby, Montana, which would help to avert similar flooding in the future. This went to the IJC in early 1951, but the dam—which would inundate parts of both Montana and British Columbia—raised difficult issues. No one questioned the potential benefits of the Libby Dam, but the very prospect of these benefits raised the thorny question of who was to get what. For example, if flooding land in British Columbia gave certain advantages to American residents living downstream, in terms of flood protection and/or energy supply, what formula could be used to determine these downstream benefits? What could be considered reasonable compensation? The issue of downstream benefits would cause much heated discussion during the lengthy negotiations surrounding the development of the Columbia River. In the short term, however, difficulties in resolving the issue led to the shelving of the Libby proposal in 1953 (Bloomfield and Fitzgerald 1958, 190-95; Swainson 1979, 45–51).

The creation of the International Joint Commission was not the only unique feature of the Boundary Waters Treaty in 1909. The treaty also specified the right of the upstream riparian to control the waters of rivers, a right inserted into the treaty at the insistence of the American negotiators and over the objections of Canada. This became significant when the chair of the Canadian section of the IJC, General McNaughton, put forward proposals for diverting rivers in the Columbia River Basin before they crossed into the United States. A number of Americans viewed this possibility with alarm, as such diversions would potentially undermine American plans for the Columbia, which assumed upstream storage in Canada.[19] McNaughton's proposal and the American reaction helped to stimulate a flurry of legal papers, debating the finer points of international river law, in particular the meaning of a specific clause in the Boundary Waters Treaty (American Society of International Law 1956; Martin 1957; Watkins 1957; White 1957; Bourne 1958a, 1958b; Cohen 1958; Bourne 1959).

Shortly after the International Joint Commission received the American proposal for the Libby Dam, the Canadian mining-smelting conglomerate that ran the Trail, BC, smelter on the banks of the Columbia, submitted another. This company—Cominco—wanted to build a dam on the Pend d'Oreille River, close to its junction with the Columbia and downstream from Trail, adjacent to the American border. The proposed Waneta dam and reservoir did not encounter the same opposition that greeted the Libby proposal and in 1952 the IJC gave its approval (Bloomfield and Fitzgerald 1958, 196; Swainson 1979, 48–49). The Waneta plant began producing power in late 1953. The addition of Waneta's considerable generating capacity created an energy surplus for Cominco, far more than it needed for its smelter operations or what could be absorbed by other regional consumers. The company worked out a tentative deal to sell its surplus power to a Washington utility, but the Canadian federal government's objections prevented the sale from going ahead. The export of electricity from Canada was a sensitive topic, and something long opposed by the federal government.[20]

The premier of British Columbia had no such misgivings. His view was to embrace any project that encouraged regional development. For example, when an American company, the Kaiser Aluminum and Chemical Corporation, proposed building a dam on the Columbia in 1954, Premier Bennett was immediately interested. As with the Libby proposal, the Kaiser plan would lead to the flooding of certain parts of southeastern BC, but it would also provide some tangible benefit for the government. The Kaiser interests proposed constructing a dam on the Arrow Lakes, at its expense, and then passing on to the province 20 percent of the power generated by the plant. To the provincial government, this seemed a relatively easy way to earn a tidy annual sum. Not only that, but the plan called for no expenditure on the government's part. In the autumn of 1954, Premier Bennett and his Minister of Lands and Forests, Bob Sommers, happily announced that the provincial government had signed a tentative agreement with Kaiser.[21]

The provincial government soon discovered that others did not share its sunny assessment of the Kaiser plan. An avalanche of criticism descended upon Victoria from the press, the opposition, and the federal government. These critics argued that

> it would be economic folly for Canada to accept [the proffered Kaiser deal], since the very cheap power generated downstream in the United States as a result of the Canadian storage would be used by the corporation to manufacture aluminum which, being produced within the protective tariff walls of the United States, would therefore be highly competitive with the Canadian aluminum manufactured at Kitimat, British Columbia. . . . in the broader context of the economy of British Columbia and of Canada as a whole, it became clear that [the Kaiser plan] would expose British Columbia's aluminum industry to damaging competition and would affect Canada's export trade (Bourne 1969).

Although Bennett was not a person easily dissuaded from pursuing his chosen path, Ottawa's opposition was hard to ignore. A major battle between the two levels of government was not long in coming.

Canada's political history is often little more than the history of federal-provincial relations writ large. The fate of the Kaiser plan—indeed, the whole twenty-year history of the Columbia River Treaty negotiations—illustrates this point only too well. From Ottawa's perspective, the Kaiser plan was not only antithetical to the national interest but also threatened to short-circuit the lengthy study of the Columbia River's future begun by the International Joint Commission a decade earlier. The Liberal government of Prime Minister St. Laurent swiftly introduced a bill—the International Rivers Improvement Act—that effectively killed the Kaiser proposal. Passed in the summer of 1955, the act stipulated that no one could construct a dam on an international river without a federal license, that is to say, without the approval of Ottawa (Swettenham 1969, 249-50).

The B.C. government, and Premier Bennett in particular, did not take kindly to Ottawa's action. As far as Bennett was concerned, the rivers of British Columbia, and any plans for their development, were a provincial affair; he saw the federal government's new law as infringing on provincial jurisdiction. Bennett and his colleagues also took exception to the none-too-subtle hints that they were an inexperienced bunch prepared to sell out the province's interests at the blink of an eye. For the moment, the provincial government knew that it had been beaten—unfairly beaten, too, it would claim—but it was prepared to wait for other opportunities. These were not long in coming.

Although the next project to make headlines in British Columbia did not concern the Columbia River Basin, it was to influence the progress of the treaty negotiations.[22] This project seems to have been sparked by a reception at the province's London House in the spring of 1956, when the representative of a Swedish industrialist learned of the tremendous potential of northeastern British Columbia.[23] Before too long, Axel Wenner-Gren sent two associates to the province, where they met the premier. The outcome of these meetings was a proposal from Wenner-Gren to embark on a massive development project, formally detailed in a letter of agreement between the government and Wenner-Gren, signed in the autumn of 1956 (Sherman 1966, 209–30; Robin 1973, 207–12).

When Wenner-Gren's ambitious scheme became public the following spring, it attracted a good deal of attention, little of which was positive. The opposition and a number of the urban newspapers denounced the plan alternatively as a massive give-away of the province's resources and/or as an unattainable pipe dream with little chance of success. Whatever the motives of the provincial government—notably those of Premier Bennett, firmly in charge of the government's energy policy—the

Wenner-Gren proposal focused attention on the enormous energy potential of the Peace River. As the discussions of the Columbia River's future became more and more drawn-out and convoluted, an impatient Bennett considered what might be achieved in northeastern British Columbia. Bennett was not a person who welcomed delays or dissent, and he knew development in the Peace River district would not be hampered by jurisdictional squabbling with the federal government.[24]

In 1958 Wenner-Gren formed the Peace River Power Development Company to inaugurate a large hydroelectric development in northeastern B.C. (Swainson 1979; Wedley 1986, 247-310). The company's plans soon encountered a major stumbling block. No one questioned the vast hydroelectric potential of the northern river system, but essential to the success of the company's project was a guaranteed market for the considerable energy that it was going to generate. To raise the necessary capital, long-term energy contracts had to be in place, tangible evidence for would-be investors of the plan's financial feasibility. The private utility, B.C. Electric, was the province's leading energy company and thus either it or its parent company, the B.C. Power Corporation, would be the obvious customer for Peace River power. Although quite interested in the Peace River project—it was represented on the board of the Peace River Power Development Company— B.C. Electric had already made plans for its future power supply. When pushed by Bennett, the company flatly refused to sign any long-term contract to purchase Peace River power. But the premier was not easily dissuaded from his plans for northern development. Increasingly he had come to see the Peace River power project as a keystone of his government's development strategy, something that dovetailed perfectly with his vision of the province's future as well as providing a lever with which to apply pressure on the federal and U.S. governments, in terms of the ongoing Columbia River talks.

★ ★ ★

As Bennett embarked on his plans for Peace River development, the federal government was becoming more interested in pushing ahead with Columbia River development. Prime Minister St. Laurent appeared to have lost patience with the International Joint Commission and in the spring of 1956 he and President Eisenhower agreed to move the discussions over the Columbia into the diplomatic arena (Swettenham 1969, 268–70; Swainson 1979, 68–74). At the same time, the federal and provincial governments began to discuss the issue of Columbia River development on what seemed a more constructive note.[25]   Before these developments produced any significant results, however, a federal election brought down the St. Laurent government, replacing it with a Conservative government led by John Diefenbaker. Even before the election, it had become clear that the Conservatives were keen to see progress with the ongoing discussions over Columbia River development.[26] Diefenbaker's first Throne Speech, delivered by the Queen in October 1957, underlined the new government's commitment to the project. "Where our predecessors had engaged in talk, my government was prepared to act," claimed Diefenbaker in his memoirs. " . . . We put forth a tremendous effort to obtain an agreement that would be fair to Canada and at the same time would develop the great power potential of the Columbia."[27] Nine months later Eisenhower was in Ottawa and brought up the subject in discussions with Diefenbaker, suggesting that "the time had come for the two of us to put some pressure on the different agencies then considering the problem so as to make sure that something was being done. He indicated that he was under the impression that the experts, and particularly General A. G. L. McNaughton, were busy finding obstacles rather than solutions" (Diefenbaker 1976, 155–56).

Within a few months, the pace of discussions on the Columbia accelerated although this does not appear to have been a direct consequence of increased political pressure.[28] The Canadian and American sections of the International Joint Commission reached a measure of agreement on how to proceed—largely due to the fact that the American section now accepted the principle of downstream benefits—and Ottawa responded by pressing the IJC on how precisely to determine and apportion those benefits. In January 1959 a formal request from both the American and Canadian governments asked the IJC for "its recommendations concerning the principles to be applied in determining: (a) the benefits which will result from the cooperative use of storage of waters and electrical interconnection with the Columbia River System; and (b) the apportionment between the two countries of such benefits more particularly in regard to electrical generation and flood control."[29] In March of the same year the International Columbia River Engineering Board (1959) produced its long-awaited report, and at the end of the year the IJC submitted its report, "Principles for Determining and Apportioning Benefits from Cooperative Use of Storage of Waters and Electrical Interconnection within the Columbia River System," to the Canadian and American governments.[30] The stage was set for the formal negotiation of the Columbia River Treaty. On 25 January 1960, Prime Minister Diefenbaker announced in the House of Commons that "negotiations between Canada and the United States for the co-operative development of the Columbia River system are to commence in Ottawa on Thursday, 11 February."[31] In just under a year, Diefenbaker and Eisenhower would formally sign the treaty, on 17 January 1961, one of the last official acts of the outgoing U.S. president. Never one for modest understatement, Diefenbaker proclaimed his hope that "in the years ahead this day will be looked back on as one that represents the greatest advance that has ever been made in international

Figure 1: The three "Plans of Development" contained in the Report to the International Joint Commission, prepared by the International Columbia River Engineering Board.

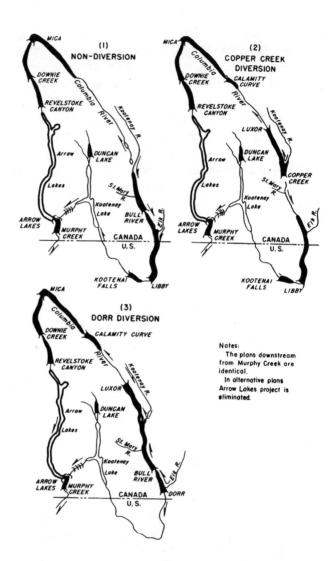

relations between countries."[32] This turned out to be wishful thinking: for the next three years, the treaty—and indeed the larger issue of Canadian-American relations—provoked bitter public controversy in Canada.

* * *

The negotiators of the Columbia River treaty had the International Columbia River Engineering Board's *Water Resources of the Columbia River Basin: Report to the International Joint Commission* to assist them in their discussions. This very extensive report outlined three possible development schemes, two of which proposed diverting the headwaters of the Kootenay River north into the Columbia (the Copper Creek diversion and the Dorr Diversion), and a third which instead proposed constructing the Libby dam further south on the Kootenay River, with no diversion of its waters into the Columbia (Figure 1). All three schemes also proposed a dam below the southern tip of the Arrow Lakes, although the International Columbia River Engineering Board suggested that two very different dams were possible, a "High Arrow" dam (which would result in significant flooding around the Arrow lakes) and a "Low Arrow" dam, which would lead to far less environmental disruption around the lakes. But the Board failed to make any recommendation as to which of the three plans was the best: "no one plan of development can be selected as representing the optimum use of the sites and water resources."[33] That job was left to the treaty negotiators, although they did have the International Joint Commission's "Principles for Determining and Apportioning Benefits from Cooperative Use of Storage of Waters and Electrical Interconnection within the Columbia River System" to guide them in their deliberations. The negotiators finally settled on a scheme which would involve no diversion of the Kootenay, the construction of the Libby Dam as well as three dams in Canada, at Mica, Duncan Lake and—most controversially—a High Dam for the Arrow Lakes. These provisions in the treaty signed on 17 January 1961 were criticized by various people and for several reasons. However it soon became apparent that one of the treaty's greatest weaknesses was Premier Bennett's unwillingness to accept its terms.

Diefenbaker had signed the Columbia River Treaty somewhat precipitously in the final days of the Eisenhower administration, perhaps to avoid having to deal with John F. Kennedy (L. Martin 1982, 181-211; Robinson 1989, 165-69; Gloin 1998; T. Nemeth 2002). A few days before the formal signing at the White House, Premier Bennett had conveyed his misgivings about the treaty to Ottawa, although the Conservative government felt that it had had a firm assurance of Bennett's support from his cabinet minister, Ray Williston.[34] A shrewd political strategist, Bennett knew he now had a significant advantage in the ongoing debate over the Columbia, which at one level pitted the federal government against its provincial counterpart. Certainly Bennett knew that the signatures of the Canadian prime minister and the U.S. president on the treaty would amount to very little if his government was unwilling to accept its conditions. The federal government could negotiate any treaty it liked, but no dams could be built on B.C.'s rivers without the approval of Victoria. "Seldom in Canada's history," noted one Vancouver commentator, "has a Canadian Prime Minister so trustingly placed his neck in a headlock by a provincial premier as Mr. Diefenbaker appears to have done to Premier Bennett's obvious strategic advantage."[35] Following the treaty's signing, the U.S. Senate quickly ratified the treaty, with only one vote opposing it.[36] Approval of the treaty in Canada was going to take far longer and it would prove much more contentious.

One of the reasons for the difficulties of the treaty was the volatility of the federal government. After being elected in 1958 with a huge landslide, Diefenbaker and his Conservatives lost their majority in an election held in June 1962; in a second election held less than a year later, in April 1963, the Liberals returned to power in Ottawa under Prime Minister Lester B. Pearson. But it was W.A.C. Bennett in British Columbia who had the greatest impact on the ongoing discussions of the treaty.

Bennett's first move caught everyone by surprise (including members of his cabinet). In one of his most controversial and unexpected acts, the premier announced in the provincial legislature on 1 August 1961 that he intended to take over both B.C. Electric and the Peace River Power Development Company.[37] With hindsight, the move makes a good deal of sense, but at the time it provoked denunciation and disbelief. Since the nationalization of B.C. Electric was a long-standing goal of the left-leaning opposition in the province, the plan passed without a dissenting vote in the legislature. On other fronts, the response was harshly critical. The financial press hurled insults and threats at Victoria, but Bennett appeared unperturbed. This wasn't about ideology, he explained, it was about development, about the province's future. He went on to point out that for years people had accused his government of being subservient to B.C. Electric. He had always denied this, and now here was proof of his independence. As premier, he would let no one—not Ottawa, not big business, not unions, not the opposition—stand in the way of his government's plans, which would benefit all British Columbians. Ever the affable populist, he smiled blandly amidst the fury and let the storm break over him.[38]

The bill that effectively nationalized B.C. Electric in August 1961 did not extend to the other major private utility in the province, West Kootenay Power. This was the utility that supplied Bennett and his constituents in Kelowna with power and this in itself may help to explain why the company avoided the fate of B.C. Electric and the Peace River Power Development Company. People in the Okanagan were well aware of the fact that they enjoyed far cheaper power than consumers in the Lower Mainland, who were serviced by B.C. Electric. If nationalization brought with it higher rates, Bennett's chances of re-election might have been threatened.[39] But another reason is probably more important in explaining Bennett's decision to allow West Kootenay Power to continue as an independent utility, outside of the new B.C. Hydro system (Robin 1973, 230-2; Mitchell 1983, 303-5; Jamieson 1962; Keate 1961; Irving 1962). Bennett had no wish to end private power generation in the province and that was not his motive in nationalizing B.C. Electric. Rather, the move reflected a larger political objective: he wanted to see both the Columbia and the Peace developments go ahead, what he liked to refer to as his "two river policy." This was, unquestionably, a grandiose scheme and one near and dear to Bennett's heart. "It is the largest and brightest concept anywhere today," he boasted, ". . . too big for little persons to grasp."[40] While many obstacles stood in the way of Bennett's two river policy, West Kootenay Power was hardly one of them; success did not require its nationalization. Personal memoirs—such as those of Robert Bonner, a cabinet minister in the Bennett government—tend to support this analysis:

> *We felt we had no alternative but to nationalize the B.C. Electric to make it a vehicle for electrical energy policy. . . . there was a growing realization that the Two River Policy was probably beyond the capacity of private interests to deal with, and when we started to become more heavily involved with the Columbia, there was no way that we could control the pressure on negotiations without having the Peace development under our control as well. It was an evolution of analysis which gradually brought us into the picture: we had to have both the Peace and the Columbia.[41]*

Bennett's ostensible justification for the controversial move in nationalizing B.C. Electric involved a feud with Ottawa over taxation: in effect, an attempt to win public support through reviving an old provincial theme of "fight Ottawa." Bennett had argued on other occasions—although no one took him very seriously—that if the federal government did not return to the province the tax revenues that Ottawa collected from private utilities, he would have no option but to nationalize the utilities, out of fairness to consumers. Publicly-owned utilities paid no income tax and Bennett claimed that this gave them an unfair advantage. After all, consumers who relied on electricity from a private utility were obliged—through their power bills—to assist the company pay its taxes. However, the focus on taxation was largely window-dressing; as David Mitchell noted in his biography of Bennett, "this was clearly more a rationalization than an explanation" (Mitchell 1983, 307).

As controversy raged in British Columbia over the government takeover of B.C. Electric, Bennett continued to participate in the Columbia River negotiations. Bennett had no objections to a Columbia River treaty. Indeed, the treaty was essential to his two river policy. His concern was with the treaty's bearing on his overall energy strategy: he wanted to ensure that it did not hinder the development of the Peace. Two things, in particular, caused him anxiety. The first was the federal government's policy regarding electrical sales to the United States: its opposition to long-term contracts involving the export of electricity went back many years. This policy posed a potentially fatal threat to the Peace River project. If Columbia River power could not be sold in the U.S., its only other market would be the Lower Mainland. However, Bennett knew that the Vancouver region could not absorb power from both the Columbia and the Peace. It followed that failure to sell the Columbia power to the U.S. would render the Peace River project redundant, and that was something that Bennett was not prepared to accept. The other issue that concerned him was the treaty's financial arrangements: how much money was to flow into his government's purse. As far as Bennett was concerned, any money earned by B.C.'s rivers belonged to the province. Not only did he expect the money from the sale of the Columbia's downstream benefits, he also had very definite ideas about the price. The instability of the federal government would help Bennett to achieve most of what he was after.

The June 1962 federal election ended Diefenbaker's majority in the House of Commons. His minority government after the election relied on the support of the federal Social Credit party to remain in office. Four months later, Ottawa formally abandoned a long-standing policy, with the announcement in the Throne Speech that "large-scale, long-term contracts for the export of power surpluses . . . should now be encouraged" (Sherman 1966, 260–2; Worley 1971, 243; Swainson 1979, 232). One knowledgeable journalist reported at the time that the rumor in Ottawa was that federal Social Credit MPs were keeping the Tories in power so that they could ratify the Columbia treaty.[42] However, the Conservative government did not survive long enough to carry this out. A second federal election, in the spring of 1963, ousted Diefenbaker and his Conservatives. Lester B. Pearson and the Liberals formed a minority government, one that was to prove a good deal more durable than Diefenbaker's.

Several months later, on 30 September 1963, Bennett and Social Credit were re-elected in British Columbia, after a hard-fought campaign that centered on "that complex and difficult subject hydro-electric power."[43] Feeling vindicated by the results of the election, Bennett confidently began serious discussions with the new federal government over possible changes to the Columbia River treaty. Unlike several members of Diefenbaker's Cabinet, Pearson and his colleagues did not have an entrenched position on the still-unratified treaty. Soon talks were underway with the Americans as well. The Canadian side was led by Paul Martin, Pearson's Minister of External Affairs and a skilled negotiator, who worked closely with several senior B.C. cabinet ministers (P. Martin 1985). The result of these discussions was a protocol—in effect, a revised treaty—signed on 22 January 1964 by Prime Minister Pearson and President Lyndon Johnson, with the enthusiastic support of Bennett and his government. As the Victoria *Daily Colonist* observed, "Ottawa composed the music but Bennett wrote the words. It is a major political triumph for the Bennett government and particularly for the Premier" (Worley 1971, 244). As the quotation suggests, the treaty gave Bennett what he wanted. The province received cash for the downstream benefits, money that would be used to pay for the construction of the three treaty dams in British Columbia. Since one of these, the Mica Dam, would also generate considerable electricity, Bennett made his well-publicized claim that the province would obtain free power from the Columbia, "And nothing is freer than free, my friend" (Mitchell 1983, 324).

If the treaty was a triumph for Bennett—he spoke of it as his greatest achievement—others regarded it as a sellout or worse.[44] Opposition came from many quarters, although it was nearly unanimous in the Kootenays, the region that would feel its impact most heavily, in the short as well as in the long term. The most obvious—and controversial—result was the plan to flood the Arrow Valley with the construction of a "High Arrow" dam. Jack McDonald, an electrical engineer who lived in the region, spoke out against the plan:

> In making the financial assessment of High Arrow the value of the unspoiled Arrow Lakes valley to the people of this province has been completely ignored. It has apparently been assumed in all the calculations that its value before High Arrow is nil. If this natural resource is considered in the light of its true value to the present and future generations in its unspoiled state the figure would far outbalance the paltry sum for which we are being asked to destroy it. We should keep in mind that a resource such as the Arrow Lakes cannot be assessed in mere dollars and it should not be sacrificed unless an undisputable benefit is to be gained. High Arrow does not meet this requirement.[45]

Richie Deane, like McDonald, was an electrical engineer employed by Cominco and he too opposed the High Arrow dam. Deane presented a thoughtful critique to the House of Commons' External Affairs Committee when it considered the final version of the treaty in the spring of 1964.[46] By that time, however, the negotiations were effectively over: the federal government was unwilling to alter the terms of a document which had been so long in the making. Kootenay residents were left with no choice but to live with the treaty's consequences, even though their views had rarely been taken into account during the lengthy process that culminated in the final agreement of 1964. Only the lone voice of Bert Herridge, the region's Member of Parliament in the House of Commons, reminded the federal government of the extent of the local opposition to the treaty.[47]

Herridge was hardly alone: a number of Canadians questioned the wisdom of exporting energy to the U.S. market.[48] They regarded the relationship between the two countries as unequal and one in which the interests of the U.S. would inevitably dominate. The report of the 1957 Royal Commission Canada's Economic Prospects had raised concerns about the American role in the Canadian economy. A groundswell of Canadian nationalism magnified by a growing widespread distaste for American foreign and domestic policies (notably with the Vietnam War abroad and racial conflict at home) added considerable weight to this concern (Azzi 1999, 125–39, Purdy 1968). While the official mind of the Canadian government remained ambivalent—thus even Walter Gordon, economic nationalist and subsequently Minister of Finance under Mike Pearson, participated in the negotiation of the free-trade style Autopact (Anastakis 2005; Donaghy 2002, 41–54; Azzi 1999, 125)—there was little doubt about the public mood. For example, when a report commissioned by the Canadian and American governments appeared in 1965, detailing "Principles for Partnership" between them, it was greeted with outrage in Canada.[49]

The scholarly literature on the Columbia River treaty frequently takes its subject to be an exercise in rational decision making. These studies tend to characterize the treaty as a model of "co-operative development," a phrase that featured in the treaty's formal title ("Treaty Between Canada and the United States of America Relating to Cooperative Development of the Water Resources of the Columbia River Basin"). But if one examines the specific obligations assumed by Canada as a consequence of the treaty, "co-operative development" seems to mean implementing longstanding American plans for the river. Thus Canada was to provide upstream water storage for U.S. dams and regulate the flow of this stored water so as "to achieve optimum power generation downstream in the United States of America." In addition, Canada would not reduce the Columbia's flow in any way that would detract from the flood control and hydro-electric power benefits, nor would it divert waters from the Columbia River basin without American consent. Canada also agreed to the flooding of its territory that would come with the construction of the Libby Dam in Montana.[50] Since the Canadian negotiators were neither supine nor stupid, the question arises as to why they agreed to these terms. Political expediency—in part, the need to bring closure to a very lengthy process—is part of the answer. Twenty years is a lengthy period to negotiate any treaty, and this was an agreement between two nations with much more than the Columbia River watershed in common. Those in Canada who did the actual negotiating knew that they were dealing with people who had a clear and carefully developed agenda for the Columbia River. In addition, it was plain by late 1962 that the American side was growing impatient, nuclear power appeared to be emerging as a serious alternative to other forms of electrical generation, and the Canadian government was anxious to secure better access to the American market for Alberta's oil and gas (T. L. Nemeth 2007). And relations between the two countries were under considerable strain after the Diefenbaker government refused to support the American government militarily at a key point during the Cuban missile crisis (Haydon 1993).

In January 1963, senior officials urged the federal government to move quickly and agree to an amended treaty, warning that failure to do so might prevent any treaty being signed.[51] It was too late, however: the Diefenbaker government was in crisis and two weeks later the minority government collapsed. The subsequent election, in early April 1963, brought the federal Liberals to power (Muirhead 2007; Beck 1968, 351–73; Thompson 1990, 87–97). The new government was determined to appear bold and forceful, and with close links to the Kennedy administration, it was anxious to "mend fences" with Washington. No one at the federal level had a particular interest in the treaty, beyond wanting to see it signed and any outstanding issues settled.

The B.C. government, on the other hand, was very interested in the treaty, and happy to see the federal government acquiesce to its demands about financing, the payment of the downstream benefit, and other matters. Bennett, the dominant force on the provincial political scene, was committed to the provision of infrastructure and development projects throughout the province. His 'two river policy' would bring construction jobs in the short term and energy revenue over the longer term. Others argued that the treaty's provisions for the sale of downstream benefits effectively prevented the province from enjoying a very real economic advantage. Thus Dal Grauer, the head of B.C. Electric, opposed the Columbia River deal as well as Bennett's two river policy, arguing that

> The potential power in British Columbia is from a number of sources and will therefore be at differing costs. It would seem to be sound public policy to keep the cheapest power for British Columbian use and to stimulate economic development here, rather than to export the least expensive power. This point seems to have particular relevance because the cheapest future power seems undoubtedly to be the downstream benefits of the Columbia River, and there might be some pressure from the United States to buy Canada's share at the American generating plants where it is produced, rather than returning it to Canada. To succumb to such pressure would, in my opinion, be a tragedy because most of British Columbia, unlike the states of the Pacific Northwest, has never had the stimulus of really cheap power. Now that the wheel of fortune in this respect is at last spinning in our direction, we should make every effort to take advantage of it.[52]

Such arguments failed to impress the provincial government. Nor did the government consider the environmental impacts of the Columbia and Peace river projects. Those people displaced by the treaty dams received negligible compensation and support.

The literature on the Columbia River Treaty ignores the treaty's role in integrating the energy economies of Canada and the United States (Clark-Jones 1987, 60–1). This integration was hardly confined to energy, of course: the Canadian and American economies grew ever more closer throughout the post-war era, culminating in the Canada-U.S. Free Trade Agreement of 1988. However, the inevitability of this integration was (and is) hotly debated. Neo-liberal economists and political scientists tended to see the trend to continentalism as a reflection of the inexorable logic of technology and trade. For nationalists of all political stripes—from the Conservative intellectual, George Grant, whose *Lament for a Nation* became an unlikely best-seller in early 1965, to the left-wing ginger group, members of the NDP "Waffle"—such an outcome could not be reconciled with their vision of an independent Canada, hence their opposition to all its manifestations.[53]

The Columbia River Treaty laid the basis for a continental power network and by doing so, effectively closed off any possibility for a national power grid within Canada, an idea that had been discussed through the late 1950s and early 1960s.[54] As its supporters noted, there were sound reasons for establishing a national grid, although the more limited goals of some provincial governments—notably British Columbia—raised obstacles that ultimately ended any hope for such a national project. The continentalist assumptions that underpinned the treaty also reflected an implicit if unimaginative economic strategy, as Grauer's comments above imply. This stands in stark contrast to the industrial strategies pursued by provincial utilities in Ontario and Quebec, as well as the more general assumptions that informed the postwar province-building projects in Alberta and Saskatchewan (Nelles 1974, 248–9; Niosi 1981, 100; Richards and Pratt 1979; Tieleman 1984).

Notes

I am grateful to the University of Alberta for a Killam Research grant which made it possible for me to visit archives in Ottawa, Vancouver, and Victoria. Thanks too to Sarah Carter and other members of the Western Canada reading group for helpful comments on an earlier version of this chapter.

1. See, for example, the booklet published by the Canadian government (produced by External Affairs), intended to explain the benefits of the treaty to a skeptical Canadian public: *The Columbia River Treaty and Protocol: A Presentation* (Ottawa: Queen's Printer, 1964). The phrase also features in the treaty's formal title, "Treaty between Canada and the United States of America relating to Cooperative Development of the Water Resources of The Columbia River Basin."

2. The treaty also dictated measures for flood control. In addition, there was a subsidiary issue involving the utility of hydropower generally, compared to that of thermal power. The latter source was less able to respond to sudden peaks in energy demand.

3. More precisely, the money was for the downstream benefits of the Canadian dams. This was a controversial topic and led to a good deal of negotiation. Although the American side was at first reluctant to acknowledge the principle of downstream benefits, by the end of 1959 (as noted below) it appears to have done so.

4. Comments on this report can be found in Billington and Jackson 2006, 155–56, and in Norwood 1980, 44–45.

5. These comments are from the cover letter by Harold L. Ickes, Secretary of the Interior and Chair of the National Resources Committee, included in *Regional Planning, Part I—Pacific Northwest* (Washington: United States Government Printing Office, 1936), iii (the cover letter was dated 21 April 1936). On the genesis and significance of this report, see McKinley 1952, 158–60, although it's worth noting that McKinley himself played a significant role in the drafting of the report (see Pitzer 1994, 441, note 14).

6. The Canadian entity I am referring to is the provincial crown corporation, BC Hydro and Power Authority, created in March 1962 when the recently nationalized B.C. Electric was merged with the BC Power Commission. President Eisenhower and Prime Minister Diefenbaker had signed the Columbia River treaty in January 1961, one of the last acts of the American president before leaving office.

7. The Army Corps of Engineers duly produced an eight-volume report (known as the "308 Review Report"), *Columbia River and Tributaries, Northwestern United States. . . .*, 81st Congress, 2nd Session, 1950. House Document No. 53 (Washington: U.S. Government Printing Office, 1952). According to a chronological list of events relevant to the Columbia River Treaty compiled by the federal government in Canada, the U.S. Senate Committee on Commerce adopted a resolution 24 September 1943, asking the Army Corps of Engineers to undertake a comprehensive survey of the Columbia River Basin in the United States. (This was listed as the first item in the chronology, with the 1944 reference to the IJC as the second item.) See E. Davie Fulton Fonds, MG 32 B11, Vol. 38, folder 62-35-2, "I.J.C. Columbia River—Negotiations," Library and Archives Canada, Ottawa.

8. Reference from the Canadian and United States governments to the International Joint Commission (Canadian Note), 9 March 1944, reprinted in Department of External Affairs and Northern Affairs and National Resources April 1964, 17. Cf. the comments in Swainson 1979, 40–41 and Johnson 1966. The IJC was created by the terms of the Boundary Waters Treaty (1909) and consists of three Canadian and three American members, co-chaired by a representative from each country. In the years since its formation it has dealt with numerous issues referred to it by either or both of the two countries. On the IJC's role and history, see Spencer et al. 1981 and Bloomfield and Fitzgerald 1958.

9. Comments on the genesis of this report in Swainson 1986.

10. Similarly, Oregon Senator Richard L. Neuberger noted in 1955: "At the time of the Columbia River reference, in 1944, the United States had numerous existing and planned projects and the elements of a long-range plan for development which even included the use of continued flows and storage of Canadian water" (Neuberger 1955, 15).

11. Under the terms of the Treaty of Washington (a treaty between the U.K. and the U.S.), the United States effectively acknowledged the legitimacy of the Canadian presence in North America, following the country's creation in 1867 and its subsequent absorption of the British-controlled territories across the prairies as well as the Crown Colony of British Columbia. Studies of the treaty and its context include Smith 1941; Stacey 1955; Myers 1978; Messamore 2004.

12. For a description and an analysis of this important commission, see Gordon 1977, 59–70; Azzi 1999, 34–65. As Azzi notes, "Before the Gordon Commission [i.e., the 1957 Royal Commission on Canada's Economic Prospects], Canadian economists did not consider foreign investment a proper topic for debate. For them it went without saying that governments should not limit the free flow of capital across national borders. The Gordon Report was seminal, triggering considerable political and academic discussion of the subject. The report gave Gordon's concerns over foreign investment an air of legitimacy" (op. cit., p. 57).

13. The pipeline debate is described in Kilbourn 1970, 111–33; McDougall 1982, 56–77; Nemeth 2007, 31–46. H. R. Norman's death remained controversial into the 1980s, when two biographies appeared, with very different approaches to their subject. A subsequent official inquiry into Norman's suicide left no doubt as to his innocence: see Lyon Fall 1991 (the inquiry's report) and Whitaker and Marcuse 1994, 402–25. Many at the time regarded Norman as a victim of an American witch hunt; Lyon for example refers to "The wave of anger that swept Parliament and [the Canadian] nation. . . . rage over a crude, cruel violation of Canada's sovereignty, and shock at the loss of one of our best and brightest." Cf. the comments in Martin 1982, 175.

14. On Diefenbaker's election, see Meisel 1960, 570–72; Newman 1963; Stursberg 1975, as well as the comments on the 1957 and 1958 election campaigns in Beck 1968, 291–328.

15. Six months later, in a background memo ahead of a meeting of the U.S.-Canadian Joint Committee on Trade and Economic Affairs, an unnamed American official referred to "the present atmosphere of radical nationalism in Canada as championed by the Conservative Government." (Paper Prepared in the Office of British Commonwealth and Northern European Affairs," 2 January 1959, op. cit., p. 746).

16. For the relevant constitutional arrangements, see Burton 1972, esp. 102–4; as Burton points out, "there is considerable scope for conflicts of jurisdiction in respect to water resources" (103). This contrasts with the U.S. jurisdiction, where federal authority over navigable rivers has been confirmed by the Supreme Court; see for example its decision in the Cowlitz River case: U.S. Supreme Court, *City of Tacoma v. Taxpayers of Tacoma*, 357 U.S. 320 (1958). Thanks to Bill Lang for drawing my attention to this case.

17. The following analysis of B.C. politics draws on several excellent contemporary accounts, notably Sherman 1966; McGeer 1972; and Hodgson 1976. Two unpublished studies are also useful: Tieleman 1984 and Wedley 1986. As John English noted (*BC Studies* No. 63 [Autumn 1984]: 73–74), Mitchell 1983 should be read with a good deal of caution, although it provides an excellent account of how Bennett himself felt his actions ought to be interpreted.

18. Although it would be inaccurate to describe Bennett as a Social Credit ideologue, he was very much a fiscal conservative. Those who have sought to explain Bennett's behavior during the treaty negotiations have tended to ignore this point. For a persuasive account that describes the Bennett government's fundamental aversion to debt, see Carlsen 1961, esp. pp. 162–82. In his 1983 biography of Bennett, Mitchell also stresses the significance of this issue, noting that "[d]ebt reduction was the prime economic objective of the Social Credit government during this period [the 1950s]; it was a virtual obsession of the premier" (277; more generally, see 272–84).

19. In 1955, for example, anxiety over the possibility of Canadian diversion led the chair of the Senate Committee on Interior and Insular Affairs to send Senator Richard L. Neuberger on a fact-finding mission to B.C. See the telegram and letter appended to Neuberger's subsequent report (Neuberger 1955, Appendix 1, 36–37). Neuberger would go on to play a significant public role in the treaty negotiations; for an account of his actions, see Wagner 1974 as well as Neuberger 1957,; 1958).

20. See for example, the lengthy discussions in Canada, House of Commons, *Debates*, Vol. IV, 1925 (19 May 1925), pp. 3339–90; House of Commons, *Debates*, Vol. V, 1925 (15 June 1925), pp. 4250–90; House of Commons, *Debates*, Vol. V, 1933 (17 May 1933), pp. 5129–33; and House of Commons, *Debates*, Vol. II, 1955 (25 February 1955), pp. 1523–42. On both the general topic and on the specific issue of Cominco's effort to export Waneta power, note the comments in Grauer 1961; Sherman 1966, 245–6; Allum 1988, 76–81; Froschauer 1999, 72–5.

21. See Swainson 1979, 57–65, Sherman 1966, 211–14. David Mitchell incorrectly asserts that the Kaiser proposal concerned the Mica Dam, north of Revelstoke (Mitchell 1983, 285). As Swainson points out (p. 57), former U.S. Senator Clarence Dill likely played a key role in these discussions, at least during the preliminary stages. Cf. Irish 1994, 346–59.

22. This could well have been intentional. For example, writing in 1961, Carlsen suggested: "It is quite conceivable that the [B.C. government's] negotiations with the Wenner-Gren interests in connection with the Peace [River] were, in part at least,

motivated by a desire to accelerate Canada-U.S. agreement re the Columbia (Carlsen 1961, 146)." Cf. *The Globe and Mail*, 30 January 1959, quoted in Swettenham 1969, 271.

23. I have taken this date from Sherman's 1966 account (214–15); Martin Robin (1973), on the other hand, has the meeting take place two years earlier (207–12).

24. In fact, the federal government felt that the Peace River project required a federal license, although BC Hydro never bothered to apply for one; see Swainson 1979, 217–18, 404.

25. Note for example the comments of Jean Lesage, Minister of Northern Affairs and National Resources, in the House of Commons, 26 March 1957, in *House of Commons Debates*, 2684–85. Cf. Swainson 1979, 70–71.

26. See for example the speech in the House of Commons by Conservative MP Davie Fulton, 2 April 1957, *House of Commons Debates*, 2986–87.

27. Diefenbaker 1976, 156. In this same volume, Diefenbaker reprints the relevant passage from the Throne Speech, 16–17.

28. According to McNaughton, the retirement of Len Jordan as head of the American section of the IJC was key; see Swettenham 1969, 271–75. Other sources substantiate the view that Jordan was seen as an impediment to resolving the issue of downstream benefits; see for example Swainson 1979, 70.

29. Quoted in Canada, Department of External Affairs, *The Columbia River Treaty: Protocol and Related Documents*, 39. Cf. the useful "Columbia River Negotiations—Selected Chronology: 1943–1964," in Canada,Department of External Affairs, *The Columbia River Treaty and Protocol: A Presentation* (Ottawa: Queen's Printer, 1964), 21–25, esp. 23.

30. The "Report of the International Joint Commission on Principles for Determining and Apportioning Benefits from Cooperative Use of Storage of Waters and Electrical Interconnection within the Columbia River System," dated 29 December 1959, is reprinted in Canada,Department of External Affairs, *The Columbia River Treaty: Protocol and Related Documents*, 39–55.

31. Reprinted in Canada, Department of External Affairs, *The Columbia River Treaty: Protocol and Related Documents*, 57. Cf. the comments in Swettenham 1969, 279–99.

32. See the account of the treaty signing in Robinson 1989, 167; the quotation is from Martin 1982, 180. Cf. Diefenbaker's comments in the House of Commons, 18 January 1961, *House of Commons Debates*, 1159.

33. "Abstract of Report to the International Joint Commission on Water Resources of the Columbia River Basin," in *The Columbia River Treaty: Protocol and Related Documents*, 37; earlier the report noted that "each of the three plans achieves about the same degree of water resource development, particularly with respect to hydro-electric power" (ibid., 26).

34. This issue is contentious. Donald Fleming (one of Diefenbaker's cabinet colleagues), records in his memoirs how the two B.C. representatives in the federal government (Davie Fulton and Howard Green) were enraged by Bennett's volte face, which—according to Fleming—they could never forgive, see (Fleming 1985, 259–75). An uncritical biography of Williston, not surprisingly, offers a rather different account: see Willston and Keller 1997, 192–3. Cf. Swainson's comments, (Swainson 1979,180–5; Sherman 1966, 238–42); and the contemporary account by Bruce Hutchison (Hutchinson 1961, 15, 60, 62–3).

35. The Vancouver publisher, Howard Mitchell, quoted in Sherman 1966, 242.

36. See especially *Columbia River Treaty: Hearing before the Committee on Foreign Relations, United States Senate, Eighty-Seventh Congress, First Session on Ex. C*, 87th Congress, 1st Session (8 March 1961). It is clear that committee members were quite happy with

the treaty; there's much congratulatory rhetoric, lauding the treaty as a demonstration of good will between the two nations and so on. There is also discussion of the possibility (or threat) of the Columbia's diversion in Canada, which—with the ratification of the treaty—would no longer be an option. Cf. Swettenham 1969, 299.

37. Useful contemporary accounts include Keate 1961 and Irving 1962, while the premier's somewhat expansive memories of these events may be gleaned from Mitchell 1983, 286–327. Cf. Sherman 1966, 243–51.

38. For an indication of his private anxiety during this period, however, see Mitchell 1983, 306–15.

39. This view was common among senior management within West Kootenay Power; see for example, the comments of M. H. Mason, "Legal Considerations: Cominco and West Kootenay Power," typescript, 1976, pp. 3–4 (copy in West Kootenay Power's Trail office). Similarly the late Jack McDonald—an electrical engineer with a lifetime's experience observing the company's career—repeated this analysis in conversation with me.

40. Quoted in Sherman 1966, 245. Bennett was right about the projects' size, however, which was such that leading American officials seriously doubted that money could be found to finance them both at the same time (Swainson 1979, 262).

41. Quoted in Mitchell 1983, 305. Ray Williston echoed Bonner's comments in an extensive series of interviews conducted in October 1975; see typescript of Interview 1375, Tape 11, Track 1, pp. 15–18, and Tape 11, Track 2, pp. 1–5, held at the B.C. Archives, Victoria.

42. See Newman 1962, 2–3; note also the comments in Robin 1973, 239–40 and Sherman 1966, 260. Swainson provides contrary evidence, although his source is scarcely an impartial one: (Swainson 1979, 316).

43. Sherman 1966, 272. The 1963 provincial election brought into sharp focus the issues surrounding the nationalization of B.C. Electric and the controversy over the Columbia River Treaty. For accounts of the campaign, see Sherman, op. cit., 271–77; (Robin 1973, 240–9, Mitchell 1983, 317–21, Swainson 1979, 259–63).

44. On Bennett's pride in the Columbia treaty—the 1964 version—see the comments in Sherman 1966, 281 and Mitchell 1983, 325. For a very different analysis, from two of the treaty's harshest critics, see Higgins 1973 and McNaughton 1963.

45. A copy of this letter survives in the E. Davie Fulton fonds, MG 32 B11, Vol. 38, 62-35-2 "I.J.C. Columbia River. Negotiations", Library and Archives Canada; it is dated 22 August 1962, and was sent to federal cabinet ministers Davie Fulton and Howard Green, as well as to area M.P. Bert Herridge and the provincial cabinet minister Ray Williston. McDonald also argued in the letter that the Mica Dam would provide far more significant water storage and flood control benefits than the High Arrow, and that the latter dam was planned simply as a stopgap until the Mica was built. For the rest of his life, McDonald remained critical of the High Arrow and published a book on the controversy not long before his death, *Storm over High Arrow: The Columbia River Treaty (A History)* (1993).

46. Richie Deane, "The Columbia Treaty and High Arrow Dam," typescript submission to the External Affairs Committee of the House of Commons, 30 March 1964. For evidence of the effectiveness of Deane's submission, see Swainson 1979, 275. For the protests and views of others in the Kootenays who objected to the treaty's provision of the High Arrow dam, see Waterfield 1970, J. W. Wilson 1973, Wilson and Conn 1983, McDonald 1993, and Loo 2004.

47. Throughout many years, Herridge spoke on Columbia River issues in the House of Commons. His speech on 24 May 1963 is typical of his grasp of the issues, as well as making clear his opposition to the final version of the treaty; see *House of Commons Debates*, 24 May 1963, 258–61. Note also the comments of his biographer, (Hodgson 1976, 203–25).

48. The level of grass roots opposition to the Columbia River Treaty can be gauged by the voluminous collection of pamphlets and other ephemera in the James G. Ripley fonds held at the University of Victoria Archives, Victoria, B.C.

49. Heeney and Merchant 1965. This study (dated 28 June 1965) arose from a meeting between U.S. President Johnson and Prime Minister Pearson in January 1964, where they "discussed at some length the practicability and desirability of working out acceptable principles which would make it easier to avoid divergencies in economic and other policies of interest to each other." (From the joint communiqué dated 22 January 1964, and quoted in Heeney & Merchant, *Canada and the United States*, Annex B, p. 1.) For the report's reception, see Donaghy 2002, 32, 44, and 132; Azzi 1999, 127–8, Keenleyside 1981, 500–2; and A. Heeney 1972, 190–4.

50. These provisions are stipulated in Articles II, IV, XII (4), and XIII (1) of the treaty. The quotation in the text is from Annex A of the treaty, "Principles of Operation," No. 6, "Power." (I have relied on the text of the treaty reprinted in Department of External Affairs and Northern Affairs and National Resources April 1964, where this quotation comes on p. 147.) The Annex goes on to acknowledge that subsequent power development at Mica (in Canada) will alter this provision. Note also the summary of the treaty in the *McGill Law Journal* Vol. 8, No. 3 (1961–1962): 216–18.

51. See the memo from R. G. Robertson, A. E. Ritchie, J. F. Parkinson, and G. M. MacNabb, 22 January 1963, marked "Confidential" and headed "Memorandum for the Minister of Justice," as well as the 32-page report prepared by W. D. Kennedy, G. J. A Kidd, G. M. MacNabb, and P. R. Purcell, dated 11 January 1963 (both documents are in E. Davie Fulton fonds, Vol. 38, 62-35-2 "I.J.C. Columbia River. Negotiations," MG 32 B11). The second document concludes with an Appendix entitled "Memorandum on Implications of Treaty Delay." The purpose of this appended memorandum, its authors explained, was "to discuss the implications which would appear to follow if the proposal or some further modification of it is not accepted by Canada before the Spring of 1963." They went on to explain that "The argument of this memorandum is based in part on the position taken by the U.S. delegation at the meeting of December 1962 when it was stated as follows:

"In January it will be two years since the Treaty was signed. . . . the need for a prompt decision as to whether this Treaty is going to be implemented has become increasingly urgent."

The memorandum concludes that "*It is therefore quite possible that failure to ratify the Columbia River Treaty in the near future would not only involve the loss of the downstream benefits, which are a major and very economic resource in themselves, but could also mean the loss of the immense renewable resource that the Treaty makes economically feasible on the Columbia River in Canada. These two resources amount to approximately 4 million kilowatts of dependable capacity*" [emphasis in the original].

52. (Grauer 1961, 283). Grauer's views were not unique; Davie Fulton made a similar point in a speech in September 1961: "It is axiomatic if we dispose of our cheapest source of power south of the border [by selling the Columbia's downstream benefits] we will have to consume more expensive power at home. . . . The higher cost of power means a higher

cost of production. It means an impaired competitive position. It is a hindrance to the industrial expansion that I have referred to, and that is so necessary to our future prosperity. It also means an obstacle in the way of the diversification of our economy and of job security." (Copy in W. A. C. Bennett Fonds, F-55, Container 58, file F-55-37-0-18, "Power Development (28-9)," held in Simon Fraser University Archives, Burnaby, B.C.)

53. On the reception of Grant (1965), see Azzi 1999, 125–7. On the Waffle, see Bullen 1983. In addition, the journal *Studies in Political Economy* (No. 32 [Summer 1990]: 173–201) has various reflections on the twentieth anniversary of the formation of the Waffle by members of that group (Reg Whitaker, Mel Watkins, John Smart, Rianne Mahon, and Pat Smart).

54. See the reference in *Hansard*, House of Commons Debates, 19 March 1962, 1923; (Froschauer 1999, 212–5, Swainson 1979, 209–10). Note also the argument in Nemeth, 2002, 191–215.

Works Cited

Allum, James Robert. *Science, Government and Politics in the Abolition of the Commission of Conservation, 1909–1921.* M.A. thesis, Trent University, 1988.

American Society of International Law. "The Diversion of Columbia River Waters: Proceedings, Regional Meeting, American Society of International Law." Seattle: Institute of International Affairs, 1956.

Anastakis, Dimitry. *Auto Pact: Creating a Borderless North American Auto Industry, 1960–1971.* Toronto: University of Toronto Press, 2005.

Azzi, Stephen. *Walter Gordon and the Rise of Canadian Nationalism.* Montreal: McGill–Queen's University, 1999.

Beck, J. Murray. *Pendulum of Power: Canada's Federal Elections.* Scarborough, Ontario: Prentice-Hall of Canada, 1968.

Billington, David P., and Donald C. Jackson. *Big Dams of the New Deal Era: A Confluence of Engineering and Politics.* Norman: University of Oklahoma Press, 2006.

Bloomfield, L. M., and Gerald F. Fitzgerald. *Boundary Waters Problems of Canada and the United States (The International Joint Commission 1912-1958).* Toronto: The Carswell Company, Ltd., 1958.

Bourne, Charles B. "Diversion: An International Problem." *Pacific Northwest Quarterly* 49, no. 3 (July 1958): 106–9.

———. "The Columbia River Controversy." *Canadian Bar Review* 37, no. 3 (September 1959): 444–72.

———. "The Columbia River Diversion: The Law Determining Rights of Injured Parties." *U.B.C. Legal Notes* 2 (1958): 610–22.

———. "The Development of the International Water Resources: The 'Drainage Basin Approach'." *The Canadian Bar Review* 47, no. 1 (March 1969): 82.

Bullen, John. "The Ontario Waffle and the Struggle for an Independent Socialist Canada: Conflict within the NDP." *Canadian Historical Review* 64, no. 2 (June 1983): 188–215.

Burton, Thomas L. *Natural Resource Policy in Canada: Issues and Perspectives.* Toronto: McClelland and Stewart, 1972.

Carlsen, Alfred Edgar. *Major Developments in Public Finance in British Columbia, 1920 to 1960.* Ph.D. thesis, University of Toronto, 1961.

Clark-Jones, Melissa. *A Staple State: Canadian Industrial Resources in Cold War.* Toronto: University of Toronto Press, 1987.

Cohen, Maxwell. "Some Legal and Policy Aspects of the Columbia River Dispute." *Canadian Bar Review* 36, no. 1 (March 1958): 25–41.

Department of External Affairs and Northern Affairs and National Resources. *The Columbia River Treaty and Protocol—A Presentation.* Ottawa: Queen's Printer, April 1964.

Diefenbaker, John G. *One Canada: Memoirs of the Honourable John G. Diefenbaker, Vol. 2, The Years of Achievement, 1957–1962.* Toronto: Macmillan of Canada, 1976.

Donaghy, Greg. *Tolerant Allies: Canada and the United States, 1963–1968.* Montreal: McGill–Queen's University Press, 2002.

Dulles, Allen. "Memorandum From Secretary of State Dulles to President Eisenhower," 3 July 1958, reprinted in United States, Department of State, *Foreign Relations of the United States, 1958–1960, Western European Integration and Security, Canada,* Vol. VII, Part 1. Washington, D.C.: U.S. Government Printing Office, 1958–1960.

Ferguson, G V. "Likely Trends in Canadian-American Political Relations." *Canadian Journal of Economics and Political Science* 22, no. 4. November 1956.

Ficken, Robert E., and Rufus Woods. *The Columbia River, and the Building of Modern Washington.* Pullman, Wash.: Washington State University Press, 1995.

Fleming, Donald. *So Very Near: The Political Memoirs of the Hon. Donald M. Fleming.* Toronto: McClelland and Stewart, 1985.

Froschauer, Karl. *White Gold: Hydroelectric Power in Canada.* Vancouver, B.C.: University of British Columbia Press, 1999.

Gloin, Kevin J. "Canada-U.S. Relations in the Diefenbaker Era: Another Look." In *The Diefenbaker Legacy: Canadian Politics, Law and Society since 1957*, edited by Donald C. Story and R. Bruce Shepard, 1–14. Regina: Canada Plains Research Centre, 1998.

Gordon, Walter. *A Political Memoir.* Toronto: McClelland and Stewart, 1977.

Granatstein, J. L. "Mackenzie King and Canada at Ogdensberg, August 1940." In *Fifty Years of Canada–United States Defense Cooperation: The Road from Ogdensberg*, edited by Joel L. Sokolsky and Joseph T. Jockel, 9–29. Lewiston, Idaho: Edwin Mellen Press, 1992.

Granatstein, J. L., and R. D. Cuff. "The Hyde Park Declaration 1941: Origins and Significance." *Canadian Historical Review* 55, no. 1 (March 1974): 59–80.

Grant, George. *Lament for a Nation: The Defeat of Canadian Nationalism.* Princeton: Van Nostrand, 1965.

Grauer, A. E. Dal. "The Export of Electricity from Canada." In *Canadian Issues: Essays in Honour of Henry F. Angus*, edited by R. M. Clark, 248–85. Toronto: University of Toronto, 1961.

Hamlet, A. "The Role of Transboundary Agreements in the Columbia River Basin: An Integrated Assessment in the Context of Historic Development, Climate, and Evolving Water Policy." In *Climate and Water: Transboundary Challenges in the Americas*, edited by H. Diaz and B. Morehouse, 263–90. Dordrecht: Kluwer Academic Publishers, 2003.

Haydon, Peter T. *The 1962 Cuban Missile Crisis: Canadian Involvement Reconsidered.* Toronto: The Canadian Institute of Strategic Studies, 1993.

Heeney, A. D. P., and Livingston T. Merchant. *Canada and the United States: Principles for Partnership.* Ottawa: R. Duhamel, Queen's Printer, 1965.

Heeney, Arnold. *The Things that Are Caesar's: Memoirs of a Canadian Public Servant.* Edited by Brian D. Heeney. Toronto: University of Toronto Press, 1972.

Higgins, Larratt. "The Alienation of Canadian Resources: The Case of the Columbia River Treaty." In *Close to the 49th Parallel Etc,*, edited by Ian Lumsden, 223–40. Toronto: University of Toronto Press, 1973.

Hodgson, Maurice. *The Squire of Kootenay West: A Biography of Bert Herridge*. Saanichton, B.C.: Hancock House, 1976.

Hutchinson, Bruce. "The Great Columbia River Foul-Up." *MacLean's*, 74, no. 11 (June 3, 1961): 15, 60, 62–3.

Igartua, Jose E. *The Other Quiet Revolution: National Identities in English Canada, 1945–71*. Vancouver, B.C.: University of British Columbia Press, 2006.

International Columbia River Engineering Board, *Water Resources of the Columbia River Basin: Report to the International Joint Commission,* seven volumes. Ottawa: The Board. 1959.

Irish, Kerry E. *Clarence Dill: The Life of a Western Politician*. Ph.D. thesis, University of Washington, 1994.

Irving, John A. "Bennett's Design for BC Progress." *Saturday Night*, 77 (May 26, 1962): 18–24.

Jamieson, Stuart. "Power in B.C." *Canadian Forum* 41, no. 496 (1962): 35–6.

Johnson, Ralph W. "The Canada–United States Controversy over the Columbia River." *Washington Law Review* 41 (1966): 711–12.

Keate, Stuart. "The Smile of the Tiger: Why Bennett Took Over B.C. Power." *Saturday Night*, 76 (September 16, 1961): 11–14.

Keenleyside, Hugh L. *Memoirs of Hugh L. Keenleyside, Vol. 1——Hammer the Golden Day*. Toronto: McClelland and Stewart Ltd., 1981.

Kilbourn, William. *Pipeline: TransCanada and the Great Debate: A History of Business and Politics*. Toronto: Clarke, Irwin & Co, 1970.

Krutilla, John V. *The Columbia River Treaty: The Economics of an International River Basin Development*. Baltimore: Johns Hopkins Press, 1967.

Lang, William L. "What Has Happened to the Columbia? A Great River's Fate in the Twentieth Century." In *Great River of the West: Essays on the Columbia River*, edited by William L. Lang and Robert C. Carriker, 144–67. Seattle: University of Washington Press, 1999.

Litt, Paul. "The Massey Commission, Americanization, and Canadian Cultural Nationalism." *Queen's Quarterly* 98, no. 2 (Summer 1991): 375–87.

———. *The Muses, The Masses, and the Massey Commission*. Toronto: University of Toronto Press, 1992.

Loo, Tina. "People in the Way: Modernity, Environment, and Society in the Arrow Lakes." *BC Studies*, no. 142/143 (Summer/Autumn 2004): 161–96.

Lyon, Peyton V. "The Loyalties of E Herbert Norman." *Labour/Le Travail*, no. 28 (Fall 1991): 219–59.

Martin, Charles E. "The Diversion of Columbia River Waters." *Proceedings of the American Society of International Law* 51 (1957): 2–8.

Martin, Lawrence. *The Presidents and the Prime Ministers: Washington and Ottawa Face to Face: The Myth of Bilateral Bliss, 1867–1982*. Toronto: Doubleday Canada, 1982.

Martin, Paul. *A Very Public Life, Vol. 2—So Many Worlds*. Toronto: Deneau, 1985.

Massell, David. " 'As though there was no boundary': The Shipshaw Project and Continental Integration." *American Review of Canadian Studies* 34, no. 2 (Summer 2004): 187–222.

McDonald, Jack. *Storm over High Arrow: The Columbia River Treaty (A History)*. Rossland, B.C.: privately printed, 1993.

McDougall, John N. *Fuels and the National Policy*. Toronto: Butterworths, 1982.

McyGeer, Pat. *Politics in Paradise.* Toronto: Peter Martin Associates, 1972.

McKinley, Charles. *Uncle Sam in the Pacific Northwest: Federal Management of Natural Resources in the Columbia River Valley.* Berkeley: University of California Press, 1952.

McNaughton, A. G. L. "The Proposed Columbia River Treaty." *International Journal* 18, no. 2 (Spring 1963): 148–65.

Meisel, John. "The Formulation of Liberal and Conservative Programmes in the 1957 Canadian General Election." *Canadian Journal of Economics and Political Science* 26, no. 4 (November 1960): 570–72.

Messamore, Barbara J. "Diplomacy or Duplicity? Lord Lisgar, John A Macdonald, and the Treaty of Washington, 1871." *Journal of Imperial and Commonwealth History* 32, no. 2 (May 2004): 29–53.

Mitchell, David. *W. A. C. Bennett and the Rise of British Columbia.* Vancouver: Douglas & McIntyre, 1983.

Muirhead, Bruce. *Dancing Around the Elephant: Creating a Prosperous Canada in an Era of American Dominance, 1957–1973.* Toronto: University of Toronto Press, 2007.

Myers, Phillip E. *Mask of Indifference: Great Britain's North American Policy and the Path to the Treaty of Washington.* Ph.D. thesis, University of Iowa, 1978.

Nelles, H.V. *The Politics of Development: Forests, Mines and Hydro-Electric Power in Ontario, 1849–1941.* Toronto: Macmillan of Canada, 1974.

Nemeth, Tammy. "Consolidating the Continental Drift: American Influence on Diefenbaker's National Oil Policy." *Journal of the Canadian Historical Association, New Series* 13 (2002): 191–215.

Nemeth, Tammy Lynn. *Canada–United States Oil and Gas Relations, 1958–1974.* Ph.D. thesis, University of British Columbia, 2007.

Neuberger, Richard L. "Power Struggle on the Canadian Border." *Harper's Magazine,* December 1957: 42-9.

———. *Study of Development of Upper Columbia River Basin, Canada and United States: Report to the Chairman of the Senate Committe on Interior and Insular Affairs Submitted by Senator L. Neuberger of Oregon.* Washington, D.C.: Government Printer, 1955.

———. "Sternest Crisis in 111 Years." *The Star Weekly Magazine,* March 29, 1958: 2–4.

Newman, Peter C. "Backstage in Ottawa: Who's where and why in the great Columbia River Debate." *MacLean's* 75, no. 23 (November 17, 1962): 2–3.

———. *Renegade in Power: The Diefenbaker Years.* Toronto, 1963.

Niosi, Jorge. *Canadian Capitalism.* Toronto: James Lorimer and Co., 1981.

Norwood, Gus. *Columbia River Power for the People: A History of the Policies of the Bonneville Power Administration.* Portland, Ore: Bonneville Power Adminstration, 1980.

Perras, Galen Roger. *Franklin Roosevelt and the Origins of the Canadian-American Security Alliance, 1933–1945: Necessary, but Not Necessary Enough.* Westport, Conn.: Praeger, 1998.

Pitzer, Paul C. *Grand Coulee: Harnessing a Dream.* Pullman: Washington State University Press, 1994.

Purdy, Al, ed. *The New Romans: Candid Canadian Opinions of the U.S.* Edmonton: M. G. Hurtig, 1968.

Richards, John, and Larry Pratt. *Prairie Capitalism: Power and Influence in the New West.* Toronto: McClelland and Stewart Ltd., 1979.

Robin, Martin. *Pillars of Profit: The Company Province, 1934–1972.* Toronto: McClelland and Stewart, 1973.

Robinson, H. Basil. *Diefenbaker's World: A Populist in Foreign Affairs.* Toronto: University of Toronto Press, 1989.

Sherman, Paddy. *Bennett*. Toronto: McClelland and Stewart, 1966.

Smith, Goldwin. *The Treaty of Washington, 1871: A Study in Imperial History*. Ithaca, N.Y.: Cornell University Press, 1941.

Spencer, Robert, John Kirton, and Kim Richard Nossal. *The International Joint Commission Seventy Years On*. Toronto: University of Toronto, 1981.

Stacey, C. P. "Britain's Withdrawal from North America, 1864–1871." *Canadian Historical Review* 36 (September 1955): 185–98.

Stursberg, Peter. *Diefenbaker: Leadership Gained, 1956–62*. Toronto: University of Toronto Press, 1975.

Swainson, Neil A. *Conflict Over the Columbia: The Canadian Background to an Historic Treaty*. Montreal: McGill–Queens University Press, 1979.

———. "The Columbia River Treaty—Where Do We Go From Here?" *Natural Resources Journal* 26, no. 2 (Spring 1986): 243–46.

Swettenham, John. *McNaughton, Vol. 3, 1944–1966*. Toronto: Ryerson Press, 1969.

Thompson, Robert N. *The House of Minorities*. Burlington, Ontario: Welch Pub. Co., 1990.

Tieleman, H. William. *The Poltical Economy of Nationaliation: Social Credit and the Takeover of the British Columbia Electric Company*. M.A. thesis, University of British Columbia, 1984.

Tomblin, Stephen G. "W. A. C. Bennett and Province-Building in British Columbia." *BC Studies*, no. 85 (Spring 1990): 45–61.

United States of America. *Columbia River and Minor Tributaries*, 73rd Congress, 1st Session—House Document No. 103, Vols. I & II. Washington: Government Printing Office. 1934.

Voeltz, Herman C. "Genesis and Development of a Regional Power Agency in the Pacific Northwest." *Pacific Northwest Quarterly* 53, no. 2 (April 1962): 65–72.

Wagner, J. Richard. "Congress and United States–Canada Water Problems: Senator Neuberger and the Columbia River Treaty." *Rocky Mountain Social Science Journal* 11, no. 3 (October 1974): 51–60.

Waterfield, Donald Cresswell. *Continental Waterboy: The Columbia River Controversy*. Toronto: Clarke, Irwin, 1970.

Watkins, Ernest. "The Columbia River: A Gordian Knot." *International Journal* 12, no. 4 (Autumn 1957): 250–61.

Wedley, John R. *Infrastructure and Resources: Governments and their Promotion of Northern Development in British Columbia, 1945–1975*. Ph.D. thesis, University of Western Ontario, 1986.

Whitaker, Reg, and Gary Marcuse. *Cold War Canada: The Making of a National Insecurity State, 1945–1957*. Toronto: University of Toronto Press, 1994.

White, Gilbert F. "A Perspective of River Basin Development." *Law and Contemporary Problems*, Spring 1957: 157–87.

White, Richard. *The Organic Machine: The Remaking of the Columbia River*. New York: Hill and Wang, 1995.

Willston, Eileen, and Betty Keller. *Forests, Power, and Policy: The Legacy of Ray Williston*. Prince George, B.C.: Caitlin Press, 1997.

Wilson, J. W., and Maureen Conn. "On Uprooting and Rerooting: Reflections on the Columbia River Project." *BC Studies*, no. 58 (1983): 40–54.

Wilson, James Wood. *People in the Way: The Human Aspects of the Columbia River Project*. Toronto: University of Toronto Press, 1973.

Worley, Ronald B. *The Wonderful World of W. A. C. Bennett*. Toronto: McClelland and Stewart, 1971.

The Columbia River Treaty: Managing for Uncertainty

James D. Barton and Kelvin Ketchum

Overview

The Columbia River Treaty is an international agreement established for the cooperative management of the Columbia River system in Canada and the United States. Since it was signed in 1961, the treaty has served as an excellent example of transboundary cooperation and governance in the face of uncertainty. This chapter will provide a brief background on the treaty, describe the many aspects of uncertainty involved in managing and governing the operation of the Columbia River system, and highlight examples of how the treaty has successfully facilitated the management of this uncertainty.

Treaty Background

The Columbia River, the fourth largest river on the continent as measured by average annual flow, generates more hydropower than any other river in North America. While its headwaters originate in British Columbia (B.C.), only about 15 percent of the 259,500 square miles of the Columbia River Basin is actually located in Canada. Yet the Canadian waters account for about 38 percent of the average annual volume, and up to 50 percent of the peak floodwaters that flow by The Dalles Dam on the Columbia River between Oregon and Washington. In the 1940s, officials from the United States and Canada began a long process to seek a joint solution to the flooding caused by the Columbia River and to the postwar demand for greater energy resources. That effort culminated in the Columbia River Treaty, an international agreement between Canada and the United States of America for the cooperative development of water resources regulation in the upper Columbia River Basin. It was signed in 1961 and implemented in 1964. The treaty has served as a model of international cooperation since 1964, bringing significant flood damage reduction and power generation benefits to both countries.

One of the major reasons for the treaty was to improve flood risk management in the Columbia River Basin. In 1948, a spring flood caused major damage all along the river and its tributaries in both countries, particularly to the cities of Trail, B.C. and Vanport, Oregon. Vanport, the second largest city in Oregon at that time, was completely destroyed. The flood displaced thirty thousand people from their homes and caused more than fifty deaths. The magnitude of the flood event served as a trigger for action and added a sense of urgency to international discussions of flood control. The United States and Canada collaborated to identify a preferred method—a coordinated development plan—that would address Columbia River Basin flooding and meet the region's increasing demands for energy. The treaty created two entities to implement the Treaty—a U.S. Entity and a Canadian Entity. The U.S. Entity, created by the President, consists of the Administrator of the Bonneville Power Administration (BPA) and the Northwestern Division Engineer of the U.S. Army Corps of Engineers. The BPA Administrator serves as chair of the U.S. Entity and is responsible primarily for hydropower operations and issues under the treaty while the Corps Northwestern Division Engineer is responsible for flood risk management under the treaty. The Canadian Entity, appointed by the Canadian Federal Cabinet, is the British Columbia Hydro and Power Authority (BC Hydro).

A main component of the treaty called for Canada to develop reservoirs in the upstream reaches of the Columbia River Basin sufficient to provide 15.5 million acre-feet of live storage. To do this, Canada built three dams: Duncan (1968), Hugh Keenleyside, also referred to as Arrow (1969), and Mica (1973). The treaty also allowed the United States to build Libby Dam on the Kootenai River, a tributary of the Columbia River, in Montana. Construction on the Libby Dam began in 1966 and was completed in 1973; the reservoir impounded by the dam is Lake Koocanusa, which backs up forty-two miles into Canada. Together, these four treaty dams more than doubled the storage capacity of the Columbia River Basin at the time.

The treaty also requires the United States and Canada to prepare annually an Assured Operating Plan for the operation of Canadian treaty storage six years in advance of each operating year. The Assured Operating Plan is developed to meet flood control and power objectives, the only recognized purposes for project operation when the treaty was signed. (Note that consumptive uses, such as drinking water and irrigation, are not constrained by the treaty.) Sharing the benefits of cooperative water management was an integral part of the treaty's design. The principle applied in the treaty was to share these benefits equally. Thus, for flood control, Canada was paid 50 percent of the estimated value of U.S. flood damages prevented through September 2024. In exchange for providing and operating the treaty storage projects for power, Canada also received an entitlement to one-half of the estimated additional downstream power benefits generated in the United States. Canada initially sold its share of this power to a group of U.S. utilities for a period of thirty years, an agreement that expired in 2003, after which the Canadian Entitlement power from downstream benefits was fully delivered to the Province of B.C. The initial lump-sum payments—from the U.S. government for flood control and from the U.S. utilities for downstream power benefits—helped to fund the construction of the three treaty dams in Canada.

Sources of Uncertainty in the Columbia River System
There are multiple sources of uncertainty involved in managing the Columbia River system. These may be related to the hydrology, climatology, or technical aspects of the system, or environmental aspects, socio/political aspects, or economic aspects, among others. From a hydrologic perspective, there is significant uncertainty and variability in runoff and streamflows, for example. The natural streamflows in the Columbia River at the Canada-U.S. border can range from 14,000 cubic feet per second (cfs) to 555,000 cfs. The variation in the seasonal and annual inflows is much larger than for many other major river systems such as the Mississippi or St. Lawrence. While some of these runoff variations on the Columbia can be predicted in advance based on snowpack measurements, there remains a substantial amount of uncertainty in both the volume and timing of the runoff. The Columbia system also has relatively limited storage available to manage these variations in streamflow compared to other major river systems in North America. For example, on the Columbia River system the total reservoir storage volume is about 20 percent of the average annual runoff volume. In comparison, on the Colorado River system the reservoir storage volume is about 300 percent of the average annual runoff and on the Missouri River system the reservoir storage volume is about 200 percent of the average annual runoff.

There is also significant uncertainty in terms of climatology. Hydrologists continue to study a number of river basin, ocean, and atmospheric parameters in order to improve the forecast accuracy for seasonal runoff volumes. One such parameter is the Southern Oscillation Index (SOI), an indicator of El Niño or La Niña conditions; the SOI has been used, with some success, to improve runoff predictions in parts of the Columbia River basin. The specter of climate change, and its impacts on snowpack and runoff conditions throughout the basin, also looms large. Scientists agree that the uncertainty around climate-change predictions is substantial.

From a technical perspective, there is uncertainty as far as the conditions of the aging infrastructure in place to manage and monitor the river. The aging infrastructure provides challenges such as potential outages and unavailability of spillway gates, power generating units, and other infrastructure. For example, the power plant at Mica Dam is now over thirty years old, and major rehabilitation of the generating units is now necessary and underway. Many power plants and other infrastructure in the U.S. are also undergoing rehabilitation.

Given the size of the basin there are a relatively limited number of data-collection sites to provide information such as precipitation, snowpack, temperature, and reservoir inflow. Government cutbacks and observer retirements have had an impact on the density of data- collection sites that affects the ability to monitor and manage the river. In addition, automated stations are gradually replacing many manual weather stations. While automated stations have the advantage of more frequent reporting, data verification for automated stations occurs less frequently and data continuity suffers due to longer station outages.

Since the time that the treaty was signed there have been significant changes in terms of environmental aspects. For example, many different species of fish and wildlife are now listed under the U.S. Endangered Species Act as threatened or endangered. In Canada, special efforts have been taken since the early 1990s to improve the spawning flow regime for several fish species. In addition, many water quality requirements have been added since the signing of the treaty. This increasing emphasis on environmental sustainability has resulted in significant changes and uncertainty in planning and operating the various reservoirs in the Columbia River system.

The socio-political landscape in the two countries has also changed considerably over time. For example, when the treaty was originally signed the primary water management focus was on consumptive uses, flood risk management, and power generation. Over time, as social and political interests have changed, there has been more emphasis on other areas such as recreation (water- and land-based), environmental protection, navigation, and others. The recent emphasis on renewable energy and energy self-sufficiency is another good example of the changing socio-political perspective. Stakeholders in the Columbia basin are much better informed now than when the treaty was signed, and stakeholder engagement is a key aspect of water management planning and operations throughout the basin. B.C.'s Water Use Planning (WUP) process is a good example of stakeholder engagement for the Canadian reservoirs. Similar processes are in place in the U.S., such as the Columbia River System Operation Review (SOR) conducted in the 1990s and the recently signed Columbia Basin Fish Accords.

There is also a great deal of uncertainty and variability from an economic and market perspective. Electrical power demand is inextricably linked to overall economic well-being in both countries. Higher oil and gas prices increase the cost of energy from thermal power plants, resulting in increased demand for power from renewable sources such as hydroelectric plants. A significant increase in wind generation in the U.S. also has implications for the operation of the hydro system. Recent world events have demonstrated how quickly economic conditions can change.

There are many other sources of uncertainty that could be highlighted; but, as these example demonstrate, management of uncertainty is a key consideration in operating the Columbia River system. The treaty was ratified nearly forty-five years ago, and all of the treaty reservoirs have now operated for over thirty-five years. During this time, Columbia system operations have been planned and managed under treaty principles, and the treaty has provided a successful framework for adapting to the various sources of uncertainty. The following section will describe some specific examples of how the treaty has been effective in facilitating change and adapting to uncertainty.

How the Treaty Has Helped Manage Uncertainty

The treaty has provided a governance structure to manage many different types of uncertainty. It involves a robust process for planning project and system operations well in advance that reserves the flexibility to readily adapt to changes that may occur once plans have been developed. One of the basic tenets of the treaty is to allow for adequate advance planning, ensuring that each country has the ability to adapt to the required changes. For example, each year an Assured Operating Plan (AOP) is prepared for the operating year six years in the future. This is done to allow each country to better understand the capabilities of the Columbia system under the treaty and to provide enough time for advance planning to make any changes that may be needed, e.g., adding new power generating resources. In addition to the AOP that is prepared six years in advance, a Detailed Operating Plan (DOP) is prepared each year, for the succeeding operating year, to allow for any updates that may have occurred since the AOP was prepared. During the actual operating year, the Treaty Storage Regulation (TSR) study, based on the DOP as well as current hydrologic conditions, defines the entities' storage and draft rights for Canadian treaty reservoirs. In some cases, the entities may choose to agree on a Supplemental Operating Agreement (SOA) to vary from TSR operation when benefits (e.g., fisheries, power, or other) to both countries are expected. There are also weekly and more frequent coordination calls as needed to adapt to changing conditions.

These planning and coordination processes that are built into the treaty provide the tools and flexibility to adapt to changing conditions. In terms of hydrologic and climatological uncertainty, the treaty has worked well in adapting to changing conditions and requirements. During the period since treaty ratification, there has been a wide range of different hydrologic conditions and the system has been operated effectively to adapt to these changing conditions. For example, in 1996–97, there was a significant high water event that would have resulted in major damage in both countries without treaty reservoir regulation. In that year, operation of treaty storage reduced the peak Columbia River discharge at Trail by about 50 percent. On the U.S. side, the Corps has estimated flood damage savings of approximately $200 million for that event. Conversely, there have been some periods of low runoff (e.g., 1992–94, 2001) and the treaty principles have provided the flexibility to adapt well to those as well.

The treaty has also allowed for adaptation to technical uncertainty. Under the treaty there is a Hydrometeorological Committee that monitors data-collection requirements, inflow forecasting procedures, and related information. Since the ratification of the treaty there have been many changes in the technological and data-collection requirements associated with the operation of the river system. The treaty has had the flexibility to readily adapt to these changes. For example, runoff forecasting models are updated on a regular basis and new techniques are implemented when value is demonstrated. There have also been issues related to dam safety at various projects in the system–the treaty planning and coordination process has facilitated international cooperation on these and other technical issues.

There are numerous examples where significant changes have occurred in the environmental/social operating procedures and requirements for the Columbia River system, both in Canada and the U.S. For example, in the U.S. there are thirteen anadromous fish species that have been listed as threatened or endangered since the treaty was signed. There are also many species of fish in Canada, such as rainbow trout and mountain whitefish, which have particular operating requirements. Through the flexibility inherent in the treaty, both the U.S. and Canada have been able to adapt river operations to meet the requirements of the various species. In addition, both countries have been able to use unilateral operating flexibility and/ or SOAs in order to improve recreational values for stakeholders. Finally, many different water quality and other related requirements have been met through international planning and coordination processes related to the treaty.

From an economic perspective, the treaty has also facilitated significant economic development in both countries. Power plants at, and/or benefitting from, treaty dams in Canada have generating capacity of about 4,500 MW. In the U.S. the generating capacity at Libby and on the main stem Columbia dams totals about 21,000 MW. Socio-political and economic conditions have changed considerably since the time that the treaty was ratified in 1964, but the treaty has allowed for these changes. Forecasts of domestic electrical demand are included as a direct input into the treaty studies. Electricity market prices are not used directly in treaty studies; however, the two entities have implemented SOAs at various times to modify treaty obligations and create additional energy value for both countries.

Summary

The Columbia River Treaty is a good example of international cooperation on water resources development and management between two nations. It has been studied by many international scholars and practitioners interested in transboundary water management. This is particularly impressive given the complex nature of the Columbia River Basin. The system is comprised of over one hundred water resource projects operated by many different public and private entities. It is operated as a coordinated system in terms of power planning and flood risk management. Although many changes have occurred in the goals, objectives, and requirements placed upon this complex river system since the treaty was signed in 1964, the treaty has the flexibility to readily adapt to changes and uncertainty. As part of its basic structure, the treaty provides a robust governance structure that incorporates the changing needs and demands of the broad spectrum of stakeholders and users throughout the system who depend on the many different outputs of this river system. Based on the strong record of success over the past forty-five years of experience in operating within the treaty framework, we believe that the treaty principles can continue to provide the right foundation needed to build upon the future of water resources development in the Columbia River Basin.

The Columbia River: Operation under the 1964 Treaty

Anthony G. White

The Columbia River system is a massive system that affects one province in Canada and seven states in the U.S. Because of the importance of this river—in supplying hydroelectric production, navigation and irrigation, as well as the fact that the system crosses an international boundary—proper coordination of all dams in the system is vital. The state, federal, and provincial organizations involved in and affected by the Columbia River system have worked to develop a coordinated management plan to ensure all uses of Columbia River dams—hydroelectric generation, flood control, irrigation, recreation, tribal culture, and fish and wildlife protection—are adequately met.

The Columbia River System

The Columbia is the fourth largest river in North America when measured by average annual flow (about 198 million acre-feet at its mouth). However, when measured using flow multiplied by change in elevation, the Columbia is the most powerful river in North America because much of the river flows from Rocky Mountain lakes and tributaries. Only 15 percent of the 259,500 square miles of the Columbia River Basin is in Canada, but between 30 and 38 percent of the average annual flow, and up to 50 percent of the peak flood flows, come from Canada.

The basin drains central British Columbia, as well as western Montana, Idaho, large portions of Oregon and Washington, and smaller areas of Nevada, Utah, and Wyoming. The Columbia River is characterized by heavy, sustained flows during the spring, with peak flow typically occurring in mid-June. Historical flows at the U.S.-Canada border vary from 14,000 cubic feet per second (cfs) to more than 550,000 cfs. Average flow at the U.S. dams ranges from 109,500 cfs at 6,809-MW Grand Coulee, the most upstream U.S. project, to 184,900 cfs at 1,089-MW Bonneville, the project furthest downstream.

Before 1968, the total usable system storage in the Columbia River Basin was only about 13 million acre-feet, or 6 percent of the average annual flow (perhaps as much as 18 million-acre-feet if all irrigation, stock pond, and private impoundments were tallied). By comparison, other major U.S. rivers—such as the Missouri and Colorado—have two to three times as much storage as average annual flow. This lack of storage caused significant flood control and hydropower management problems at the eleven mainstem federal and municipal dams on the Columbia River in the U.S. (Bonneville Power Administration 2001).

The 1964 Columbia River Treaty is the culmination of centuries of diplomacy between the United States and Canada (or Great Britain). A brief overview of the treaties preceding it sets the stage for the geographic boundary between the two countries that would divide the Columbia River Basin.

Treaty of Paris (1783). This treaty was the formal end of the Revolutionary War and recognition by Great Britain of the thirteen colonies to be free, separate, and independent states. Initial boundaries based upon very inaccurate surveys and maps of North America were set forth, and led to later confusion (Treaty of Paris 1783).

Anglo-American Convention of 1818. Also called the Treaty of 1818, this treaty extended the boundary between the U.S. and British North America from the Lake of the Woods, along the 49th Parallel (latitude) to the Stony (later "Rocky") Mountains. It resolved long-standing boundary issues, settled fishing rights disputes, allowed joint settlement of the Oregon Territory (the Columbia District of the Hudson's Bay Company) while deferring resolution of legal and territorial issues, and gave territory around the Red River to the U.S. while ceding the northern-most tip of the Louisiana Purchase (now in Alberta) to Britain (Anglo-American Convention 1818).

The Oregon Treaty (1846). Some sixty-seven years after the creation of the United States, Americans had steadily colonized the regions west of the Mississippi River. President James Polk's offer to extend the 49th Parallel through the Oregon Territory was rejected by Great Britain, leading political expansionists of the "manifest destiny" school to reach out for annexation of the area up to the northern boundary of 54 degrees, 40 minutes North ("54-40 or fight"), which was also the southern boundary of Russian America, and would have included the entire Columbia Basin (Pletcher 1973).

Secretary of State (later President) James Buchanan negotiated and concluded this treaty shortly after the outbreak of the Mexican-American War. It was this treaty that split the Columbia Basin, leaving about 15 percent of the basin area in what was to become Canada and British Columbia (The Oregon Treaty 1846).

The Boundary Waters Treaty of 1909. In the interests of providing a way to prevent and resolve disputes between the U.S. and Canada, in 1909 a treaty was signed between the U.S. and Great Britain to establish principles and mechanisms to help resolve disputes and to prevent future ones, primarily those concerning water quantity, water quality, and navigation along the boundary between Canada and the United States. "Boundary waters" in this treaty are defined as the waters from main shore to main shore of the lakes and rivers and connecting waterways, or the portions thereof, along which the international boundary between the

United States and the Dominion of Canada passes, including all bays, arms, and inlets thereof (excluding tributary waters or the waters of rivers flowing across the boundary).

Perhaps more importantly, this treaty established the International Joint Commission (IJC), a group of three Americans and three Canadians to address and resolve water boundary disputes, slated to play a major role in creation of the Columbia River Treaty.

In the Columbia Basin, the IJC at the request of Canadian and U.S. governments created in 1938 a Kootenay Lake Board of Control to manage an IJC Order controlling lake levels and protecting agricultural lands both upstream and downstream of the lake. While Grand Coulee dam was being finished, the IJC in 1941 created a Columbia River Board of Control to monitor Lake F.D. Roosevelt elevations (behind Grand Coulee), which under some conditions could back Columbia River water up into Canada at the border one to two feet.

The IJC in 1946 issued an Order creating the Osoyoos Lake Board of Control to monitor reconstruction and operations of the Zosel Dam, originally built as a log storage pond but now used for recreation, irrigation, and domestic water supply (Boundary Waters Treaty 1909).

Finally the Statute of Westminster (1931) established a level of equality between the Commonwealth's self-governing nations and Great Britain, and henceforth allowed Canada to execute its own treaties without requiring approval from London.

The Columbia River Treaty (1964)

When the Grand Coulee power plant came into service in 1942, there was little immediate demand for the electricity. But the low-cost power available in the Northwest was soon used to attract manufacturing industries supporting the World War II effort, including the creation of support communities such as Vanport, Oregon, just north of Portland, which housed shipyard workers. Aluminum smelters and other heavy industries were also attracted to the Pacific Northwest to aid in wartime production.

The economy and population of the Pacific Northwest grew rapidly during the postwar period. This growth spurred the construction of a number of federal and nonfederal dams on the Columbia mainstem and its tributaries.

However, these new projects did not and could not adequately address flooding concerns since they provided little storage. In 1944, the United States and Canada had asked the IJC to investigate development of Columbia Basin water resources in Canada. The Commission established the International Columbia River Engineering Board to conduct technical studies in the basin, an effort that proceeded comparatively slowly until 1948.

In May of 1948, a spring flood caused major damage from Trail, British Columbia, to Vanport, Oregon. Vanport, the second largest city in Oregon at that time, was completely destroyed. The flood displaced 30,000 people from their homes and caused an estimated more than fifty deaths. The magnitude of the flood event served as a trigger for action and added a sense of urgency to international discussions of flood control. The United States and Canada collaborated to identify a preferred method—a coordinated development plan—that would address Columbia River Basin flooding and meet the region's increasing demands for energy.

U.S. President Eisenhower and Canadian Prime Minister Diefenbaker signed the Columbia River Treaty (CR Treaty) in 1961 to complete three large storage dams in Canada between 1968 and 1973, and to permit the building of Libby Dam if the U.S. so desired. The treaty was finally ratified by the countries in 1964, and came into force for an indefinite period of time (but a minimum of sixty years) on September 16, 1964.

Tendrils of the Boundary Waters Treaty remain in the Columbia River Treaty. The operation of the Kootenay River, with respect to Libby Dam, must comply with an IJC order from the 1930s governing Koortenay Lake. Either country may (but are not required to) use the IJC as a dispute resolution mechanism, whereas Article XVII seems to state that the Boundary Waters Treaty (and therefore the IJC) has no direct jurisdiction in the Columbia Basin as long as the Columbia Treaty remains in effect: "Upon termination of this [CR] Treaty, the Boundary Waters Treaty, 1909, shall, if it has not been terminated, apply to the Columbia River Basin, except insofar as the provisions of that [Boundary Waters] Treaty may be inconsistent with any provisions of this [CR] Treaty which continues in effect" (Columbia River Treaty 1964).

Operation under the Treaty
A U.S. Entity and a Canadian Entity are required by the treaty to be appointed to carry out the treaty's duties and obligations. The U.S. Entity, created by the president, consists of the administrator of the Bonneville Power Administration (chair) and the Northwestern Division engineer of the U.S. Army Corps of Engineers (member). The Canadian Entity, appointed by the Canadian Federal Cabinet, is the utility British Columbia Hydro and Power Authority (BC Hydro).

Each Entity appoints personnel to monitor and review policy matters (coordinators), administrative and negotiation issues (secretaries), and operations—the Columbia River Treaty Operating Committee, consisting of U.S. and Canadian sections of four persons each. Each Entity also appoints a two-person section of the Hydrometeorological Committee, responsible for establishing and maintaining a network of streamflow and snow reporting stations for forecasting river flows and volumes.

The Operating Committee and support personnel prepare each year an Assured Operating Plan for Canadian storage projects, estimating the most likely range of river operation six years in advance; a Determination of Downstream Power Benefits document which calculates the Canadian Entitlement (electric power owed Canada in return for building the three storage dams); and a Detailed Operating Plan for the next year, refining the operation of the Assured Operating Plan from five years prior. The Operating Committee reviews and maintains operating plans for flood control and the operation of Libby Dam, and prepares and presents any special reports required by the Entities.

The Columbia River Treaty also established the Permanent Engineering Board (PEB), required by the two federal governments to monitor and report on the results being achieved under the treaty. Additionally, the board is directed to assist in reconciling differences concerning technical or operational matters that may arise between the Entities. The U.S. secretaries of Army and Energy each appoint a PEB member and an alternate. The governments of Canada and British Columbia also each appoint a Canadian member and alternate (Canadian and U.S. Entities, December 2008).

Because of this treaty, management of the Columbia River is as much about international relations as it is about public administration within the U.S. With the construction of Grand Coulee and the Duncan, 1,792-MW Mica, and Keenleyside storage dams in Canada, the shape of the natural runoff can be modified. This involves storing water in high-flow periods and releasing it during low-flow periods to meet power, flood control, fisheries, and other needs (including irrigation, navigation, recreation, and municipal use).

The Columbia River is known for its extensive hydroelectric development. Grand Coulee and seventeen other dams with hydro facilities on the Columbia River and lower Snake River (the Columbia's largest tributary)—as well as tributary projects 525-MW Libby, 428-MW Hungry Horse, 42.6-MW Albeni Falls, and 402-MW Dworshak— provide nearly 28,500 MW of capacity and 90,000 gigawatt-hours of electricity production each year (See Figure 1, Introduction; Postlewait and White, 2005). Table 1 lists these projects, in order from upstream to downstream, and major tributaries.

The River Management Team

A number of diverse organizations are involved in managing water in the Columbia River. In addition to hydroelectric project owners and operators, these include organizations in charge of forecasting, making recommendations for dam operation, and marketing power produced by the projects.

Several organizations contribute to the operation of the Columbia River system through their efforts to collect and analyze data for water supply, including: the

U.S. Bureau of Reclamation; British Columbia Hydro and Power Authority (BC Hydro); the U.S. Army Corps of Engineers (Corps); National Weather Service River Forecast Center; the U.S. Department of Energy's Bonneville Power Administration (BPA); the Columbia River Treaty Hydrometeorological Committee (consisting of members of BPA, the Corps, and BC Hydro and mandated under the treaty to develop an international water supply data analysis system); the U.S. Geological Survey; and the U.S. Natural Resources Conservation Service (Figure 1).

Among the data and forecasting needed to predict ultimate flows and volumes in the Columbia River Basin are: rainfall, temperature, runoff, snow pack, soil moisture conditions, required spill, generator availability, volume expectation, fish hatchery operations, refill probability, and water withdrawals for consumptive uses such as irrigation or municipal use.

Project Owner/Operators
Seven different organizations own hydro projects along the Columbia River. They operate these projects under agreed-upon rules to optimize each organization's benefits with regard to one another, under contracts and agreements to optimize power made available under the Columbia River Treaty.

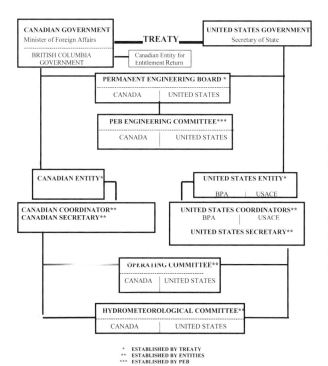

Figure 1.
Note: other organizational charts have the PEB and Entities on the same line; neither is defined by the CR Treaty as being "over" the other in an organizational context.

The Corps operates nine main-stem and three tributary projects. Most of these, built between 1938 and 1975, were constructed to produce hydropower and provide for navigation. Other purposes for these dams include flood control, recreation, irrigation (the Bureau of Reclamation pumps water from several Corps facilities as well for irrigation), and fish and wildlife protection and enhancement.

Reclamation operates Grand Coulee and Hungry Horse dams for multiple uses, including hydropower, flood control, and irrigation of the Columbia Basin Project lands. Like the Corps, Reclamation also operates other dams in the Columbia River Basin that have additional purposes, including irrigation, flood control, hydropower, recreation, and fish and wildlife protection and enhancement.

The public utility districts of Douglas, Chelan, and Grant counties in Washington State operate the five nonfederal U.S. projects in Table 1 under licenses issued

Table 1: Dams with Hydroelectric Facilities on the Mainstem Columbia River

Dam	Owner	Rated MW	First Power
Columbia			
Mica	BC Hydro	1,792	1977
Keenleyside (Arrow)	Columbia Basin Trust (Columbia Power)	185	2002
Revelstoke	BC Hydro	1,980	1984
Grand Coulee	Bureau of Reclamation	6,809	1941
Chief Joseph	Army Corps of Engineers	2,457	1955
Wells	Douglas County PUD	840	1967
Rocky Reach	Chelan County PUD	1,287	1961
Rock Island	Chelan County PUD	622	1933
Wanapum	Grant County PUD	1,038	1963
Priest Rapids	Grant County PUD	907	1959
McNary	Army Corps of Engineers	993	1953
John Day	Army Corps of Engineers	2,160	1968
The Dalles	Army Corps of Engineers	1,813	1957
Bonneville	Army Corps of Engineers	1,089	1938
Lower Snake			
Lower Granite	Army Corps of Engineers	810	1975
Little Goose	Army Corps of Engineers	810	1970
Lower Monumental	Army Corps of Engineers	810	1969
Ice Harbor	Army Corps of Engineers	635	1962
Other Major Tributaries			
Hungry Horse	Bureau of Reclamation	428	1952
Libby	Army Corps of Engineers	525	1975
Albeni Falls	Army Corps of Engineers	43	1955
Dworshak	Army Corps of Engineers	402	1973

by the Federal Energy Regulatory Commission. The projects were created and are operated primarily for hydropower but provide other significant benefits for navigation, fisheries, and other purposes.

BC Hydro is a provincial crown corporation that operates the Duncan, Mica and Keenleyside projects for the Columbia River Treaty purposes of flood control and power; 1,980-MW Revelstoke is a run-of-river project that is not part of the treaty. BC Hydro also operates these projects for several other purposes, including recreation, fish and wildlife protection, and navigation. The 185-MW Arrow Lakes Hydro plant (at Keenleyside) is owned by the Columbia Basin Trust.

The U.S. Department of Energy's Bonneville Power Administration (BPA), created by Congress in 1937 as part of the Department of Interior, markets all surplus power produced at Corps and Reclamation projects in the Columbia River Basin. In addition, BPA repays the debt incurred to build the federal dams on the river, is a member of the federal group of agencies that manages and operates day-to-day flows and storage, and finances much of the federal fish recovery efforts in the Basin and the activities of the Northwest Power and Conservation Council, using revenues from hydroelectric generation.

Other Organizations

As mentioned, the Columbia River is operated for multiple purposes. The following parties, related to many of those purposes, maintain an interest in the operations: The National Oceanic and Atmospheric Administration (NOAA) National Marine Fisheries Service (NMFS) and the U.S. Fish and Wildlife Service (FWS) have responsibilities under the U.S. Endangered Species Act to make recommendations for implementing actions that will reduce the effects of hydro operations on threatened or endangered species. For example, when the Corps, Reclamation, and BPA—collectively called the Action Agencies—attempt to optimize operations to benefit anadromous and resident fish, NOAA's NMFS advises them of when the fish are moving and when specific operations need to begin at which dams. The Action Agencies are the implementing agencies per the recommendations listed in the Federal Columbia River Power System (FCRPS) "Biological Opinions" issued by FWS in 2000 and NMFS in 2004 (and under consideration in 2009). The Biological Opinions cover operations that include the fourteen federal projects in Table 1.

The US Environmental Protection Agency makes recommendations and advises the operators of all dams in the system on water quality, especially with respect to dissolved gases and water temperature.

The Northwest Power and Conservation Council (an interstate compact of the states of Idaho, Montana, Oregon, and Washington created under a 1980 federal law) makes recommendations to the Corps, Reclamation, and BPA for the operation of the power system and implementation of energy conservation programs. In addition, the council develops fish and wildlife mitigation plans under operations of the FCRPS.

Native American Indian tribes have historic and tribal treaty rights to fish the Columbia River and its tributaries. The tribes provide advice and recommendations to federal operators individually or through organizations like the Columbia River Inter-Tribal Fish Commission (CRITFC) and the Upper Columbia United Tribes (UCUT), as well as proposing ways to preserve cultural/historic artifacts.

State agencies, particularly those involved with fish and wildlife and water quality, actively make recommendations to the federal operators. These include the Oregon Department of Fish and Wildlife, Montana Department of Fish, Wildlife and Parks, Idaho Fish and Game Department, and the Washington Department of Fish and Wildlife (Postlewait and White 2005).

Federal and nonfederal hydropower project operators coordinate their operations in a number of arenas. Following are two examples.

The Columbia mainstem hydro owners/operators work together under a contract called the Pacific Northwest Coordination Agreement. This work is overseen by a group called the Northwest Power Pool, a voluntary organization of major electricity generators in the Northwest, British Columbia, and Alberta, founded in 1942. This U.S.-only agreement provides the coordination contemplated in the Columbia River Treaty to deliver power and flood control benefits in the U.S., through annual determination of firm energy load-carrying capability, monthly regulation to set draft points, and determination of interchange and in-lieu energy.

The Columbia River Treaty Operating Committee is made up of representatives of the Corps, BPA, and BC Hydro. The committee communicates almost daily and meets bi-monthly to develop plans and implement operations of the Canadian treaty projects (Bonneville Power Administration 2001).

Conclusion

The geology and physical geography came first—the Columbia River is as powerful as it is because of the composition of the earth underneath it, where it arises in the Rocky Mountains, its elevation drop, and the breadth of area of the basin which it drains. Political geography as defined by political agreements and treaties divides the river resource itself, and provides the opportunities and limits on who can exploit the resource at what place, by what means, and within what time frame.

The Columbia River Treaty's heritage reaches back to the founding of the United States, and that heritage informs us as to what is possible and permissible. In particular, the building of dams on one nation's land that impacts, positively or negatively, another nation is a matter for negotiation.

The treaty represented the culmination of over twenty years' worth of study, calculation, and negotiation, from 1944 to 1964. Involved were the IJC, the

International Engineering Committee, the U.S. State Department and Canadian Department of Foreign Affairs and International Trade, plus the technical, economic, and political individuals and organizations they saw fit to involve. During this period, the scope and subject matters of the treaty were defined, studied, and refined.

At the national level, the U.S. president and Canadian prime minister agree to and sign treaties, and present them to, respectively, the U.S. Senate and Canadian Parliament for ratification. As with all bilateral or multilateral treaties, the Columbia River Treaty could have been amended from the first day, September 16, 1964 of implementation— and yet it remains un-amended forty-seven years later. Such amendments would have to be agreed to by both nations, and the process is the same as adopting a new treaty: executive departments proposing, the executive agreeing to and signing, and the legislative body ratifying. Interpretations of existing treaties can often be made by the executive departments themselves, evidenced by an exchange of diplomatic notes, or in the usual method of challenges in the judiciary.

The Entities appointed by their respective governments to carry out the treaty, as required by that treaty itself, focus on the two treaty-defined benefits, increased firm power production and flood control. There have been no major floods in the Columbia Basin since 1948; increased firm power production on the U.S. portion of the Columbia mainstem since 1968 with the completion of the first Canadian treaty dam has been divided equally between the two countries with each gaining about 500 average megawatts of energy and 1200–1300 megawatts of capacity.

Other purposes of the multipurpose Columbia have not been ignored; interests such as irrigation, navigation, recreation, tribal rights, fish and wildlife, municipal use, and water quality are addressed by the Entities and their Operating Committee in an operational context, consistent with the treaty's power and flood control requirements. This system of extensive coordination and cooperation, on both sides of the international border, allow the Entities to carry out their assigned duties and responsibilities while at the same time responding to other competing interests for how the Columbia is operated.

Works Cited

Anglo-American Convention of 1818. *Treaties and Other International Agreements of the United States of America, 1776–1949.* V12. TS 112.

Bonneville Power Administration. April 2001. "The Columbia River System: Inside Story Second Edition." *Bonneville Power Administration* www.bpa.gov/power /pg/columbia_river_inside_story.pdf.

Boundary Waters Treaty of 1909. *Treaties and Other International Agreements of the United States of America, 1776–1949.* V12. TS 548.

Columbia River Treaty. 1964. *Treaties and Other International Acts Series* No. 5638.

Canadian and U.S. Entities. *Annual Report of the Entities, 2007–2008.* Portland, Ore.: U.S. Army Corps of Engineers, December 2008.

Oregon Treaty. 1846. *Treaties and Other International Agreements of the United States of America, 1776–1949.* V12. TS 120.

Pletcher, David M. 1973. *The Diplomacy of Annexation: Texas, Oregon, and the Mexican War.* Columbia: University of Missouri Press.

Postlewait, Lori A. and Anthony G. White. 2005. "Managing the Columbia: Making Optimum Use of a Massive River System." *Hydro Review.* September.

Treaty of Paris. 1783. *Treaties and Other International Agreements of the United States of America, 1776–1949. V12* TS 104.

Changes in Empowerment: Rising Voices in Columbia Basin Resource Management

Barbara Cosens

Introduction

The growing demand for a public voice in natural resource decision making discussed in Chapter 4 has little application if local capacity to participate is lacking. In the Columbia River Basin both the empowerment and capacity of basin communities to participate has grown substantially since negotiation of the 1964 treaty, suggesting that the public is prepared to participate and even likely to demand participation in any decision on whether to and how to modify the treaty. Factors contributing to and defining this rise in empowerment and capacity include: (1) legal recognition of the treaty rights of certain Native American tribes to participate in the harvest and management of Columbia basin fisheries within the United States; (2) establishment of the Northwest Power and Conservation Council in the United States in 1980; (3) Constitutional recognition of the rights of First Nations in Canada in 1982; and (4) legislative recognition of the Columbia Basin Trust in Canada in 1995. The two essays in this chapter illustrate some of those voices. The following paragraphs describe the legal and legislative changes that have given rise to that empowerment.

The Rights of Native Americans

The pulse of the Columbia River is defined by the annual migrations of anadromous fish—fish spawned in fresh water that migrate to the ocean as smolts and return as adults to their natal streams to repeat the cycle. The importance of Columbia basin fisheries to Native American tribes is reflected in oral tradition and ceremonies. Of greatest significance was the annual return of the chinook salmon (Landeen 1999). The tribes gathered with other northwest tribes at Celilo Falls on the Columbia River to harvest the salmon supply that would carry them through winter (Landeen 1999). A federal district court opinion noted "[t]hese fish were vital to the [northwest] Indian diet, played an important role in their religious life, and constituted a major element of their trade and economy. Throughout most of the area salmon was a staple food and steelhead were also taken, both providing essential proteins, fats, vitamins, and minerals in the native diet" (*United States v. Washington* 1974). Salmon provided the primary protein source, but its importance to the tribe did not end with food supply. "Salmon" plays a major role in Native American mythology as shown by this excerpt from "The Maiden and Salmon" (Hines 1999):

> *And now Salmon came up the river after making a phenomenal recovery to life. "I go now to have revenge." He came up the river. He would swim along for a while; then, he would go ashore to walk along, up the valley. While he was thus walking, he saw a lodge with smoke wafting from it. "Let me just go in." He entered noiselessly ['xu-l']. There sat an old man spinning; it was Spider. Salmon said to him, "Why are you spinning, old man?"*
>
> *"Oh just to sew my clothes," he replied. But Salmon knew well enough what he was doing, that he was making a fishnet. The old man had told him this, because from the very beginning he had identified him, by smell, as Salmon.*
>
> *Salmon went outside and said to all the salmon, "You will swarm past here, all of you salmon. You will come to the old man; you will thus take pity on him."*

The life cycles of Columbia basin fisheries were used to mark time:

> *8. Then cam Hesu'al (Ha-soo-ahl), the time when the hesu (eels) move to the upper tributaries. (Hesu was a favored fish in the Nez Perce diet).*
> *9. Next came Qoyst'sal (Khoy-tsahl), the season of the run of the blue back salmon (k'ohyl-ehkts) in the upper tributaries. . . .*
> *11. Then came Nat'soxliwal (Nah-t' sohkh-le-wahl), the time when the nat'sox (chinook salmon) return to the upper rivers, ready to journey to the spawning streams.*
> *12. August was Wawama'ayqll'al (wa-wam-aye-k'ahl), the time when the chinook salmon reach the canyon streams and fishermen move to the upper rivers.*
> *13. September was Piq'uunm'ayq'al (Pe-khoon-mai-kahl), the season when the fall salmon run upstream and when the fingerlings journey down river.*
> (Landeen 1999).

The spiritual, cultural, and subsistence reliance of the northwest tribes on Columbia basin fisheries led to the inclusion of what has been interpreted to be highly significant language in a series of treaties negotiated by Isaac I. Stevens, then territorial Governor of Washington Territory, with various northwest tribes south of the 49th parallel at the council of Walla Walla in 1855 (Josephy 1965). This language can be found, for example, in Article 3 of the Nez Perce Treaty: "[t]he exclusive right of taking fish in all the streams where running through or bordering said reservation is further secured to said Indians; as also the right of taking fish at all usual and accustomed places in common with citizens of the Territory" (Treaty between the United States of America and the Nez Perce Indians 1855).

The language stating that the right is "in common with citizens of the Territory," was interpreted by Judge Boldt of the U.S. District Court, Washington, in 1974, to entitle treaty tribes to up to 50 percent of the harvestable fish that pass (or would pass absent harvest en route) the usual and accustomed fishing places (*United States v. Washington* 1974; *Washington v. Washington State Commercial Passenger Fishing Vessel Association* 1979). At the time of the council in 1855, non-Indian fishing in the area was minor; however, once canneries made large-scale commercial fishing possible, non-Indian harvest began to present major competition for the fish (United States v. Washington 1974; Washington v. Washington State Commercial Passenger Fishing Vessel Association 1979). Yet the ruling recognizing the legal right of Native American's equal access to fish would not come until years later and over a decade after the Columbia River Treaty was finalized. In affirming the District Court, the Ninth Circuit Court of Appeals interpreted the right of treaty tribes "in common with citizens of the Territory," as analogous to a co-tenancy, stating:

> [C]otenants stand in a fiduciary relationship one to the other. Each has the right to full enjoyment of the property, but must use it as a reasonable property owner. A cotenant is liable for waste if he destroys the property or abuses it so as to permanently impair its value . . . By analogy, neither the treaty Indians nor the state on behalf of its citizens may permit the subject matter of these treaties to be destroyed. (*United States v. Washington* 1975).

In 1977, in the wake of these decisions, the four tribal governments implicated, the Nez Perce, Confederated Bands of the Yakama Nation, Confederated Tribes of the Umatilla Indian Reservation, and Confederated Tribes of the Warm Springs Reservation, formed the Columbia River Inter-Tribal Fish Commission (CRITFC) to unite the efforts of the four tribal governments to renew their sovereign authority in fisheries management (CRITFC 2010). This legal recognition of rights combined with the capacity building reflected in the scientific and policy work of CRITFC, has elevated the status of the four tribes to co-managers of salmon in the U.S. portion of the Columbia River basin.

In addition to the tribes participating in CRITFC, the five upper Columbia tribes in the United States have joined together on various resource issues of common concern forming the Upper Columbia United Tribes (UCUT). The primary common issue among the Coeur d'Alene Tribe, the Kalispell Tribe of Indians, the Spokane Tribe of Indians, the Kootenai Tribe of Idaho, and the Confederated Tribe of the Colville Reservation is the blockage of their lands from anadromous fish migration by Grand Coulee Dam. In 2005, UCUT and its member tribes entered a memorandum of understanding with Bonneville Power Administration recognizing the sovereign role of the tribes in management of, among other things, fish and water resources.

Eight populations of salmon that make use of the waters of the Columbia basin at some time during their lifecycle are listed by NOAA Fisheries under the Endangered Species Act (NOAA Fisheries 2010). Current listings of salmon species found in the Columbia Basin: Snake River Sockeye (endangered), Upper Willamette River Chinook (threatened), Lower Columbia River Chinook (threatened), Upper Columbia River spring-run Chinook (endangered), Snake River fall-run Chinook (threatened), Snake River spring/summer-run Chinook (threatened), Lower Columbia River Coho (threatened), Columbia River Chum (threatened) (Final Listing Determinations for 16 ESU's of West Coast Salmon, 70 Fed. Reg. 37160, 37193 (June 28, 2005)). Upper Columbia steelhead are listed as threatened. Initial listing (70 Fed. Reg. 52630 Sept. 2, 2005) designated Upper Columbia Steelhead as endangered. Upgrading to threatened was pursuant to court order of June 18, 2009. As an extreme example of the decline of anadromous fish, the annual run of Snake River Sockeye, known for the nine-hundred-mile journey up the Salmon River to the tributaries to Redfish Lake to spawn, has dwindled to a few hundred:

> *The residual form of Redfish Lake sockeye, determined to be part of the ESU in 1993, is represented by a few hundred fish. Snake River sockeye historically were distributed in four lakes within the Stanley Basin, but the only remaining population resides in Redfish Lake. Only 16 naturally produced adults have returned to Redfish Lake since the Snake River sockeye ESU was listed as an endangered species in 1991. All 16 fish were taken into the Redfish Lake Captive Propagation Program, which was initiated as an emergency measure in 1991. The return of over 250 adults in 2000 was encouraging; however, subsequent returns from the captive program in 2001 and 2002 have been fewer than 30 fish.* (70 Fed. Reg. at 37179).

It should be noted that Idaho Department of Fish and Game data indicate that the numbers provided in the quote are incomplete. Their data indicate historic distribution in five lakes in the Stanley Basin, and return of eighty-four fish in the period from 2001—2005 (Robertson 2006). Either data set shows a dramatic reduction.

Although a remarkable life history, the salmon's lengthy migration exposes the species to numerous threats and jurisdictions. Along with habitat destruction, competition with hatchery fish and over harvesting, among the chief threats are blockage of migratory routes and slowing of water flow by hydropower dams (CRITFC 1996; Ruckelhaus 2002). Thus, Native American tribes may seek expansion of the purposes of operation of dams, optimized under the 1964 treaty for hydropower and flood control, to include consideration of the impacts on anadromous fish.

The States

As described by Mouat in Chapter 1, the province of British Columbia played a large role in the final stages of treaty negotiation. The same cannot be said for the states on the U.S. side of the border. Primary development of the Columbia River had been financed at the federal level, the states lacking the capacity for major river development. At the time of treaty ratification, a major battle between public and private power was playing out on the Snake River and would end with private power receiving the support it needed to prevail from the Eisenhower Administration (Brooks 2009). However, as described by Vogel in Part IV, Chapter 1, the development resulting from the Columbia River Treaty actually made regional capacity building possible.

In 1980, the Northwest Power Act (Pacific Northwest Electric Power Planning and Conservation Act 1980) authorized an interstate compact approved by the legislatures of Idaho, Montana, Oregon, and Washington giving the four states a greater role in decision making with respect to electric power and fish and wildlife in the Columbia River basin. The resulting Northwest Power and Conservation Council is made up of two political appointees from each state, has legal and technical staff, and is funded through power revenues from Bonneville Power Administration. The council has three primary objectives: (1) to develop a 20-year electric power plan that will guarantee adequate and reliable energy at the lowest economic and environmental cost to the Northwest, (2) to develop a program to protect and rebuild fish and wildlife populations affected by hydropower development in the Columbia River Basin, and (3) to educate and involve the public in the Council's decision-making processes (Northwest Power and Conservation Council).

The Act requires all actions of Bonneville Power Administration (BPA) to be consistent with the council's electric power plan (Pacific Northwest Electric Power Planning and Conservation Act 1980; § 839b(d)(2)). In contrast, the fish and wildlife program is intended to be based on input from states, tribes, and federal agencies, and to complement their activities, but it is not the role of the council or BPA to reconcile the fish and wildlife program with hydropower operations in the basin (Pacific Northwest Electric Power Planning and Conservation Act 1980; § 839b(h)). As part of its efforts to develop a fish and wildlife program and to involve the public, the council undertook a sub-basin planning process on fifty-eight tributary watersheds and main-stem segments in order to identify and prioritize habitat restoration opportunities. This was completed in 2005 (Northwest Power and Conservation Council). That effort is analyzed by Heikkila and Gerlak in Part IV Chapter 2. The council's positive efforts to involve the public in habitat restoration decisions, juxtaposed with the absence of any connection between these efforts and BPA decisions on power, have led to an informed but frustrated public (see e.g., Blumm 2002). The fact that the entity which must generate hydropower consistent with the council's energy plan and one of the U.S. entities appointed to operate the treaty are connected (BPA and the BPA administrator respectively) merely provides an informal means of communication and coordination but not one that must be utilized. The council has no direct role in treaty implementation or in any decision to modify the treaty, but its formation and efforts to involve and educate the public signal an increase in capacity within the basin to seek a role in the decision-making process.

Neither the tribal organizations nor the council should be viewed as providing a unified voice in treaty decisions for their respective members. Upstream and downstream tribes and states frequently have conflicting interests. Nevertheless, these organizations have greatly increased the knowledge and capacity of their members to weigh in on treaty issues. This capacity did not exist in 1964.

First Nations and Canadian Communities in the Basin

The capacity of First Nations to participate in decision making has also increased in the Canadian portion of the basin, although possibly not to the degree of tribes in the U.S. portion. Britain granted Canada full sovereignty in 1982 when the Canadian Constitution was patriated (thereby eliminating the need for an act of the British Parliament to amend the constitution). The Constitution Act recognized aboriginal and treaty rights, including rights acquired through land claim agreements, of aboriginal people in Canada (Constitution Act, Part II, Section 35 1982). As noted by Garry Merkel in the essay to follow, this formal constitutional recognition is expected to elevate the status of First Nations in providing input on any decision by Canada regarding the treaty.

In addition, the Columbia Basin Trust has become a major player on the Canadian side of the basin. Its formation is described on its Web site:

> Despite the significant changes that occurred across the Columbia Basin as a result of the Treaty, there was a lack of consultation with residents. The people of the Basin came together in the early 1990s to press the Province of BC for recognition of the injustice of this situation. Local governments coordinated their efforts (at the regional district and tribal council levels and in partnership with elected officials) under the formation of the Columbia River Treaty Committee, in order to approach and negotiate with the Province (Columbia Basin Trust 2008).

Negotiations were successful and, in 1995, Columbia Basin Trust was established. A binding agreement was also established which resulted in the following for the residents of the basin through Columbia Basin Trust: (1) $276 million to finance power project construction; (2) $45 million, with the trust used as an endowment; and (3) $2 million per year from 1996 to 2010 for operations (Columbia Basin Trust 2008).

The trust has participation from First Nations and other communities in the Columbia River basin in Canada. The recognition of the trust by the Province of British Columbia in 1995, and the knowledge and capacity built by the trust through its substantial funding, suggests that the people of the basin in Canada will not be excluded from future treaty decisions. The involvement of the trust in hydropower development and partial funding through sale of the Canadian Entitlement also suggests that its input may be more nuanced than its roots in seeking redress from harms caused by implementation of the treaty would suggest.

In the following essays you will see two perspectives on the impact of the Columbia River Treaty on indigenous peoples and the role sought in any review of the treaty. Garry Merkel, as both a First Nation member and board chair of Columbia Basin Trust, outlines the legal and legislative changes recognizing First Nation rights in Canada and leading him to conclude that no reconsideration of the treaty will go forward without a First Nation role. Former Coeur d'Alene judge and tribal member Mary Pearson gives us a view of the legal changes in the United States that suggests both increasing empowerment in Columbia River fisheries management and decreasing recognition of tribal sovereignty at the federal level. Her passionate argument born of frustration gives us a glimpse of the impact these changes have on the lives of the River People.

Works Cited

Blumm, Michael. *Sacrificing the Salmon: A Legal and Policy History of the Decline of Columbia Basin Salmon* (Den Bosch, The Netherlands: Book World Publications 2002).

Brooks, Karl B. *Public Power, Private Dams: The Hells Canyon High Dam Controversy* (Seattle: University of Washington Press 2009).

Columbia Basin Trust. 2008. Web site at http://www.cbt.org.

Columbia River Inter-Tribal Fish Commission (2010). Web site at http://www.critfc.org/text/work.html.

Columbia River Inter-Tribal Fish Commission. "'Wy-Kan-Ush-Mi Wa-Kish-Wit' Spirit of the Salmon:" The Columbia River Anadromous Fish Plan of the Nez Perce, Umatilla, Warm Springs, and Yakama Tribes." Columbia River Inter-Tribal Fish Commission (1996). http://www.critfc.org/text/trptext.html.

Constitution Act, Part II, Section 35 (1982). http://www.solon.org/Constitutions/Canada/English/ca_1982_html.

Hines, Donald M. *Tales of the Nez Perce.* (Fairfield, Washington: Ye Galleon Press 1999).

Josephy, Alvin M, *The Nez Perce Indians and the Opening of the Northwest.* (New York: Mariner Books 1965).

Landeen, Dan, and Allen Pinkham. *Salmon and His People, A Nez Perce Nature Guide.* (Lewiston, Idaho: Confluence Press 1999).

NOAA Fisheries, Northwest Regional Office. *ESA Salmon Listings* (July 1, 2010). http://www.nwr.noaa.gov/ESA-Salmon-Listings/.

Northwest Power and Conservation Council. Web site at http://www.nwcouncil.org/.

"Pacific Northwest Electric Power Planning and Conservation Act." *Northwest Power Act, Pub. L. No. 96-501, 94 Stat. 2697* (1980).

Robertson, Cindy, interview by Barbara Cosens. *Fisheries Biologist* (January 19, 2006).

Ruckelhaus, Mary H., Phil Levin, Jerald B. Johnson, and Peter M. Kareiva. "The Pacific Salmon Wars: What Science Brings to the Challenge of Recovering Species." *American Review of Ecological Systems* v. 33 (2002).

"Treaty between the United States of America and the Nez Perce Indians." 12 Stat. 957 (June 11, 1855).

Upper Columbia United Tribes. Web site at http://www.ucut.org/index.ydev.

United States v. Washington. 384 F.Supp. 312 (D.Wash., 1974).

First Nations

Garry Merkel

The Columbia River is immense and deeply engrained in the minds, hearts, souls, and livelihoods of countless historic and contemporary cultures. The earliest accounts of these relationships arise from the indigenous peoples, generically referred to as Native Americans or tribes in the United States or First Nations or aboriginal in Canada.

The Columbia River Treaty was one in a string of events on the Columbia River that completely re-shaped the contemporary social framework Most Columbia River tribes and First Nations were deeply impacted by the treaty by loss of salmon and losses or significant changes in resident fish populations, hunting areas, gathering areas, and spiritual sites. In many situations the treaty completely and negatively re-shaped these societies and communities.

When the treaty was first established the Canadian crown had few known legal obligations to First Nations. First Nations were generally not aware of the process and had no influence on the decision-making process. Since that time Canadian courts have (a) recognized that aboriginal rights exist (a constitutionally protected priority right to exclusively use and occupy the land in a manner consistent with communal practises); (b) established that only the crown can infringe on aboriginal rights but the infringement has to be shown justifiable; (c) established very high consultation standards for potential infringements on aboriginal rights; and (d) required the crown to adequately accommodate any justifiable impacts to aboriginal rights.

Canada and British Columbia will both have to meet these requirements with Columbia River First Nations with respect to any discussions regarding the treaty.

Canadian governments and First Nations are still working out ways to deal with these emerging legal responsibilities. However, based on existing practice, we can expect that a distinct process will be established between Canadian governments and Columbia River First Nations for any treaty discussions that will almost certainly include: (a) a government to government orientation;(b) resources to ensure informed First Nations involvement; (c) significant First Nations influence on the local outcomes of the discussions; and (d) some type of accommodation for First Nations.

All of these elements point towards a significant First Nations role in treaty discussions on the Canadian side—not necessarily a veto position but almost certainly an opportunity to have a much more significant First Nations role now and into the future of the Columbia River Treaty.

The River People and the Importance of Salmon

Mary L. Pearson

*[L]ife evolved around the salmon that found water. That was a gift from Mother Nature
… to continue the relationship with mother earth and grandfather spirit, and to benefit
from these mighty salmon, we must come together once again. Water and especially the
salmon within it, guided our spirituality and ceremonies and it gave us life.* (Seyler
2000)

Introduction

The region now known as the Pacific Northwest has been populated by Indian
tribes for thousands of years. Many of these tribes lived on and around the rivers
that defined their entire culture. The people depended on the salmon in these rivers
for 60 percent of their diet. Salmon was, and is, the center of their religion, their
culture, their economy, and their lives. They call themselves River People and the
big river, the Columbia, was called the Path of Life.[1] It is this river and the people
who lived and continue to live near the Path of Life and its tributaries with which
we are concerned. It is the River People who have been and will be impacted by
the U.S. and Canadian treaty that I will discuss. This chapter will explain who the
River People are, the importance of the salmon to them, and why the tribes should
be sitting at the table talking about the salmon when the United States and Canada
talk about the treaty.

The "River Tribes" were many. They included the three bands of the Nez
Perce;[2] the three bands of the Umatilla;[3] the three bands of Warm Springs;[4] and
the fourteen bands that make up the Yakama Nation.[5] These tribes, known today
as the Yakama Nation, the Confederated Tribes of the Umatilla, the Warm Springs
Confederated Tribes, and the Nez Perce Tribe, are also called the Columbia River
Treaty Tribes because they have treaty-protected rights to fish on the mid-Columbia
River downstream of where The Dalles Dam now sits at what used to be Celilo
Falls near the Dalles, Oregon.[6]

Other River People include the Confederated Tribes of the Colville Reservation
made up of twelve bands;[7] the Spokane Tribe of Indians,[8] and the Coeur d'Alene,
Kootenai, and Kalispell tribes. All of these tribes have been affected by the building
of dams in violation of the 1846 treaty between the two countries—the U.S. and
Canada. The River Tribes have especially been affected by the building of the
Grand Coulee Dam on the edge of the Colville Reservation. Once this dam was

built without fish passage(s), the upriver stocks for the Columbia, including the tributaries that supported Native life for centuries were cut off all the way to the headwaters of the river. The Spokane and other upriver tribes such as the Coeur d' Alene, Kalispell, and Kootenai had already been cut off by the dams installed first in the City of Spokane, then at Post Falls, and later at the Nine Mile Falls Dam nine miles from Spokane near the Spokane Reservation in their ceded area. The importance of the salmon to the River People cannot be overemphasized.

Brother Salmon and the First Foods Ceremony

The salmon was the most important commodity to the River People at the time of the coming of the Europeans. It still is. Indian people and the salmon lived in harmony with the balance of nature. The water was free, food was plentiful, and the rhythms of the season were spent gathering food for the next year.

The River People depend upon the fish in the Path of Life, also known as the River That Gives Us Life, and always gave thanks to the Creator in the First Foods Ceremonies. The River People revered the salmon and that reverence was shared by the recognition that salmon was not only central to their diet, but also to their religion. In the First Foods Ceremony, salmon holds the place of honor at the table. Elizabeth Woody, a Warm Springs poet, described the First Foods Feast in the following way:

> Together, tribal members and Salmon weave a unique cultural fabric designed by the Divine Creator. What the mind cannot comprehend, the heart and spirit interpret. The result is a beautiful and dignified ceremonial response to the Creator in appreciation for the willingness of Nature to serve humankind (Woody et al. 2003).

The ceremony may have varied from tribe to tribe, but for many of the Columbia River tribes that follow the Washat or Seven Drums Religion, the River People shared this ceremony. Salmon is first in the place of honor on the table to show how important it is. It is followed by three other foods: meat (elk or deer), roots and then berries. These four foods were the staples of life for the River People and contained a diet rich in calcium, iron, Vitamin C, healthy oils, and minerals.

For those upriver tribes who lost salmon nearly seventy years ago, the First Foods ceremonies have withered and died or for the most part been replaced by Christianity. Although the Coeur d'Alene, Kalispell, and Spokane tribes still retain a few longhouses for ceremonies (such as funerals) and their winter dances, called Medicine Dances, for the most part traditional religion has almost disappeared along with the Salmon Ceremony, and the Elders. This is a tremendous economic, cultural, and religious loss. Only recently, in the spring of 2005, and again in 2006, the Coeur d'Alene Tribe conducted a First Foods Feast, the first in many, many years.

Fishing Sites

There were two main fishing sites where the people gathered several times a year and where they fished together. One was located at Celilo Falls on the lower Columbia River and the other at Kettle Falls further north. The upriver tribes frequented both.

Celilo

Fishing at Celilo Falls or Wyam, its Indian name, has been described as the most important fishing site for the four treaty tribes. It is on the mid-Columbia river. It was one of the longest continuously inhabited communities in North America and is estimated to have existed for twelve thousand years (Woody et al. 2003). It is the traditional site for salmon fishing by the River People. Other tribal people also came to Celilo each year. They came from the coast and from the Interior Salish; from the Colville and the Spokane tribes; from the Flathead and Coeur d' Alene, where they joined the Yakama, the Tygh, Wasco, and Paiute, along with the Nez Perce, the Cayuse, and Walla Walla People.

Celilo is where a large set of staircase waterfalls provided a natural herding point for salmon traveling upstream. It was a place where the tribes also gathered to trade fish for dentalium shells and other items not available inland. They gathered to race horses and bet on the outcomes. They came to find marriage mates and to cement intertribal relationships. The annual trip to Celilo provided needed resources and it was lost forever when Bonneville Dam was built.[9]

Kettle Falls

The other important fishing site to the River People, particularly the Upriver People, was Kettle Falls. This site had been in existence for more than nine thousand years. It became very important to the surrounding tribes of the Colville, Lakes, Spokane, Sanpoil, Kalispell, Flathead, Chewelah, and the Kutenai about 2,600 years ago.[10] David Thompson, the English mapmaker, explorer, and historian, wrote of seeing large (20' by 60') cedar plank sheds filled with drying salmon at Kettle Falls in June 1811. In the mid-1800s, Father Pierre Jean DeSmet described fishing with spears and using J-Baskets and catches of as many as 3,000 salmon in a day there. It is one of the more important historical and cultural locations on the 1,243 miles of the Columbia River.

Others describe being able to take up to eight hundred salmon a day weighing up to forty pounds at each location (Eels 1840). This area was home to many now called the Colville Confederated Tribes. The falls were described by Thompson as "ten feet of steep slope, [broken] in places" (Kershner 2006). The village centered there was thousands of years old and people gathered to mine for quartzite to make tools as well as to gather salmon with runs beginning in June and running through summer.

Additionally, Kettle Falls remained an important fishing site for many other bands than the local Colville and Lake tribes, i.e. the Okanogan, Sanpoil, Spokane, Kalispel, Flathead, Coeur d' Alene, Kootenai, even the Nez Perce and Palouse (Kershner 2006). It has been described as "the center of civilization for this part of the river" (Chance 1986). In 1939, the federal government placed the entire population of four hundred making up Kettle Falls on trucks and moved them to a new, much higher site three miles away. It was several more years before rising water from the Grand Coulee Dam backed up and covered the falls and the old town with thirty feet of water, destroying not only the town but the salmon run that had spawned it (Kershner 2006).

Dam Building Impacts on the River People
The building of dams on the Columbia River and its major tributaries, particularly the Snake River in Idaho, has had a devastating impact, both mental and physical, on thousands of River People. Physical losses were caused by the loss of nourishment, economic independence, and purpose. Mental and emotional losses were more complicated because of the integral relationship between the people and salmon described earlier.

B. D. (Before the Dams), Indians in the Northwest living on or near the Columbia River and its tributaries were economically independent and healthy, both physically and emotionally. The salmon was plentiful. It was used for survival, in ceremony, and it was traded and shared with others who came from great distances.

With the building of Coulee Dam on the edge of the Colville Reservation, twenty-one thousand acres of the tribes' prime bottomland where the people had lived for thousands of years were flooded. The hunting, farming, and root gathering disappeared overnight. Some tribal burial grounds were moved at the last minute with no time for ceremonies or proper care for the remains. The Bureau of Reclamation hired a Spokane funeral home before the flooding to help dig up the bones of the Colville Tribe's loved ones and move the bodies to higher ground (Kershner 2006). However, Reclamation waited too long. The water began to rise before the graves could all be removed. Between six hundred and eight hundred graves from thirty-three different cemeteries were affected (Kershner 2006). The graves removed in 1941 were estimated at 1,027 and in 1979 as 1,388. The pain of these losses to the rising waters has had a tremendous impact on the River People.

The same thing happened to the Spokane people. The Bureau delayed so long in asking anthropologists from the University of Washington to catalog what was going to be covered by water that the workers had to scramble out of the way of the rising water and leave without cataloging much of the work. These devastating losses of the River People's ancestors can be compared to the "ethnic cleansing" that took place in Bosnia by actions of the Serbs a few years ago (Kershner 2006).

The impacts of pain from accumulated losses over the salmon and riverbank areas have been succinctly described by a Yakama psycho-social nursing specialist:

> [T]here's a huge connection between salmon and tribal health. Restoring salmon restores a way of life. It restores physical activity . . . mental health . . . improves nutrition and thus restores physical health . . . it's a big deal. It allows families to share time together and builds connections between family members. It passes on traditions that are being lost. If the salmon come back, these positive changes would start (Kershner 2006 quoting Chris Walsh).

This description of the importance of salmon to tribal health is critical for non-Indians and their governments to understand in order for the change that needs to take place. Fish passages may not be the only answer but it is one that we can all live with and it restores life and health to a large part of the original people of this land.

The Numbers

In 1805, Thomas Jefferson, the second president of the United States, sent Meriwether Lewis and William Clark to explore the Northwest to see if it was worth settlement. When they came west they visited various tribes along the way, first spending a little time with the Nez Perce who provided them with fish (and kept them from starving) when they were nearly out of supplies. The explorers wrote about Wishram, a place where they stopped, on what is now the Washington side of the Columbia River and reported seeing one million salmon at that location (Brown 2005).

There were Spokane People also at Wishram when Lewis and Clark visited there. Still by the year 1825 when Fort Spokane was moved to Fort Colville, the Spokane People were taking as many as seven to eight hundred salmon a day at the fish barrier upstream from the Fort (Brown 2005). There was no shortage of fish anywhere on the river. Today, the Spokane must go to Chief Joseph Dam, located below Grand Coulee or drive to a fish hatchery to gather salmon for funerals and other ceremonies.

The numbers of salmon began to decline with the coming of the Europeans and they continue to decline (Brown 2005), with the most severely affected species being the Snake River salmon that is required to run the gauntlet of eight dams on the Columbia and Snake rivers to get back to Idaho. It is estimated that the one million salmon seen by Lewis and Clark in 1805 is the number that made it back to Idaho and eastern Oregon.

Declines worsened with dam building, starting in 1890 with the dam on Monroe Street at Spokane Falls, in Spokane, Washington, followed in 1906 by another dam built at what is now Post Falls, Idaho. Both dams were on the Upper Spokane River in Eastern Washington, both were hydroelectric, assuring that the city of Spokane had electricity and streetlights before San Francisco, California or Portland, Oregon. However, the dams cut off the salmon that used to return to what is now Lake Coeur d'Alene, the headwaters of the Spokane River, depriving the Kalispell, Coeur d'Alene, Kootenai, and Flathead tribes of salmon from locations they had fished for centuries—and yet there was no provision to provide electricity to the Indians (Ray 1980).

The first critical impact to the Spokane tribal fishery was in 1909 with the building of a dam at Nine Mile Falls close to the big fishing site on the Little Spokane River in Washington. The Nine Mile Falls Dam was built without fish ladders, cutting off the salmon from that location east. A bigger impact began with non-Indian commercial fishing and the impacts reached genocidal proportions to the Columbia River peoples' cultures with the building of the main stem Grand Coulee and Bonneville Dams in the late 1930s and early 1940s.

Grand Coulee Dam, in particular, was responsible for more than 50 percent of the loss of salmon in the Columbia and its tributaries. Grand Coulee Dam was directly responsible for the killing of a culture that depended upon salmon as the core of its religious practices and its primary food source. The devastating impact of Coulee Dam can only be explained in terms of genocide, as the dam's destruction of the salmon runs directly ended the ability of the Columbia River People to fish for salmon, which in turn damaged their cultural existence and diminished their source of sustenance. At one point in time, the salmon population was so decimated that the only fishing allowed to the River People was for ceremonies (Columbia River Inter-Tribal Fish Commission n.d).

By 1994, the number of salmon returning to Idaho and eastern Oregon was down to seventeen hundred, a record low, and in 1994 not one single spring Chinook returned to the Grand Ronde River, on the edge of the Umatilla Reservation in northeast Oregon. The ocean fishery declined 80 percent in the four-year period from 1988 to 1992. Coho, the second most abundant fishery, declined by 95 percent between 1976 and 1993, and wild salmon declined by 80 percent (Columbia River Inter-Tribal Fish Commission n.d). However, in June 2008, the Columbia River Inter-Tribal Fish Commission announced a forecast of fifty-two thousand upriver Chinook with an estimated nine thousand fish for the tribal fishery (Columbia River Inter-Tribal Fish Commission 2008). This was announced by the commission, but there is no explanation of the source or how much is due to hatchery stock or why it has taken seventeen years for this number to come back.

In a little more than one hundred and fifty years, the dams have turned what had been desert and empty space in the Northwest into a high-tech region with a gross national product that ranks tenth in the world (Columbia River Inter-Tribal Fish Commission 2008). They gave the Pacific Northwest the cheapest electricity in the nation and marketed it to the population centers in California, leaving the River People to pay the price. The federal government earns $400,000,000 per year on electricity from the Grand Coulee Dam, situated half on the Colville Indian Reservation, where its Indian residents pay double the cost for electricity of that paid by non-Indians on the other side of the river in Grant County.

Federal Breaches
When the United States was a newly formed country, it entered into a treaty with England in 1846, which guaranteed open navigation on the Columbia River. It created the Northwest Territory and provided that:

> *[It] being understood that all the usual portages along the line thus described shall, in like manner, be free and open. . . .[It] being, however, always understood that nothing in this article shall be construed as preventing, or intended to prevent, the government of the United States from making any regulations respecting the navigation of the said river or rivers not inconsistent with the present treaty."[11]*

This language not only envisioned an open waterway between the two countries, it guaranteed it. The United States was well aware of the importance of fish and fishing to the Columbia River People. In 1845 Captain Charles Wilkes, a member of the U.S. Exploring Expedition, wrote about Kettle Falls and the Indians he saw there from the Colville, Spokane, Kalispell, Kootenai, and Coeur d'Alene people, and he wrote of the importance of the salmon and how many he saw, nine hundred to a thousand.[12] The language about authority to make regulations respecting the river or rivers does not contemplate the building of dams. It promises open waterways. The guarantee that the river(s) would always be open to navigation also meant the river would always be open to fishing. This provision has been breached many times over by both countries. With the 1846 treaty in place, a wave of pioneers from the east headed west where they could find free land and a new beginning. They also found many Indian tribes, which meant that the United States was required to negotiate treaties numbering in the thousands to make way for the new settlers.

Treaty making began in the 1850s and the U.S. president again sent two men to do the job: Joel Palmer was sent as governor for the Oregon Territory and Isaac I. Stevens as governor for the Washington Territory. Both governors began negotiating treaties up and down the coasts of their respective territories and then moved their interests inland to the tribes on the Big River.

The treaties were negotiated and entered into between sovereign nations. The tribes insisted that they be guaranteed fishing rights because of their dependence on and relationship with the salmon. For a people whose bodies depended upon a 60 percent salmon intake, the demand for guarantees of continued rights to fish is understandable.

The treaties contained the following promises:

> [t]he right of taking fish at all usual and accustomed places, in common with citizens of the Territory, and of erecting temporary buildings for curing them; together with the privilege of hunting, gathering roots and berries, and pasturing their horses and cattle upon open and unclaimed land. [13]

Almost immediately, the new citizens of the United States tried to deny the River People the rights they had retained to fish the Columbia River and its tributaries. Tribes began to be excluded from their "usual and accustomed" fishing grounds by the expansion of businesses and white settlers. First, the Indians were crowded out, then they were fenced out, then the new states began to require licenses for them to fish. When the Indians resisted, the states began to arrest and jail them. This attack on Indian treaty rights has resulted in millions of dollars being spent on litigation by all of the parties involved. It has been a constant battle for a way of life that has now been drastically changed for the people and for the salmon upon which they depend.

It was no different for the upriver tribes of the Colville, Spokane, and Coeur d'Alene people who made contracts with the United States after Congress banned the making of treaties in 1871.[14] The contracts were formalized as agreements, statutes, or executive orders (Cohen 1982). Executive Order tribes, however, were not recognized as having "hunting rights" until 1975, in *Antoine v. Washington*, when the United States Supreme Court held that the phrase "the hunting rights of Indians in common with other persons would not be taken away or abridged"[15] meant that the Executive Order tribes had retained the right to hunt and fish, as had the Columbia River tribes in the Columbia River treaties, and that agreements, like treaties, once ratified by Congress, become "the supreme law of the land" (Cohen 1982).

Contrary to the diminished notion of sovereignty articulated by the U.S. Supreme Court in recent years, treaties were made with people recognized to have the status of sovereign nations. Furthermore, the states in the west were not in existence at the time the federal government entered into treaties with Indian tribes. Tribes have never been considered the equivalent of states. The treaties and later executive orders remain enforceable contracts between these sovereigns. Treaties have a special place in the United States Constitution and the fishing provisions are considered to be "the supreme law of the land" on an equal footing with the Constitution.[16]

In these supreme documents, which the Indian tribes still consider sacred promises, the federal government used specific treaty language to protect those things that the tribes deemed most important, i.e., the ability and freedom to continue to feed themselves the mainstay of their diet, salmon. The U.S. federal government has breached its solemn promises made in the 1846 treaty with England and the treaties with the tribes by building dams with no fish passages in the Indians'"usual and accustomed" fishing sites and in covering those treaty-protected fishing sites with water from the dams. At this time 232 genetically unique Pacific salmon and steelhead are extinct and at least six species were added to the Endangered Species List in 1999.[17] By destroying these species of fish, the U.S. is also destroying the cultures that rely on their continued existence. The U.S. government has knowingly destroyed the number one food source for thousands of River People in a manner that can only be described as acts of genocide.[18]

Congress began passing federal legislation that breached the River Peoples' treaties as early as 1920, when it passed the Federal Power Act that was the precursor to the building of dams on the River That Gives Us Life. The Act was amended and used again and again to deprive the River People, who had lived along the Spokane, Snake, and Columbia rivers, of most of their food supply and their economic stability. Congress then enacted legislation to purportedly mitigate those losses by building twenty-four of the twenty-six mitigation fish hatcheries below Bonneville Dam when the area above Bonneville is where the "usual and accustomed" fishing sites of the Indian tribes are located, thereby deliberately depriving the River People of mitigation.[19] This was not surprising since the man who was charged with overseeing and protecting Indians, their land, and their resources, and who was acting as secretary for the Department of Interior at that time, Harold Ickes, made the following statement in 1947:

> *The Department [of the Interior] feels that the Columbia River fisheries should not be allowed indefinitely to block the full development of the other resources of the river . . .* The overall benefits to the Pacific Northwest *from a thoroughgoing development [meaning the erection of dams] of the Snake and Columbia are such that* the present salmon run must if necessary, be sacrificed. *"[20]*

Ickes, as head of Interior, with the ultimate responsibility to act as trustee for Indians, also oversaw the Bureau of Reclamation at the time he made this statement and Grand Coulee Dam was under consideration. No one raised this conflict of interest on behalf of the tribes.

This sentiment was repeated by former Washington State Attorney General and Senator, Slade Gorton, while under consideration for appointment to President Bush's cabinet as Secretary of the Interior:"There is a cost above which, regrettably, you let species disappear."[21] Gorton was referring to salmon but he might as well have been referring to Indians. It reflects a sentiment that if you hold out long enough for the salmon to die so will the Indians.

The U.S. government then enacted legislation that placed control of dam building at the federal level. The Federal Water Power Act created in 1920 was amended in 1935, changing its name to the Federal Power Act asserting the federal role in dam building. It was contained no provision for payment for the use of Indian water. Through this Act, the Grand Coulee Dam was built without mention of payment to the tribes. Later it was amended again, to provide specifically for protection of Indian reservations,[22] but it was too little, too late. Today, dam licenses issued within any Indian reservation require a specific finding by the commission that the license will not interfere or be inconsistent with the purposes for which the reservation was created or acquired. Whether the Federal Power Act can be used to the advantage of the tribes in getting fish passages for all of the dams on the Columbia and its tributaries remains to be seen.

The 1964 treaty that provided for the building of the four dams on the Columbia River in Canada and Montana was ostensibly for "flood control" and "other purposes" and provides for 165 million acre-feet of water storage at the three dams that are all on the Canadian side. The treaty isn't one-sided, but it does provide for "benefits to the two countries in the form of additional hydroelectricity and flood control in a manner that will make the largest contribution to the economic progress of both countries and to the welfare of their peoples of which those resources are capable." This allows the United States to use the hydropower created by the dams in Canada for the "most effective use," such obligation being determined by the downstream benefit, which can only benefit the United States. These plans provide for flood control and the production of optimum hydroelectricity absent some specific agreement between the two countries. Canada stores the water and gets its payments, but the United States appears to have the upper hand. It gets all of the electricity, which it sells to California.

The treaty does not provide for other uses, such as water for salmon or steelhead migration in the lower Columbia, or for irrigation or navigation. One reason is because Canada has been cut off from the salmon and steelhead on the Columbia River ever since the erection of Grand Coulee Dam, which was built without fish passages in 1941. A second reason is that Canada retains salmon fisheries on the Frazier River. A third reason is that there were no Indian tribes at the table during negotiations. It doesn't appear that either side considered fish ladders to restore the salmon at or above Coulee Dam which is one of the reasons for this chapter. But the purpose of the treaty is for the hydroelectric power produced by the four upriver dams and the increased hydropower generation at the downriver dams. The benefits, as we have seen, are to be accomplished "in a manner that will make the largest contribution to the economic progress of both countries and to the welfare of their peoples of which those resources are capable." Clearly, this language contemplates only a narrow purpose with no other use of the water resource appearing to be valuable in this negotiated document. Without tribal voices at the negotiation table, the value of the salmon, the way of life on the River That Gives Us Life through navigation and fishing, and other values of the river from a native cultural perspective remain silent.

Some of the River People and federal agencies entered into new accords in May 2008 for the Columbia River Basin that "deliver specific, scientifically valid biological results for the region's fish," but this does not help the upriver tribes and the Nez Perce refused to sign. This accord follows the Pacific Salmon Treaty of 1985 entered into between Canada and the United States for the purposes of cooperation, research, enhancement, and management of mutual concerns of Pacific salmon stocks. This agreement lasted until 1992 and from that date to 1998 the countries could not agree on how to amend and modify the outmoded 1985 Treaty.

It took six years of negotiations to reach an agreement that included habitat provisions versus a negotiated catch arrangement, the establishment of two funds for enhancement, one north and one south, and the creation of a Trans-boundary Panel and Committee on Scientific Cooperation. The primary reason for the treaty was to address the interception of native stocks of one country by the other. However, neither of these agreements addressed the 1964 Columbia River treaty in which the tribes were not involved. It does not address the need for fish passages past Chief Joseph and Grand Coulee, the three northern dams in Canada, and the dam on the Kootenai River in Montana, built more than a hundred years after the 1846 treaty and inherited by Canada.

When the United States negotiated the treaties with the River People, payment for the lands sold to the government did not include payment for water in the tribes' ceded territory, thus, the water has been appropriated by the United States without compensation. Water adjudications are going on all over Indian country,[23] but this issue has not been litigated on the Columbia River and most of its tributaries. Furthermore, litigation has not occurred on whether the reservation of rights to fish at usual and accustomed sites and within reservation boundaries carries an implied warranty of not only water quality, but the availability of the water to sustain the salmon. The tribes have a right to share in the profits from the dams and a right to compensation for the loss of fish, their economy, and their way of life. Failure to provide such payment was, and is, a further breach by the federal government of its contracts with the tribes.

In the two treaties between the United States and our northern neighbor—the one executed in 1846 with England creating the Northwest Territory and the 1964 treaty with Canada—the original people of both continents were not invited to the table, nor were their rights considered. In 1846, the reasons for this lack of consideration may have been because there were no treaties with the tribes at the time and in the European mindset Indians were considered to be savages, and uncivilized.[24] Although Indians did not become citizens until federal legislation was passed in 1924, there is no reason that the tribes should not be sitting at the table today. The tribes were not at the table in 1964 because they were not yet politically organized. Since that time, however, the Columbia Inter-Tribal Fish Commission (CRITFC), created in 1977,[25] and the United Columbia Upriver Tribes (UCUT), created in 1982,[26] joined by the Colville in 1999, have been politically active.

What We Can Do

It is not too late to right this wrong, to restore salmon and a way of life to the River People by recognizing the genocidal effect that the loss of salmon has had on the tribes. It can be done during this period of pre-negotiations by inviting the tribes to the table, by assisting them with finding the funds to pursue these discussions, and by setting aside old feelings and rigidity.

As possible renegotiation draws near, this is an opportune time for the two countries to issue an apology to the River People on both sides of the border that share this great river. It is time to consider the river tribes' aboriginal and treaty rights to the salmon and the water in the Columbia River, and to involve them in the negotiations. It is also time to build fish ladders from Grand Coulee Dam on up. Salmon and electricity can co-exist. Alternative sources of power are available during the time that fry need to be rushed to the ocean. There are no alternative sources of food for the River People whose mental and physical health and economic well-being have been and are being damaged. Fish passages may just be more economical, less divisive, and healthier for all than compensation to the tribes for loss of the salmon, their way of life, and for the use of their water resources.

Notes

1. The Columbia River got its present name after the purchase of the Northwest Territory from England in 1846.

2. The upriver, downriver, and Joseph bands.

3. The Umatilla included the Cayuse, Palouse, and Walla Walla tribes.

4. The Warm Springs Confederacy includes the Tygh, Wasco, and Paiute tribes.

5. The Yakama Nation includes the Kah-milt-pah, Klickitat, Klinquit, Kow-was-say-ee, Li-ay-was, Oche-chotes, Palouse, Pisquose, Se-ap-cat, Shyiks, Skinpah, Wenatshapam, Wishram and Yakama tribes.

6. Each of the four Columbia River tribes signed a treaty in 1855 that guarantees them the "right to fish at usual and accustomed places" forever.

7. Including the Colville, Nespelem, San Poil, Lake, Palus, Wenatchi, Chelan, Entiat, Methow, Southern Okanogan, Moses Columbia, and the Nez Perce Band of Chief Joseph.

8. The Spokane tribes are made up of three bands named for the upper, middle, and lower Spokane River.

9. Umatilla History document.

10. Upper Columbia United Tribes Fisheries Center, Eastern Washington University Department of Biology, Cheney, Wash. Fisheries Technical Report No. 2, (various authors) at p. 26, quoting from Bohm and Holston (1983).

11. Treaty between the United States and England, June 15, 1846.

12. Id. at 30.

13. Treaty between the United States and the Yakama Nation of Indians, concluded at Camp Stevens, Walla Walla Valley, 9 June 1855, ratified by the Senate, 8 March 1859, proclaimed by the President of the United States, 18 April 1859.

14. Act of March 23, 1871, ch. 120, § 1, 16 Stat.544, 566 (codified at 25 U.S.C. § 71).

15. *Antoine v. Washington,* 420 U.S. 194 (1975).

16. Article VI, cl. 2, Constitution of the United States.

17. The Upper Columbia Spring Chinook, also known as king, or Tyee; the Columbia River Chum or dog salmon; the Oregon Coast, Southern Oregon/Northern and central California; Sockeye is on the endangered list for the Snake River. Steelhead, as well as Chum, Coho, Sockeye, and Chinook are all on the endangered species list on the upper Columbia, threatened on the upper Willamette, and the mid-Columbia, Snake, or lower Columbia areas.

18. Title 18 U.S.C. § 1091(4). Genocide—subjects the group to conditions of life that are intended to cause the physical destruction of the group in whole or in part.

19. Enacted into Law on September 16, 1964.

20. Harold Ickes, Secretary Department of Interior. Advance Press Release for the June 1947 meeting at Walla Walla, Wash. by the Bureau of Reclamation. (Emphasis added in text).

21. *A River Lost*, Blaine Harden (W.W. Norton & Co., 1997), 237.

22. 16 U.S.C. 791–828c; Ch. 285, June 10. 1920, 141 Stat.

23. The Nez Perce Tribe have agreed to water adjudication in the Fifth District Court that will pay them $96 million in three separate funds and provide them with eleven thousand acres of land in exchange for their water. Native American Rights Fund report on Snake River Basin Adjudication, Nez Perce Tribe Water Rights. Moore, Steve, March 23, 2005.

24. In 1783, Washington himself held the attitude that "the Indians after all were little different from wolves, … both being beasts of prey, tho' they differ in shape" and were "hunted as beasts" in a "war of extermination." *The Conquest of the New World, American Holocaust*, Oxford University Press, 1992, Stannard, David E. at 119, quoting Richard Drinnon's *Facing West: The Metaphysics of Indian-Hating and Empire-Building,* 1980 (Minneapolis: University of Minnesota Press).

25. i.e., the Yakama, Umatilla, Nez Perce and Warm Springs tribes.

26. i.e., the Spokane, Coeur d'Alene, Kalispell and Kootenai, joined by the Colville in 1999.

Works Cited

Brown, John A., and Robert H. Ruby. *The Spokane Indians* (Norman: University of Oklahoma Press, 2005).

Chance, David H. *People of the Falls* (Kettle Falls Historical Center, 1986). Cited in Kershner 2006.

Columbia River Inter-Tribal Fish Commission. n.d. "The Tribal Vision for the Future of the Columbia River Basin and How to Achieve It." Web site at http://www.critfc.org/legal/vision.pdf.

Columbia River Inter-Tribal Fish Commission. Newsletter, June, 2008.

Eels, Rev. Cushing. Letter to *Missionary Herald*, February 25, 1840. Quoted in Kershner 2006.

Kershner, Jim. "For a Time a River Ran Through It," *Spokesman Review* (October 22, 2006).

Ray, Verne. *The Sanpoil and Nespelem.* (New York: AMS Reprints 1980).

Seyler, Warren. Keynote Address. Presented at the Columbia River Basin Tribal Water Quality Conference, Spokane, Washington, (November 15–16, 2000).

Woody, Elizabeth, Seth Zuckerman, and Edward C. Wolf, eds. *Salmon Nation: People and Fish at the Edge* (Portland, Ore: Ecotrust, 2003).

CHANGING VALUES

Managing Transboundary Natural Resources: An Assessment of the Need to Revise and Update the Columbia River Treaty

Matthew McKinney

INTRODUCTION

Historical Perspective

The Columbia River Basin is the fourth largest river basin in the United States, equal to the size of France (see Appendix A). It includes parts of Oregon, Montana, Idaho, Nevada, Wyoming, Utah, Washington, and British Columbia (Lang 2010; Lang and Carriker 1999). The Columbia River has ten times the flow of the Colorado River and two and one-half times the flow of the Nile River. It is one of the most hydroelectrically developed river systems in the world, with a generating capacity of more than 21 million kilowatts. There are eleven dams on the main stem in the United States and three in Canada, in addition to more than four hundred other dams for irrigation and hydropower on tributaries. While this infrastructure has generated many benefits in the form of power and flood control, many people argue that it has adversely impacted fish, navigation, irrigation, recreation, and indigenous cultures.

In 1944, planners in Canada and the United States recognized that cooperative development of the basin might generate more benefits than each country acting independently (U.S. Army Corps of Engineers and Bonneville Power Administration 2009). The planners requested that the International Joint Commission (IJC) study the feasibility of cooperative development in the Columbia Basin. From 1944 to 1959, the IJC studied a range of options to cooperatively develop and manage resources within the basin. Following years of negotiation, the governments of Canada and the United States ratified the Columbia River Treaty and Protocol in 1964.

The "Treaty Between the United States of America and Canada Relating to Cooperative Development of the Water Resources of the Columbia River Basin" (CRT) is considered one of the most far-reaching water treaties in the world (15.2 U.S.T. 1964).[1] It required Canada to build three new storage dams—Keenleyside, Duncan and Mica (referred to as The Treaty Dams)—to optimize flows for hydroelectric power and flood control in both nations. The Treaty Dams provide more constant year-round streamflows, as spring floods from snowmelt are held back and released throughout the year. In return for building the dams, Canada is compensated by the United States through two mechanisms that have efficiently transferred hundreds of millions of dollars annually. First, the United States paid Canada for half of the estimated flood control benefits provided by the Treaty Dams until 2024 (after 2024 the U.S. will pay operating costs and economic losses for any requested Canadian flood control operation). Second, the treaty set-up a system in which the United States compensates Canada for one-half of the downstream hydroelectric power benefits generated by the upstream storage dams (known as the "Canadian Entitlement"). Canada also retains the rights to use all of the power generated by the Canadian Treaty Dams.

The CRT is considered by some experts to be one of the most sophisticated transboundary natural resource treaties in the world. Unlike other international water treaties, it does not focus on allocating fixed quantities of water, but rather allocates a mix of "benefits" to each country. The primary benefits of the CRT—hydroelectric power, flood control, and compensation—were largely fixed in 1964. The governance of the Columbia River under the original CRT thereby excludes many of the values that society has found increasingly important in the intervening years, particularly the quality and quantity in-stream flows for ecosystem health, as well as legal obligations to tribes for treaty-based water and fishery resources.

The administration of the CRT is governed by what are commonly referred to as the "Entities"—including the Bonneville Power Administration and Army Corps of Engineers from the United States, and BC Hydro from Canada. A Permanent Engineer Board is responsible for the preparing and approving an Annual Operating Plan. Appendix B presents an organizational chart for the CRT.

The CRT does not have an expiration date. However, after sixty years of implementation (no sooner than 2024) the treaty can be terminated and renegotiated as long as either the United States or Canada give at least ten years notice of their intent to terminate.[2] The flood control agreement expires in 2024.[3] While these deadlines may seem to be beyond the planning horizon of most political decision makers, many professionals involved in management of the basin have started to think about how, if at all, the CRT might be revised and updated in light of all the changes that have occurred since 1964.

Purpose and Methods

The purpose of this chapter is to present and discuss the findings of an assessment of the need to revise and update the CRT. During fall 2008, five graduate students in the University of Montana's Natural Resources Conflict Resolution Program set out to interview people representing the "Entities" in the United States and Canada (i.e., those agencies with primary authority to formulate and implement the CRT), other government agencies, tribal governments, and selected scholars.

Due to time and financial constraints, the team was forced to limit the number of interviews. A more complete and robust assessment would include interviews with representatives from various interest groups and other stakeholders.

Appendix C presents a list of interviewees, and Appendix D presents the interview questions. In addition to the interviews, we reviewed scientific, legal, and other commentary on the merits of the CRT. We also provided interviewees a chance to review and comment on earlier drafts of this and to provide any additional technical information. We also presented and discussed the results of this assessment at the 2009 conference.

We are very grateful to all of those people and organizations that provided feedback. Throughout this assessment, the team was guided by the Code of Professional Conduct for the Association for Conflict Resolution—which, in essence, compels members of the team to operate as nonpartisan, impartial servants of all stakeholders and decision makers.

The following sections of this article present our findings, along with conclusions and recommendations that build on what people told us and best practices in multiparty negotiation and collaboration. We do not attribute ideas or comments to interviewees, preferring to operate on the principle that what is most important is what was said, not who said what. This principle allows everyone to consider the merits of the ideas presented, regardless of who said what. The people we interviewed represented their own viewpoints, not official positions of their organizations.

Our hope is that this assessment helps inform and invigorate discussions about the future of the CRT.

PERFORMANCE TO DATE
What Is Working?
Nearly all of the interviewees said that the CRT is working well for its intended purposes —hydroelectric power production and flood control. Many people also agreed that the technical operations of the CRT have been very successful (i.e., the combination of the operating committee, annual operating plans, and the Permanent Engineering Board). One person asserted, "Lots of things are working well. The CRT is probably one of the most successful agreements of international cooperation." Another person explained how floods have been reduced with the infrastructure of new dams and the careful management of hydropower systems. One interviewee said that the treaty "contributed hugely to the reduction of global warming by reducing the use of fossil fuels," and has also "provided the Northwest with some of the cheapest electricity in the world."

Although nearly all respondents said the treaty is working well for its original purposes, many interviewees cited various problems with the CRT. These difficulties—explained more fully below—include adverse impacts on fish and wildlife in both Canada and the United States; the loss of thousands of acres of habitat; and harm to tribal interests (particularly fishing and hunting), community interests, and farming interests in both Canada and the United States. A few respondents, however, stated that the operating team does a pretty good job of integrating fishery interests when it can.

The interviewees are somewhat split over the distribution of benefits within the CRT. Some say, "The framework allows some of the economic benefits to be divided on an equitable basis between the countries." Others said that they "would like to see a better exchange of benefits across the border."

Finally, some respondents said that no single aspect of the treaty is working well. In fact, some interviewees failed to answer the original question (*What is working well with respect to the CRT?*), and instead cited various ways in which the CRT is not working well—it does not satisfy current social and environmental needs, it does not allow for the legal rights of tribes in Canada and the United States to be met, and it does not provide sufficient opportunity for stakeholders and the public to be informed and engaged. These issues are addressed in more detail below.

Drivers of Change

According to the interviewees, the primary reason to revise and update the CRT is the changes that have taken place since ratification of the treaty. One interviewee succinctly noted that "we have moved from a time when the primary interests (hydropower and flood control) were readily quantified and generally complimentary, to a time with many more interests that are extremely difficult to quantify and often mutually exclusive." This sentiment was echoed by many of the interviewees.

The interviewees identified six specific drivers or reasons to revise and update the Columbia River Treaty: (1) ecosystem health; (2) expectations for public participation; (3) tribal interests and rights in both Canada and the United States; (4) population growth; (5) climate change; and (6) other considerations. Appendix E, Chronology of Major Events Since 1964, provides additional information on issues and decisions that influence the management of the Columbia River.

1. *Ecosystem Health*. Nearly all of the interviewees explained that issues around ecosystem health are one of the most compelling drivers to revise and update the CRT. This driver is a catch-all term for a number of specific issues identified by the respondents, including but not necessarily limited to:

(a) The emergence of ecosystem health values as reflected by a series of environmental laws passed by the U.S. Congress since 1964, including the National Environmental Policy Act of 1969 (42 U.S.C. 1964), the Clean Water Act of 1972 (33 U.S.C. 1972), and the Endangered Species Act of 1973 (16 U.S.C. 1973). The most influential of these to the CRT is the Endangered Species Act, which provides protections for endangered and threatened plants and animals (listed species) and the habitats upon which they depend.

(b) The impact of Treaty Dams and reservoir operations throughout the basin on fish species, including salmon and resident fish—particularly as these impacts influence the maintenance of commercial, tribal, and recreational fisheries.

(c) The impact of land use development, resource development (e.g., mining), and transportation infrastructure on fish resources.

(d) The importance of conserving and restoring fish and other wildlife in recognition of traditional cultural, spiritual, and legal rights of the tribes and First Nations consistent with the tenants of environmental justice (Wood 1995).

(e) The adequacy of water supply in the face of continued population growth and climate change.

(f) The degradation of water quality from point sources (e.g., industrial and municipal effluent) and nonpoint sources (e.g., urban growth, agriculture, and forestry) and the impact on fishery and other resource values.

The interviewees concluded that these and perhaps other ecosystem health issues are not adequately taken into account in the existing treaty. Some of the interviewees also noted that these ecosystem health interests are often at odds with each other—for example, upstream and downstream fisheries may need water retained or released at conflicting times.

2. *Expectations for Public Participation.* Most of the interviewees agreed that another reason to revise and update the CRT is an increased expectation for public and stakeholder involvement in future management of the system.

As explained above, most interviewees agree that the Columbia River must be managed to meet a broader and more complex set of values beyond the original focus on hydropower and flood control. These respondents explained that the best way to integrate the interests and concerns that revolve around ecosystem health, tribal rights, and recreation is to make the process of managing the Columbia River system more open, transparent, and inclusive. Several people cited the Pacific Salmon Commission as one model to improve the governance of the basin under the CRT—in large part because it provides meaningful opportunities for tribes and stakeholders to be involved in decision making and implementation (Pacific Salmon Commission 2010).

Some respondents explained that it is not only important to engage organized stakeholder groups and unaffiliated citizens in setting priorities for the system, but also to involve them in monitoring and adapting the system over time.

Not all interviewees agreed about whether or when citizens and stakeholders should be engaged. Some of the respondents suggested that the best time to engage non-governmental interests is after the entities and other key actors have a chance to work through some of the issues and propose some type of revised plan.

3. *Tribal Rights.* According to many respondents, another significant reason to revise and update the CRT is to fulfill the trust responsibilities and legal obligations of the federal government with respect to the interests, customs, and legal rights of First Nations and Native Americans (collectively referred to here as tribes). The existing operations of the treaty dams have caused further damage to what was an already compromised fishery,[4] and to which the tribes have a reserved legal right. The interviewees noted that the U.S. government has a "trust and fiduciary responsibility to the tribes on actions that affect their treaty-protected resources." These resources include salmon fisheries, other fish species, wildlife, and native plants. The tribes' legal rights have been established through a long history of legal action.[5]

In 1855, Native Americans in the Columbia River Basin signed a series of treaties with the United States which ceded most of their lands, but reserved exclusive rights to fish and hunt within their reservations as well as rights to fish in usual and accustomed places off the reservation (Center for Columbia River History). First Nations have similar rights based on Section 35 of the Constitutional Act (1982) that gave constitutional protection to the aboriginal and treaty rights of the First Nations in Canada. To exercise these rights, some interviewees explained that the tribes have had to resort to the courts. Over time, the courts have increasingly recognized these legal rights as reflected by several notable cases[6] (*Winters v. United States*,[7] *Sohappy v. Smith/U.S. v. Oregon*,[8] *U.S. v. Washington*,[9] *Settler v. Lameer*,[10] and the Haida and Taku River decisions in Canada[11]). In addition, the Pacific Salmon Treaty incorporated tribal rights as it set out to cooperatively provide recommendations to managers of Pacific salmon stocks.[12] With so many existing court decisions and treaties governing fisheries management, some interviewees made it clear that the tribes do not want fisheries management simply "incorporated" into the CRT per se. Rather, they want the governance of the basin under the treaty to meet their legally defined interests on par with other designated benefits or values of the system.

Some interviewees explained that First Nations have been displaced and otherwise negatively impacted during the creation of CRT storage reservoirs and dams in Canada. First Nations apparently lost significant hunting, fishing, and gathering land that they have historically relied on. Interviewees also explained that First Nations are concerned about the disruption of burial grounds and artifacts by the fluctuating water levels of reservoirs. These respondents went on to explain that First Nations have not been adequately compensated for the sacrifices they have made to develop the Columbia River under the CRT, and that their interests are not being represented in the current treaty.

In conclusion, some interviewees explained that the needs and interests of tribes should be reflected in any process to revise and update the CRT. Most of the interviewees concluded that tribes from Canada and the United States should have a decision-making role in any process to revise and update the CRT.

4. *Population Growth.* Some interviewees expressed concern about the potential impact of population growth on the management of the Columbia River system, and the system's ability to meet increased demands for water and energy. When the CRT was ratified, it was anticipated that some of the dams on the Columbia would provide growing populations with water and power (in fact, consumptive use is the highest priority of the CRT). However, the Columbia Basin has continued to grow at unprecedented rates (between 20 to 40 percent in urban areas since 1960) (Independent Scientific Advisory Board 2007). If populations in the lower basin continue to grow at the current rates, many interviewees agree there will be a significant increase in demand for water and power in major metropolitan areas of the Pacific Northwest.

Energy producers have voiced concerns about meeting future demands in the region (Northwest Power Planning Council 2002). Some have expressed concern about California's continued dependence on Columbia River hydropower, which could indirectly increase electricity rates as the demand for this power increases in the Northwest. Some interviewees concluded that the CRT should be revised and updated to prioritize the future water and energy needs of the basin before exporting either resource out of the basin.

5. *Climate Change.* Some interviewees said that the uncertainty and potential impacts associated with climate change (particularly its impact on future water availability) is another compelling reason to revise and update the treaty. In addition to a lack of information on climate change, interviewees suggested that changes to snowpack, temperature, and precipitation patterns would likely influence the management of the Columbia River to meet multiple interests. While the impacts of climate change are uncertain, interviewees explained there is a growing need to develop management scenarios to both mitigate and adapt to whatever impacts may emerge.

Some interviewees said that some public resource agencies have started working on this issue, but that more attention might be focused on it in the near future. Others suggested that the CRT, in its current form, could already accommodate these issues through adjustments in operating plans.

6. *Recreation.* The impact of reservoir management on recreation in and around reservoirs and the associated impact on tourism was also mentioned by some interviewees as an important driver to change the CRT. At least one interviewee expressed concern about the need to maintain consistent water levels for reservoir-based recreation. The ability to access reservoirs and associated recreational resources is impaired when reservoirs levels fluctuate to meet downstream needs and interests.

Some interviewees noted that recreation objectives are sometimes met through special operating agreements authorized by the Detailed Operating Plan, but that these objectives must be balanced against competing objectives including power.

PROSPECTS FOR THE FUTURE

People's Preferences

In light of the changes that have taken place since 1964, we asked interviewees what their preference was in terms of the future of the CRT—maintain the status quo, terminate the treaty, or revise and update some or all aspects of the treaty.

Most interviewees expressed a desire to revise and update the CRT. A frequent sentiment was that "things have changed" and there are additional considerations that were not prevalent during the 1960s negotiation, such as climate change, sensitivity to ecosystem health, consideration of fish and wildlife, expectations for public involvement, and increased pressure (socially and legally) for tribal input. In addition, many respondents feel there is potential for more equitable sharing of benefits.

While most of the interviewees agree that the treaty needs to be revised and updated, many of them also explained that they hope this can be done short of renegotiating the entire treaty. These respondents seem to embrace a principle of "keep the foundation in terms of what is working, and build on that foundation to revise and update the CRT." Some of these interviewees expressed a concern about opening Pandora's box if the entire treaty is open for renegotiation, fearing that valuable benefits might be lost. One respondent put it very clearly: "[renegotiation is] probably the best option, as it would allow consideration of many facets, and allow for broad consultation with stakeholders. However, for almost the same reasons, a new treaty is probably impossible to accomplish, given the diversity of values, and rampant self-interest." Other interviewees concluded that the only option to fully incorporate their interests would be a full renegotiation of the CRT.

A few interviewees said that letting the CRT continue as is may be the easiest and, therefore, most preferred option. These respondents explained that opening the CRT to full and complete renegotiation and involving a diversity of interest groups has the potential to "dissipate more benefits than it could possible create." However, other respondents think that the existing CRT can integrate some, if not all, of the interests not currently reflected in the CRT through various procedures built into the existing framework. (The following section presents some of these more "informal" approaches to revise and update the CRT). These options provide an opportunity to reduce the risks inherent to termination and renegotiation (e.g.. the potential for national interests to over run the interests of the Pacific Northwest).

Finally, none of the interviewees expressed a preference to terminate the treaty. Some respondents noted that termination would most likely result in a loss of the existing benefits associated with the CRT. This sentiment is captured by one respondent's answer: "The CRT cannot stay the way it is and I don't see how you can operate the river in these times without a cross-border agreement, so letting it go away is not an option." While none of the interviewees said that the CRT should be terminated, some respondents speculated that other people or agencies might believe that termination is in their best interest.

In sum, nearly all interviewees agree that the CRT should be revised and updated. The question is, how should it be revised and updated? To answer this question, the following section reviews the legal and institutional options available. We offer these options as a place to begin a conversation, keeping in mind that we are not legal experts and cannot fully explore the ramifications of these options in this chapter.

Legal and Institutional Options

Based on our research, there appear to be several options potentially available to revise and update the CRT. The purpose of this section is to simply lay out, in a preliminary way, what the legal and institutional options are to revise and update the treaty, and to thereby inform and invigorate ongoing conversations.

Option 1—Maintain the status quo: As explained earlier, the CRT has no expiration date. If the United States and Canada agree (and neither country sends the other a notice to terminate), the existing treaty could presumably stay in place. One potential complication with this option is that Canada's obligations for annual flood control operation expire after sixty years (in 2024). At this time, the United States Congress would have to authorize additional payments to Canada for providing any requested flood control measures (Bonneville Power Administration and the Army Corps of Engineers 2008).

Option 2—Terminate the treaty: A second option is to terminate the CRT by giving formal notice any time after 2014 (Article 19). If one country chooses to terminate without renegotiating some or all of the existing CRT, the governance of the Columbia River would default to the 1909 Boundary Waters Treaty (Article 15 Section 2). Under this option, each country would maintain exclusive control of the use of the river on their side of the border (Boundary Waters Treaty Article II 1909). It also means that consent from the International Joint Commission must be obtained for any change in the flow of water at the boundary (Boundary Waters Treaty Article IV 1909).

Option 3—Revise and update the treaty: The third option is to revise and update the CRT. According to our research, there are several ways to accomplish this objective. The following options are presented from most formal (and therefore, perhaps hardest) to most informal (and perhaps easiest). The CRT itself does not specify any procedures to revise or update the treaty; it only provides a procedure for terminating the treaty.

A. Renegotiate the treaty. As implied above, the existing CRT could be renegotiated after either Canada or the United States submit a notice to terminate. If both countries agree to renegotiate, then they can presumably proceed with whatever renegotiation process they determine appropriate under existing international and federal laws and customs. This option might best be referred to as a "formal renegotiation" of the CRT under the auspices of existing law and practice, which would include the United States Congress and the Canadian Parliament.

B. Negotiate a "partner treaty." A "partner treaty" could be negotiated that elaborates on and amends the CRT. This option may, however, raise questions about how to resolve potential conflicts between the CRT and the "partner treaty." In addition, since it would be a new international treaty, it would need the approval of the United States Congress and the Canadian Parliament.

C. Negotiate formal amendments. Yet another option to revise and update the CRT is to seek formal amendments. According to international law, a treaty may be amended under the same rules that govern creation of the treaty, as long as the current treaty does not prohibit this (Vienna Convention on the Law of Treaties 1969). Therefore, amendments go through the same formal diplomatic process as a formal negotiation, but do not necessarily open the whole treaty for consideration. Currently, there are no formal amendments to the CRT.

D. Negotiate and implement protocols. Another option is to engage in diplomatic discussions without the presumption of terminating and completely renegotiating the entire treaty. After the CRT was initially signed in 1961, additional negotiations about the distribution of benefits and operations were completed with a diplomatic "Exchange of Notes" resulting in the Protocol (dated 1964), which is attached to the CRT (Bankes 1996). Although this Protocol contains some significant provisions, it is viewed as consistent with the original CRT and therefore not considered a formal renegotiation needing ratification. Protocols are simply another frequently used form of international negotiation.[13] The use of this option begs the question of how far the Entities can go in revising and updating the treaty through the use of protocols before such changes trigger a formal renegotiation. Any substantial changes require consultation with the United States State Department and the Canadian Department of Foreign Affairs and International Trade (DFAIT) to authorize the agreement with a diplomatic "Exchange of Notes."

E. Incorporate new "Entities" or advisors. In the United States, an Executive Order (EO) issued by President Johnson in 1964 carried out the implementation of the treaty (Executive Order 11177 1964). The EO designated the U.S. Entities (BPA and the Corps) and the formation of the U.S. Section of the Permanent Engineering Board (PEB). This EO may be modified by the president, which may provide an opportunity to expand participation on the U.S. side. For example, the president could modify the composition of the "U.S. Entity," perhaps including the U.S. Fish and Wildlife Service, the National Oceanic and Atmospheric Administration, Environmental Protection Agency, and even tribal representatives as legally recognized sovereign nations within the United States. Alternatively, Section

204 of the existing EO states that the U.S. Section of the PEB may call upon other federal agencies to aid it in "the performance of its functions." A third option may be for the secretary of state, as the lead negotiator on behalf of the United States, to assemble a negotiating team composed of the existing Entities as well as federal, tribal, and regional governments.

F. Adjust annual operating plans. According to Article 14, Section 2(k) of the CRT, another option to revise and update the treaty is to adjust the annual operating plans. According to this section, the implementing agencies have the authority for "preparation and implementation of detailed operating plans that may produce results more advantageous to both countries than those that would arise from operation under the plans [required by the CRT]." Presumably, this means that the Operating Committee and/or the Entities have the authority to integrate what are sometimes referred to as "nontreaty interests" (i.e., those that are not currently recognized in the CRT) into the annual operating plans, as long as such actions are viewed as beneficial to both countries. The Operating Committee has also used Supplemental Operating Agreements to include objectives other than power and flood control. Whether or not these approaches to revising and updating the CRT are acceptable to all interested and affected parties is an open question. Nevertheless, the Operating Committee has apparently used this option in the past (Muckleston 2003).[14]

Options on How to Proceed

Although we did not assume interviewees would conclude that the CRT should be revised and updated, we wanted to ask various questions related to the process of revising the CRT if they did. During the interviews, we framed a series of questions around the process of renegotiating the CRT. We quickly learned, however, that the word "renegotiating" is a term of art, interpreted by many people as the formal process of terminating the existing treaty and negotiating a completely new treaty. In response to this potential confusion, we have chosen to talk about "revising and updating" the CRT, whether the process to do that is more or less formal.

Convening the Dialogue

Many interviewees said that if the CRT were renegotiated in any formal sense (in other words, if either the United States or Canada terminate the existing treaty and seek to negotiate a completely new treaty), then the conveners would be the U.S. Department of State and the Canadian Department of Foreign Affairs and International Trade. Many of the interviewees explained that this is a simple matter of law—international negotiations must start with the highest level of government. In other words, they claimed that the CRT, the law governing international treaties, and law in both the United States and Canada dictate who can convene formal negotiations over international treaties and transboundary resources.[15]

Some interviewees, however, suggested that the authority to convene a multiparty negotiation to simply revise and update the treaty could be delegated to the Entities (U.S. Army Corps of Engineers, Bonneville Power Administration, the Province of British Columbia, and BC Hydro). As explained in the previous section, one option along these lines is to amend the EO that directs implementation of the CRT and officially name other agencies and perhaps tribes as part of any formal negotiation process.

A number of other respondents said that if the CRT is revised and updated through some type of informal process as explained in the previous section, then the question of who convenes the process is a bit more open. The interviewees identified a number of options along this line:

(a) Sovereign entities, including tribes, United States, and Canada;

(b) Entities and other governmental agencies (e.g., U.S. Bureau of Reclamation, Environmental Protection Agency, National Oceanic and Atmospheric Administration, U.S. Fish and Wildlife Service, BC Ministry of Environment, Ministry of Energy, Mines and Petroleum Resources, Environment Canada, etc.)

(c) Northwest Power and Conservation Council[16] and Columbia Basin Trust—two distinctly regional entities;

(d) Permanent Engineering Board;

(e) International Joint Commission; and

(f) First Nations on both sides of the 49th parallel.

The interviewees who advocated a more informal approach to convening seemed to be open to combining the options presented above. These interviewees effectively articulated the following principle around the question of convening—create a homegrown process, one that is convened and coordinated by and for the people within the Columbia River Basin; the convening body should be viewed as credible and legitimate by all affected parties, particularly in terms of making sure all of the interests within the basin are sufficiently represented.

Representation—Who Should Be at the Decision-making Table?
A small minority of interviewees suggested that only representatives from the Entities should be at the table. An equally small number of respondents said that everybody who has an interest or stake in the basin should be included.

The majority of respondents said that any process to revise and update the CRT should provide meaningful opportunities for all decision makers, stakeholders, and citizens to influence the process and the outcomes. Realizing that it would be cumbersome to have representatives from every conceivable stakeholder group at the table, most of these individuals were quick to mention that the process to revise and update the CRT will need to be multi-faceted. That is, there should be a core negotiating group, one or more advisory groups, and multiple opportunities for public participation.

In terms of who should be in the core negotiating group, respondents identified the following options:

(a) Only sovereign entities, including the tribes, United States, and Canada (i.e., the Entities and tribal representatives, similar to the negotiation process that led to the Pacific Salmon Treaty (Pacific Salmon Commission 2010);

(b) Some combination of the Entities, tribes,[17] and:

- British Columbia, Montana, Idaho, Oregon, and Washington;

- Groups that have legal rights to resources within the basin (e.g., irrigators);

- Representatives of identifiable "communities of interest" (that is, groups of individuals that share a common interest, such as conservation, recreation, municipalities, utilities, irrigators, fisheries, etc.); and

- Multi-stakeholder, place-based groups, such as the Lower Columbia River Estuary Partnership, Clark Fork River Watershed Council, and the watershed planning groups in British Columbia.[18]

Public Participation—How to Inform and Engage Unaffiliated Citizens
Almost without fail, the interviewees stated that any process to revise and update the CRT should be open, inclusive, and transparent. The core concern here seems to be that all citizens and stakeholders should have an opportunity to (1) be informed and educated; and (2) provide input and advice. As the interviewees moved beyond this principle to clarify when and how the public should be involved, their responses were quite varied.

The respondents identified the following options with respect to when the public should be involved (at this point it is helpful to distinguish between "organized" stakeholder groups and "unaffiliated" citizens; the focus of this section is on the latter):

(a) At the beginning to help frame values, issues, options, and priorities;

(b) At key stages throughout the process, such as those defined by the National Environmental Policy Act (Council on Environmental Quality 2007);

(c) After they are sufficiently informed about the choices and consequences, and before the final plan is ratified.

In terms of how to involve the public, most of the interviewees seemed to agree on a principle that the process for public participation should be jointly designed by all of the affected parties: the Entities, other governmental agencies, tribal representatives, and both communities of place and communities of interest.

More specifically, the interviewees identified the following options on how to involve the public:

(a) Include one or more "public" representatives at the negotiating table;

(b) Convene regional (i.e., basin-wide) and sub-regional dialogues to inform and educate the public, seek their input and advice to clarify values and priorities;

(c) Coordinate separate public processes within the United States and Canada;

(d) Follow the legally required opportunities for public involvement as defined by the National Environmental Policy Act;

(e) Encourage citizens to provide input and advice through informal means such as lobbying elected officials, talking to interest groups, etc.

When asked who might be in the best position to facilitate meaningful public involvement, the interviewees identified the following options:

(a) A team of facilitators and mediators;

(b) Agencies;

(c) Universities;

(d) Communities of interest; and

(e) Place-based groups.

Process Management—The Role of Facilitation and Mediation

The responses to a question about facilitation and mediation were quite varied. On the one hand, some of the interviewees said that an impartial, nonpartisan person or team was not used during the original CRT negotiations, and it is not a common practice in negotiating international, transboundary agreements around natural resources. Nor is it legally required. Other respondents said that, while facilitation might be helpful, the Entities will not likely support such a role and the diplomats representing the different countries are (at least in theory) capable of playing this role.

On the other hand, the majority of respondents seemed to embrace something like a principle that recognizes the value of using a nonpartisan, impartial person or team to:

(a) Assess the needs and interests of other stakeholders, similar to what has been done with this assessment;

(b) Help design an inclusive, informed, transparent, and effective process to revise and update the CRT;

(c) Organize and convene regional and sub-regional dialogues and other opportunities to facilitate public involvement;

(d) Work with communities of interest and place-based groups to prepare for participation in the process to revise and update the CRT;

(e) Facilitate communication, understanding, and agreement across government agencies;

(f) Mediate, as necessary, the conversation among the core group of decision makers (whoever that is).

The interviewees who support this role also said that a facilitator or mediator must be impartial and nonpartisan, and should have some knowledge about the treaty, issues, players, and the process to revise and update the CRT. Along these lines, the respondents suggested that one or more of the following groups might play such a role:

(a) Northwest Power and Conservation Council *and* Columbia Basin Trust;[19]

(c) International Joint Commission; and

(c) University-based public policy and conflict resolution centers within the Columbia River Basin.

Information—The Need for Scientific and Public Learning

Most of the interviewees asserted that there is a huge need to promote and support both scientific and public learning. In terms of what information is needed, respondents offered the following suggestions:

(a) Clarify the process (and options) to revise and update the CRT;

(b) Identify public values, interests, and priorities – throughout the basin;

(c) Clarify legal rights (e.g., the rights of First Nations and Native Americans) and how different legal rights may conflict (e.g., endangered species vs. power and water supply);

(d) Examine the likely influence of numerous variables on what is preferable and what is doable in terms of revising and updating the CRT, such as:

- Climate change;
- Environmental laws adopted since 1964;
- Change in population;
- Demand for energy;
- Demand for water; and
- Species at risk.

(e) Map the options and consequences (costs, benefits, and trade-offs) to accommodate multiple interests and communicate this information to citizens, stakeholders, and decision-makers; and

(f) Make hydrological models and data available to citizens and stakeholders to facilitate a broad-based understanding of the trade-offs of various policies.[20]

In terms of how to gather and distribute the desired information, many of the respondents seemed to embrace the principle of joint fact finding.[21] That is, several interviewees suggested that existing information should be pooled and made publicly available.[22] Then, based on what they know, they can begin a dialogue to jointly identify what they don't know, what they need or want to know, and how they might go about learning together. The respondents suggested that this approach to gathering, evaluating, and disseminating information would increase the chances that the information is politically relevant, scientifically valid, and widely accepted.

Some of the more specific methods recommended by interviewees include:

(a) Survey to clarify public values, interests, and priorities;

(b) Scenario building to examine options, consequences, and trade-offs;

 (c) Modeling to assess the impacts of alternative operating scenarios;

(d) Research to clarify legal rights and potential conflicts among legal rights;

(e) Dialogue and deliberation to facilitate communication, understanding, and agreement on how to revise and update the CRT; and

(f) A public education campaign to raise awareness and seek informed input and advice.

To facilitate joint fact-finding, the interviewees suggested that the following resources should be utilized to help generate and distribute the desired information: tribes, universities, Northwest Power and Conservation Council, Columbia Basin Trust, students, watershed groups, agencies, and consultants.

Governance—Implementing and Adapting to Change

When asked what type of governance arrangement would be most effective to implement the treaty, many interviewees suggested that the answer would depend on the outcome of any process to revise and update the CRT. Others said that the existing system is working extremely well, particularly the Permanent Engineering Board (PEB), the Operating Committee (which meets frequently), and the annual operating plan. They said this success is due in large part to a history of trust and technical understanding. This consortium is responsible for assembling flow records, resolving differences that may arise among competing uses, and creating annual reports of accomplishments. The PEB consists of two members appointed by Canada and two by the United States.

The underlying idea or principle here seems to be build on what works and adjust accordingly (McKinney and Johnson 2009). In other words, if the treaty integrates new interests, including but not limited to tribes and ecosystem health, then the existing boards and committees designed to govern and implement the treaty should be accordingly revised. For example, if water quality interests are integrated into the CRT, perhaps is makes sense to have a representative from the Environmental Protection Agency and Environment Canada to serve on one or more of the governing bodies. Likewise, if fisheries are integrated, a representative from the National Oceanic and Atmospheric Administration (which oversees endangered species recovery) might be appointed. Some interviewees opposed the idea of incorporating new players into the governance of the CRT, preferring to clarify that such interests should be better integrated and balanced with hydropower power and flood control.

Other respondents suggested a menu of possibilities in terms of governing the system:

(a) Create a standing scientific and technical work group that can research and respond to questions and issues (i.e., fishery management) as they arise, perhaps then reporting to an entity such as the PEB;[23]

(b) Create a policy board to address value-based issues, to complement the work of the PEB and operations committee (which is focused more on the technical operations of the system);

(c) Create a consultative committee of other agencies and stakeholders to monitor, evaluate, and suggest adaptations to the operations of the system;

(d) Create a transboundary commission that would include both policy and technical components. One option here might be to amend the Pacific Salmon Commission; another is to explore the possibility of creating some type of integrated commission from the Northwest Power and Conservation Council and the Columbia Basin Trust, which seem to have more legitimacy, credibility, and capacity to integrate multiple uses. A third option identified by at least one interviewee is to create an International Watershed Board under the auspices of the International Joint Commission (International Joint Commission 2010).

(e) Clarify governance protocols in terms of how decisions are made, disputes resolved, and goals and strategies are adapted.

Success and Barriers

Interviewees identified a number of indicators of success, as well as barriers to revising and updating the CRT. The comments on success and barriers include both substantive and process issues, which are integrated throughout the following discussion.

Indicators of Success

1. Build on what is working. Most of the interviewees asserted that success should be measured, at least in part, by building on what is working: generating and distributing power, preventing floods, and reducing the use of fossil fuels. Maintaining an equitable distribution of benefits between the United States and Canada was also mentioned along these lines.

2. Prevent harm and provide more explicit benefits to tribes. Most of the interviewees feel that the CRT's impact on indigenous people must be addressed in any process to revise and update the treaty. The sentiment of one tribal representative captures the essence of this issue: "The cultural aspects of flooding

must be addressed. There are burial grounds and artifacts that are constantly being disrupted by fluctuating water levels in the reservoirs. Tribes are constantly forced to reexamine how to handle burial remains. It is like having your family dug up on a regular basis." Some of the respondents explained that tribal interests include compensation for past harms, as well as the prevention of future harms (or, more positively, the provision of benefits in the future).

3. Balance multiple uses and benefits. In addition to building on what is working, and accommodating tribal interests, most of the interviewees said that one of the most important indicators of success will be to strike a balance between the multiple uses or benefits of the river, including power generation, flood control, ecological health (such as fish and wildlife concerns), cultural interests, recreation, and other offstream uses. An obvious element of this indicator of success is the need to be responsive to laws, policies, and judicial decisions that have been adopted since 1964, including the Endangered Species Act, National Environmental Policy Act, Pacific Salmon Treaty, and judicial decisions recognizing the rights of tribes. Another aspect is to equitably balance or distribute benefits and costs upstream and downstream.

4. Develop strategies to mitigate and adapt to climate change. Most respondents said that a new and improved CRT must include strategies to mitigate and adapt to climate change. Given that there is a great deal of uncertainty about how climate change may affect the amount and timing of precipitation, these respondents explained that it is important to consider a range of scenarios, impacts, mitigation procedures, and strategies to adapt the operating system as new information becomes available. Many interviewees also said that it is important to maintain the production of hydropower as a renewable energy source that helps reduce CO_2 emissions.

In addition to successfully addressing several substantive issues, the interviewees also identified a number of process issues necessary for success.

5. The Entities should be more open and transparent. According to several interviewees, the success of revising and updating the CRT will depend on the Entities embracing, supporting, and implementing a more open and transparent process. This means, at least in part, that the Entities should not name the issues, frame the options, and design the process without the participation of other governmental agencies, tribes, organized interest groups, and unaffiliated citizens. Instead, the Entities should seek to engage all of these people and organizations as early as possible.[24]

6. Provide opportunities for all interests to be meaningfully involved. Most respondents voiced a desire for a more inclusive process, meaning that people other than those associated with the Entities should be meaningfully engaged in the process to revise, update, and implement the CRT. This should include state and provincial agencies, tribes, organized stakeholder groups, as well as unaffiliated

citizens. As discussed above, this indicator of success could be accomplished through regional roundtables, watershed groups, web-based surveys, and direct contact with organized interest groups

Barriers to Overcome

Not surprisingly, the barriers to overcome in revising and updating the CRT correspond in many ways to the indicators of success identified by the interviewees.

1. Overcome institutional inertia. Many of the respondents said that one of the biggest barriers to overcome in revising and updating the CRT is to overcome the inertia of the status quo. The confidence among interviewees varied about the degree to which the Entities might embrace and support an open, inclusive process from the get go. Some concluded that, while this is critical, it is not likely to happen. Other respondents explained that the challenge is to encourage and provide incentives for the Entities and others to move beyond self-interest and focus more broadly on the mix of benefits provided by the system. Unfortunately, as noted by the interviewees, the Entities are often unwilling to move beyond self-interest because they are bound by legislative mandates.

2. Determine who participates, when, and how. Another barrier identified by most of the interviewees is the fundamental question of who participates, when, and how – and who decides these questions. While this barrier is somewhat related to the issue of institutional inertia, most respondents seem compelled to highlight it given its fundamental nature. Most of the respondents realize that the more people, organizations, and interests engaged in the process, the more complex and harder it will be to find common ground. That said, many of the interviewees explained that this barrier can and should be overcome.

The interviewees variously identified a number of specific questions along the lines of: (1) Who is at the decision-making table? (2) Who is allowed to provide input and advice and when? (3) How will the decision-makers demonstrate that they have responded to the input and advice of individuals and groups? and (4) How will disputes be resolved among those at the decision-making table and between the decision makers and those individuals and groups providing input and advice? The overarching question here is who decides how to address these process issues?

3. Address specific substantive issues. Many of the interviewees explained that any process to revise and update the CRT must successfully address a number of substantive issues, some of which will be more difficult than others:

• Maximize power and flood control benefits while meeting the needs, interests, and legal requirements for tribes and endangered species;

• Equitably distribute costs and benefits among the two countries, tribal nations, four states, and one province;

• Quantify the costs (and appropriate compensation) to tribes in the upper basin that have lost cultural assets due to flooding;

• Clarify the impacts of dams and water quality to anadromous fish and tributaries, and determine how to mitigate the impacts;

• Resolve issues between commercial ocean fisheries and inland fishermen, given that there is not likely to be a sufficient resource to meet all of their respective interests;

• Maintain high-quality recreation and aesthetic values in upper basin reservoirs while meeting the downstream needs for fish and hydroelectric power;

• Build in flexibility and adaptability to deal with (what most interviewees assume is inevitable) climate change. Seek agreement on the scientific and technical facts associated with climate change, beginning with how much the average annual flow of the Columbia River might change over time and how future shortages should be allocated among different uses and benefits;

• Encourage the action agencies to be open and forth-coming across the border prior to giving notice and undertaking negotiations, and to engage other governmental agencies, tribes, stakeholders, and unaffiliated citizens in the design of any process to revise and update the CRT. Avoid the conventional posturing and behind-the-scenes bargaining associated with these types of multiparty negotiations; and

• Clarify how the National Environmental Policy Act and other legislation and judicial decisions passed since 1964 will influence the process to revise and update the treaty.

Conclusions and Recommendations

The findings based on the interviews speak for themselves. Building on these findings, along with best practices for multiparty, multi-issue negotiation,[25] we offer the following conclusions and recommendations. To supplement these recommendations, Appendix D provides a synthesis of the top-ten lessons learned about transboundary water management.

A. Build on the options and recommendations articulated by the interviewees. While the intent of this assessment was not to build agreement per se on the future of the CRT, the interviewees collectively articulated something like a set of principles that may be useful in future discussions related to the process of revising and updating the CRT.[26] The principles are:

(1) *Convening*: Create a homegrown process, one that is convened and coordinated by and for the people within the Columbia River Basin; the convening body should be viewed as credible and legitimate by all affected parties, particularly in terms of making sure all of the interests within the basin are sufficiently represented.

(2) *Representation*: Provide meaningful opportunities for all decision makers, stakeholders, and citizens to influence the process and the outcomes.

(3) *Public Participation*: The Entities, other government agencies, tribal representatives, and both communities of place and communities of interest should jointly design the process for public participation.

(4) *Process Management*: Use a nonpartisan, impartial person or team to help design and facilitate an inclusive, informed, and transparent process.

(5) *Information*: Engage in joint fact finding to promote and support both scientific and public learning.

(6) *Governance*: Build on what works and adjust accordingly.

B. Conduct a more complete assessment. The assessment of the Entities, other governmental agencies, and tribes is a solid beginning. However, it is incomplete. A more complete assessment should be conducted to identify the interests and concerns of organized stakeholder groups and unaffiliated citizens, and to evaluate the strengths and weaknesses of the CRT. A nonpartisan, impartial third party under the auspices of the Entities and perhaps other governmental agencies and tribes should complete this assessment.

C. Clarify the options to revise and update the treaty. The material presented here is a beginning. Perhaps a diverse team of people could more clearly articulate the menu of legal and institutional options on how to revise and update the CRT.

D. Evaluate options to involve organized stakeholders and unaffiliated citizens. As revealed above, most of the interviewees believe that whatever process is used to revise and update the CRT, it should provide more opportunities for stakeholder and public involvement. Building the menu of possibilities presented herein, perhaps a team of people could evaluate, refine, and develop options to meaningfully engage stakeholders and citizens.

E. Facilitate scientific and public learning. Many (if not most) of the interviewees concluded that it is important to identify what we know, don't know, and need to know to make informed decisions and to promote scientific and public understanding about the CRT and the social, economic, and environmental forces shaping the future of the basin. Once again, perhaps a diverse team of people could build on existing initiatives to answer these questions, and develop one or more strategies to facilitate public learning on these complex issues.

The Columbia River Basin is a special place, and the CRT is widely viewed as a model for managing transboundary natural resources. We thank all of the interviewees and reviewers of this article, and hope that it informs and invigorates attempts to foster livable communities, vibrant economies, and healthy environments in this region.

Notes

1. For a review of the literature on the formulation and implementation of the Columbia River Treaty, start with Keith W. Muckleston, *International Management in the Columbia River Systems* (Report prepared for UNESCO's International Hydrological and World Water Assessment Programme, 2003); Richard Paisley, "Adversaries Into Partners: International Water Law and the Equitable Sharing of Downstream Benefits," *Melbourne Journal of International Law* (2002): 280–300; Nigel Bankes, *The Columbia Basin and the Columbia River Treaty: Canadian Perspectives in the 1990s* (Northwest Water Law and Policy Project, 2001, available at www.lclark.edu/dept/water); John Volkman, *A River in Common: The Columbia River, The Salmon Ecosystem, and Water Policy* (Western Water Policy Review Commission, 1997); Neil Swainson, *Conflict Over the Columbia: The Canadian Background to an Historic Treaty* (Institute of Public Administration of Canada, 1979); John Krutilla, *The Columbia River Treaty, The Economics of an International River Basin Development* (Resources for the Future, 1967).

2. If the treaty is terminated and not renegotiated, the Boundary Waters Treaty of 1909 will govern the transboundary Columbia River; however, certain provisions of the treaty continue so long as the projects exist, including called-upon flood control, Libby coordination, and Kootenay diversion rights.

3. Some interviewees interpret the agreement as stating that flood control provisions will continue beyond 2024.

4. When the U.S. Senate debated the treaty in 1961 (i.e., Senate Committee on Foreign Relations, Hearing Document, March 8, 1961), they assumed there couldn't be any more salmon losses since Grand Coulee had already knocked out the runs that would have returned to the area where the Canadians were to build their dams. However, they ignored the serious detrimental impacts to the fisheries that remained and the resulting impact to the tribes' right to fish.

5. In addition to the narrative provided here, additional information on the tribes' legal rights is presented in Appendix E, Chronology of Major Events.

6. Though most of these court cases were not specifically referenced by interviewees, we feel their inclusion here reflects and refines the discussions with interviewees about legal rights of tribes in the US and Canada.

7. *Winters v. United States* (1908), which allows tribes to reserve future water needs in the amount necessary to meet the primary purpose of the reservation when established, with priority based on the date of establishment of the reservation. This means that Native Americans in the Columbia Basin have an authority to legally define their water rights.

8. *Sohappy v. Smith/U.S. v Oregon* (1969) held that the tribes were entitled to a "fair share" of the fish runs and the state is limited in its power to regulate treaty Indian fisheries (the state may only regulate when "reasonable and necessary for conservation"). Further, state conservation regulations were not to discriminate against the Indians and must be the least restrictive means. (Columbia Inter-Tribal Fish Commission)

9. In *U.S. v. Washington* (1974) the judge held that a "fair share" was 50 percent of the harvestable fish destined for the tribes' usual and accustomed fishing places and reaffirmed tribal management powers. This principle was then applied to the fisheries under *U.S. v. Oregon* in the Columbia River Basin.

10. *Settler v. Lameer* (1974) ruled that treaty fishing is a tribal right, not an individual right, and the tribes had reserved the authority to regulate tribal fishing on and off the reservations.

11. In the Haida and Taku River decisions in 2004, the Supreme Court of Canada ruled that the Crown (federal government) has a legal duty to consult and accommodate First Nations when considering an action that might adversely impact Section 35 rights (established or potential).

12. Pacific Salmon Treaty, Pacific Salmon Commission, http://www.psc.org/publications_psctreaty.htm.

13. Another example of a protocol used to incorporate additional interests is the Migratory Bird Convention (Canada), which amended a 1916 agreement between the U.S. and Canada to incorporate aboriginal practices and conservation. See Canada Legal Information Institute. http://www.canlii.org/ca/sta/m-7.01/part302240.html

14. For example, the Libby Coordination Agreement (2000), which allowed for maintenance of in-stream flow for endangered fish species in the U.S. and provided compensation to Canada for lost benefits.

15. Please note that this is the view of some interviewees and may or may not be consistent with established law.

16. Some interviewees suggested that this group is not appropriate to convene, as they have not been adequately satisfying tribal legal rights and have recently been sued by the tribes.

17. Some interviewees noted that the tribes are concerned about having fair distribution of representation (for example, the conservation, recreation, or fisheries groups have multiple representatives, where all tribes only have one). Given their sovereign role and multiple tribes involved, they would expect to have many representatives.

18. These are only some representative examples of the type of multi-party, place-based groups that might play this role. Other groups that might meet these criteria include the Lower Columbia Solutions Group and the Deschutes River Conservancy.

19. Note that the Northwest Power and Conservation Council and Columbia Basin Trust are separate entities with different missions. Although they are both regional agencies created through legislation, have some similar interests, and have direct connections with the interested public, they are not identical and they operate under different mandates and different authority. In addition, some interviewees do not view these groups as nonpartisan due to their lack of tribal representation.

20. Note that some interviewees suggested that these models and data have been viewed by the Entities as proprietary, rather than public, information.

21. For a history of the idea of joint fact finding, see Herman A. Karl, et al., "A Dialogue, Not a Diatribe: Effective Integration of Science and Policy through Joint Fact Finding," *Environment* 49(2007): 20–34; Peter S. Adler and Juliana E. Birkhoff, *Building Trust: When Knowledge from Here Meets Knowledge from Away* (The National Policy Consensus Center, undated); Matt Leighninger, *The Next Form of Democracy: How Expert Rule is Giving Way to Shared Governance … and Why Politics Will Never Be the Same* (Vanderbilt University Press, 2006); R. D. Brunner et al., *Adaptive Governance: Integrating Science, Policy, and Decision Making* (Columbia University Press, 2005); John t. Scholz and Bruce Stiftel, eds., *Adaptive Governance and Water Conflict: New Institutions for Collaborative Planning* (Resources for the Future, 2005); Gail Bingham, *When the Sparks Fly: Building Consensus when the Science is Contested* (RESOLVE, 2003); Frank Fischer, *Citizens, Experts, and the Environment: The Politics of Local Knowledge* (Duke University Press, 2000); Peter Adler et al., *Managing Scientific and Technical Information in Environmental Cases: Principles and Practices for Mediators and Facilitators* (RESOLVE, 2000); John R. Ehrmann and Barbara L. Stinson, "Joint Fact-

finding and the Use of Technical Experts," in Lawrence Susskind et. al., *The Consensus Building Handbook* (Sage Publications, 1999): 375–99; Kai Lee, *Compass and Gyroscope: Integrating Science and Politics for the Environment* (Island Press, 1995); Lawrence Susskind, "The Need for a Better Balance between Science and Politics," in *Environmental Diplomacy: Negotiating More Effective Global Agreements* (Oxford University Press, 1994): 66–79; Connie Ozawa, *Recasting Science: Consensual Procedures in Public Policy Making* (Westview Press, 1991); Sheila Jasanoff, *The Fifth Branch: Science Advisors as Policymakers* (Harvard University Press, 1990).

22. Many interviewees explained that the Northwest Power and Conservation Council maintains one of the best Web sites, including information on the history of the CRT and key players and issues, and provides newsletters and other opportunities to inform and educate the public and to seek their input and advice. See www.nwcouncil.org. Some interviewees explained that the Northwest Power and Conservation Council and/ or Columbia Basin Trust might play the role of an impartial, nonpartisan coordinator of information: the various action agencies would submit information to that the council and the trust, who would make the information available for public use.

23. A good example here is the Columbia River Fish Working Group, a joint advisory group created by Washington and Oregon to develop recommendations on a variety of Columbia River fishery-related issues facing the two states.

24. For more on designing and facilitating collaborative processes, see Appendix E: Principles of Collaboration for Natural Resources.

25. For an introduction to this topic, start with Lawrence Susskind and Jeffrey Cruikshank, *Breaking the Impasse: Consensual Approaches to Resolving Public Disputes* (Basic Books, New York, 1987); Barbara Gray, *Collaborating: Finding Common Ground for Multiparty Problems* (Jossey-Bass, 1989); Lawrence Susskind et. al. *The Consensus Building Handbook: A Comprehensive Guide to Reaching Agreement* (Sage Publications, 1999); and Julia M. Wondolleck and Stephen L. Yaffee, *Making Collaboration Work* (Island Press, 2000).

26. It should be emphasized that the framing of these principles is based on the results of the interviews, but that the interviewees themselves have not formally endorsed these principles.

Works Cited

Bankes, N. "The Columbia Basin and the Columbia River Treaty: Canadian perspectives in the 1990s," *Northwest Water, Law and Policy Project*. 1996.

Bonneville Power Administration and the Army Corps of Engineers. *Columbia River Treaty History and 2014/2024 Review*. 2008.

Center for Columbia River History. *Treaties and Executive Orders Archive*. Web site at http://www.ccrh.org/comm/river/treaties.htm

Columbia Inter-Tribal Fish Commission, A Short Chronology of Treaty Fishing on the Columbia River, Web site at www.critfc.org/text/timeline3.html

Council on Environmental Quality. *Collaboration in NEPA: A Handbook for NEPA Practitioners*. 2007.

Executive Order 11177, Sept. 16, 1964 (29 Fed Reg. 13097, 3 CFR, 1964–65 Comp.): p 243, as amended by Executive Order 12038, Feb 3, 1978 (43 CFR 4957, 3 CFR, 1978 Comp.): p 136.

Independent Scientific Advisory Board. *Human Population Impacts on Columbia River Basin Fish and Wildlife* (ISAB Human Population Report, June 8, 2007): 9.

International Joint Commission. Web site at http://www.ijc.org/rel/comm/ref1198.

Lang, William L. *Columbia River*. Center for Columbia River History. http://www.ccrh. org/river/history.htm. 2010.

Lang, William L., and Robert Carriker. *Great River of the West*. Seattle; University of Washington Press. 1999.

McKinney, M., and Johnson, S. *Working Across Boundaries: People, Nature, and Regions*. Lincoln Institute of Land Policy. 2009.

Muckleston, K. *International Management in the Columbia River System*. UNESCO, Technical Documents in Hydrology. 2003.

Northwest Power Planning Council. *Northwest Power Supply Adequacy: Reliability Study-Phase 1 Report*, Paper Number 200-4. 2002.

Pacific Salmon Commission. Web site at http://www.psc.org.

Treaty Relating to Boundary Waters and Questions Arising Between the United States and Canada, U.S.–Great-Britain. T.S. No 548. 1909.

Treaty Relating to Cooperative Development of the Water Resources of the Columbia River Basin, U.S.-Canada. 15 U.S.T. 1555. 1964.

U.S. Army Corps of Engineers and Bonneville Power Administration. *History and 2014/2024 Review*. 2009.

Vienna Convention on the Law of Treaties. 1969. Accessed at: http://untreaty.un.org/ilc/texts/instruments/english/conventions/1_1_1969.pdf

Wood, M. C. "Protecting the Attributes of Native Sovereignty: A New Trust Paradigm for Federal Actions Affecting Tribal Lands and Resources," *Utah Law Review*. 1995. 109–237.

Appendix A: List of Interviewees
Nigel Bankes, University of Calgary
Kat Brigham, Umatilla Tribe and Columbia River Inter-tribal Fish Commission
Barbara Cosens, University of Idaho
Lynette de Silva, Oregon State University
Bill Green, Canadian Columbia River Inter-tribal Fish Commission
John Harrison, Northwest Power and Conservation Council
Charles Hudson, Columbia River Inter-tribal Fish Commission
John Hyde, Bonneville Power Administration
Kelvin Ketchum, BC Hydro
Bob Lohn, National Oceanic and Atmospheric Administration
Mike Matylewich, Columbia River Inter-tribal Fish Commission (USA)
Pat McGrane, Bureau of Reclamation
Bruce Measure, Northwest Power and Conservation Council
Garry Merkel, Columbia Basin Trust
D. R. Michel, Upper Columbia United Tribes (and Colville)
Rebecca Miles, Nez Perce Tribes
Daniel Millar, Environment Canada
Keith Muckleston, Oregon State University
Tim Newton, Permanent Engineering Board
Richard Paisley, University of British Columbia
Ken Peterson, PowerEx (retired)
Bob Heinith, Columbia River Inter-tribal Fish Commission
Doug Robinson, BC Hydro
Derik Sandison, State of Washington, Department of Ecology
John Shurts, Northwest Power and Conservation Council
Marvin Wodinsky, Canadian Department of Foreign Affairs (retired)
Aaron Wolf, Oregon State University

Appendix B: Interview Questions[1]
What is your interest, role, and history related to the Columbia River Treaty?
What is working well with respect to the Columbia River Treaty?
What is the most preferred future for the Treaty? Options may include, but are not limited
  to:
Allow the treaty to expire?
  Extend the existing treaty?
  Renegotiate (i.e., revise and update) the treaty?
  If the treaty is renegotiated, how could it be improved? What is the single most
  important change from your perspective?
  What are the key issues (or drivers) that should be addressed in the renegotiation of the
  treaty?
  What do you think will be the easiest issues to address?
  Most difficult?
  Where will conflicts arise?
  Between whom and on what issues?
From a process perspective:
  Who will or should convene the negotiation?

Who should be at the table during the renegotiation?

What type of information is needed (scientific, technical, legal, etc.), and who should provide that information?

How should the public be involved?

What role, if any, might an impartial facilitator or mediator play in the design and coordination of the renegotiation process?

What type of governance arrangement will be most effective to implement the treaty? (That is, who should have decision-making authority, how should decisions be made, disputes resolved, etc.)?

What will a successful renegotiation of the treaty look like? What do you think are the opportunities or potential gains possible through such a renegotiation? For your interests? For the region as a whole?

What do you think are the barriers to a successful renegotiation of the treaty? Do you have any suggestions on how to overcome the barriers?

Is there anything else you would like to share?

Who else should we talk to?

Appendix C: Chronology of Major Events Since 1964

1964: Columbia River Treaty was implemented, delineating power and flood control benefits between the U.S. and Canada. In addition it authorized construction of a number of Canadian storage facilities to improve storage capacity in the system and maximize hydropower generation.

1965: Water Resources Planning Act, which established a Water Resources Council in the United States to be composed of cabinet representatives, including the secretary of the interior. The council was charged with maintaining a continuing assessment of the adequacy of water supplies in each region of the U.S. The council also was mandated to establish principles and standards for federal participants in the preparation of river basin plans and in evaluating federal water projects with respect to agricultural, urban, energy, industrial, recreational, and fish and wildlife needs.

1966: To protect dwindling runs of summer chinook above Bonneville Dam, the Oregon Fish Commission asked the Oregon State Police to strictly enforce the law forbidding non-Indian commercial fishing upriver from Bonneville.

1968/69: *Sohappy v. Smith* and *United States v. Oregon*. Fourteen Yakama tribal members filed suit to prevent the state of Oregon from interfering with their off-reservation treaty fishing rights. The court found that the state's authority to regulate Indian fishing for conservation purposes was limited as treaties provide tribes an absolute right to a fair share of the fish produced by the Columbia River system.

1969: National Environmental Protection Act, which requires U.S. federal agencies to examine the impacts of proposed major federal actions significantly affecting the environment.

1973: U.S. Congress passes the Endangered Species Act: "The purposes of this Act are to provide a means whereby the ecosystems upon which endangered species and threatened species depend may be conserved, to provide a program for the conservation of such endangered species and threatened species, and to take steps as may be appropriate to achieve the treaties and conventions..."

1974: *United States v. Washington*. A federal district court in the state of Washington found that Native American tribes were entitled to the opportunity to take up to 50 percent of the harvestable number of fish that can be taken. This harvestable sharing principle was also applied in *US v. Oregon* (see above).

1977: Four Indian tribes with treaty fishing rights on the Columbia River formed the Columbia River Inter-Tribal Fish Commission to coordinate fish management policies and objectives. The participants are the Nez Perce Tribe, Confederated Tribes of the Umatilla Reservation, Confederated Tribes of the Warm Springs Reservation, and Confederated Tribes and Bands of the Yakama Indian Nation.

1980: In December, the U.S. Congress approves and President Jimmy Carter signs into law the Northwest Power Act, which authorizes the four northwest states of Idaho, Montana, Oregon, and Washington to form the Northwest Power and Conservation Council (the agency was known until 2003 as the Northwest Power Planning Council) and gave the council three distinct responsibilities: 1) prepare a program to protect, mitigate, and enhance fish and wildlife, and related spawning grounds and habitat, of the Columbia River Basin that have been affected by hydropower dams, while 2) assuring the Pacific Northwest an adequate, efficient, economical, and reliable power supply, and 3) informing the public about energy and fish and wildlife and involving the public in decision making. The council met for the first time in April 1981.

1985: Pacific Salmon Treaty was ratified as a cooperative agreement between U.S. and Canada to research and enhance Pacific salmon stocks.

1988: Snake River coho salmon were considered extinct.

1991: In April, the U.S. National Marine Fisheries Service proposed to list Snake River sockeye as an endangered species. In June, the NMFS proposed to list Snake River spring/summer and fall chinook as threatened species, but declined to list lower Columbia coho on the grounds that the population was so infused with the genetic material of hatchery-bred coho that no truly wild coho remained.

1995: In May, British Columbia's Legislative Assembly approved the Columbia Basin Trust Act, which established the Columbia Basin Trust "to help create a prosperous economy with a healthy and renewed natural environment." The trust is "an autonomous and independent organization of communities," according to its literature. Through the trust, millions of dollars will flow into the Canadian Columbia River Basin from the sale of electricity in the United States—so called "downstream benefits"—made possible by the operation of storage reservoirs behind the three Canadian dams of the 1962 Columbia River Treaty.

1995-1999: Endangered Species Act Listings. Nine additional species of fish throughout Columbia Basin were listed under the Endangered Species Act.

1999: The Entities determined that some provisions of the CRT covering Entitlement delivery did not address the realities of the Pacific Northwest grid, and that new rules covering the cost of electric transmission had not been anticipated. This change was considered to be "substantial" and the United States' State Department and the Canadian Department of Foreign Affairs and International Trade (DFAIT) were consulted, and ultimately covered the agreements with an Exchange of Notes.

2000: The Entities agreed to coordinate the operation of Libby Dam with Canadian projects to self compensate Canada for losses incurred as a result of the operation of the dam. The original difference of opinion was presented to the U.S. State Department and DFAIT, but no resolution appeared to be possible, so the Entities were allowed to see if a pragmatic resolution could be developed. The idea of self-compensation allowed an agreement to be developed, without compromising the original position of either country. The agreement provides both parties with very short termination options, so there is an incentive to make it work, rather than go to a very lengthy arbitration process.

2001-2004: Salmon and steelhead returns to the Columbia River were far above recent ten-year averages. Some, such as the returns in 2003, were the highest since record keeping began at Bonneville Dam in 1938. In 2003, more than 920,000 chinook salmon were counted crossing Bonneville Dam, where the ten-year average count was 399,000. A number of factors appeared to be contributing to the increased run sizes, including improved fish passage at dams, improved spawning and rearing habitat, improved feeding conditions in the ocean, and a reduction of intercepting fisheries. In 2004, as strong runs continued, scientists at NOAA Fisheries who monitor the runs said it appeared the runs would stay high at least through 2006.

2008: The Pacific Salmon Treaty established the Pacific Salmon Commission, a bilateral body that recommends to the U.S. and Canada the ocean salmon fishing levels in Southeast Alaska and British Columbia. The United States and Canada adopted a new set of fishing regimes for Chinook, coho, chum and transboundary rivers on December 23, 2008 through an exchange of diplomatic notes. Fisheries had steep declines and there were closures of fishing seasons.

Appendix D: Top Ten Lessons for Transboundary Water Management[2]

River basins may not be a sufficient "territory of the problem." If the goal is to build livable communities, vibrant economies, and healthy environments, it may be necessary to reach beyond the river basin itself.

The initiation and implementation of transboundary water agreements depends first and foremost on political will. If sovereign entities are not willing to share power and negotiate mutual gain solutions, it will be difficult (if not impossible) to forge practical transboundary arrangements.

The development and implementation of transboundary water plans should seek to involve all the stakeholders in an open, transparent, and inclusive process. To the extent practicable, the multiple stakeholders should be allowed to help name issues, frame options, deliberate over the consequences of alternatives, and seek consensus on how to move forward.

The most effective transboundary initiatives tend to embrace multiple issues—water, land, economic development, and so on. In this respect, transboundary conversations may start by focusing on water allocation but eventually embrace a larger mix of issues and concerns.

The participants in a transboundary water negotiation should employ joint fact finding, clarifying what they know, don't know, and need to know in order to make informed decisions. Joint fact finding promotes political legitimacy, an interest-based agenda (i.e., a focus on the most salient issues), and scientific credibility.

The most effective transboundary agreements and governance arrangements are homegrown, not imposed from outside the river basin. International conventions and protocols may be necessary to create an independent framework when none exists and to otherwise inform and invigorate homegrown efforts—but such conventions and protocols are not sufficient.

Transboundary negotiations should apply widely accepted principles such as equitable use, reasonable use, sharing benefits, and doing no harm.

Universities and scholars can play a valuable role in terms of informing and invigorating transboundary water efforts; harvesting lessons learned from other experiences; and perhaps in some cases serving as an impartial, nonpartisan convener and coordinator of transboundary collaboration. An important corollary to this lesson is that scholarly studies are necessary, but not sufficient.

In the process of initiating, convening, negotiating, and implementing transboundary water agreements, it is critical to be sensitive to the relative distribution of power among the participants. In theory, hegemons should use their power to initiate, inspire, and model cooperative behavior.

To promote transparency and accountability, a multi-stakeholder group should monitor the implementation of transboundary water agreements and plans. This group should expect and plan for change—new information, issues, and players. In this respect, the governing arrangement should be flexible and adaptive.

Appendix E: Principles of Collaboration on Natural Resources

The Collaborative Democracy Network has identified at least fifty different theoretical frameworks for collaborative planning and policy making.[3] Although there is some variation among these frameworks, the following propositions constitute a coherent and widely shared group of general principles that inform collaboration on natural resources.[4] It should also be noted that, while something like these principles might inform and invigorate many collaborative processes (particularly collaborative processes convened by government entities with authority and responsibility over particular issues), many such processes are organic, homegrown efforts by citizens and non-governmental agencies. In the latter case, the collaborative process often starts with a diverse coalition of people committed to a common agenda. Other people are then brought on board as they agree with the purpose and scope of the emerging conversation.

Purpose driven. Collaboration focuses on the needs and interests of the participants—it is purpose driven. It is intentionally designed and managed in line with an agenda, ground rules, sideboards, a timetable, and a budget approved by all the parties.

Inclusive. Collaboration includes those people that are interested in or affected by the issues; those that are needed to implement any outcome; and those that might challenge the process or the outcome. Participation is voluntary. The most effective method to identify participants is to allow stakeholder groups to select their own representative.

Informed. All parties have an equal opportunity to share views and information. The process fosters mutual learning, common understanding, and consideration of a variety of options. It enables participants to jointly develop and rely on the best available information, regardless of sources. Scientific information represents one way knowing, and is integrated with other ways of knowing (e.g., anecdotal information).

Deliberative. Participants jointly name issues and frame options, thereby clarifying their underlying interests and predispositions. They respect and listen to each other, consider the rationale or reason for diverse viewpoints (i.e., the interests that underlie the positions), and seek solutions that integrate as many interests as possible.

Consensus seeking. Participants seek consensus (defined as unanimity), but accept overwhelming agreement.

Accountability. Participants—including decision-makers—strive for transparency and communicate in good faith their interests and expectations. Participants respect and work within existing laws and policies, and appropriately seek the input and advice of constituents, the public, and decision makers.

Supplemental. The product of collaboration is a recommendation, not a final decision. Decision makers do not abdicate their decision-making authority, and any proposals are vetted through formal public involvement and decision-making processes.

Implementation. The product of collaboration is a written document that participants agree to support individually. Participants seek ratification of the outcome by the people or groups they represent (if any), as well as the public and decision makers.

Adaptive Management. Given that it is impossible to estimate all of the impacts or consequences of the recommendations, policy choices/decisions are monitored so that continuous adjustments can be made.

Process Managers. Professional facilitators and mediators increase the effectiveness of collaboration. This individual or team, which should be jointly selected and/or approved by all participants, can help the stakeholders design the right process, convene meetings, resolve disputes, and coordinate implementation.

Notes to appendices

1. Some of these interview questions (such as question 3) were revised as we conducted interviews and learned more about the CRT.

2. These lessons – synthesized by Matthew McKinney -- are based on presentations and dialogue at the 5th World Water Forum (Istanbul, Turkey, March 2009), and include experiences from Europe, Central Asia, the Middle East, and the United States.

3. Learn more about the Collaborative Democracy Network at www.csus.edu/ccp/cdn.

4. These propositions are adapted from a speech presented by Professor Lawrence Susskind at *Water in the West*, Bozeman, Montana, September 2006. Also, see *Breaking Robert's Rules: The New Way to Run Your Meeting, Build Consensus, and Get Results,* Lawrence Susskind and Jeffrey Cruikshank, Oxford University Press, 2006; and Gerald Cormick et al., *Building Consensus for a Sustainable Future: Putting Principles into Practice* (National Round Table on the Environment and the Economy, 1996).

The Past and Future of the Columbia River

Paul W. Hirt and Adam M. Sowards

Introduction

This volume uses uncertainty in environmental management as an organizing principle. We address this theme and several others, including the interconnection between natural systems and social systems, the centrality of historical context in shaping river management regimes and international agreements, the importance of humility in environmental management, and the value of balancing efficiency with equity in resource allocations. We unpack all of these themes in a historical survey of the changing human relationship with the Columbia River over the past century. Our analysis reveals trends in policy development, engineering, ethics, and international relations. Our own organizing principle is, simply, that studying the past illuminates the present and helps us chart an informed course into the future. *History is instructive.*

At any given time in any given place, uncertainty is a prevailing factor. Nature presents itself as patterned disorder, whether it is climate fluctuations affecting biological production or salmon appearing in numbers higher or lower than anticipated or rivers overtopping their streams' banks and spreading across the floodplain (Orsi 2004, 165–83; Botkin 1990). But if natural systems are unpredictable, so too are human systems—perhaps to an even greater extent. Political systems are constantly evolving with policies adapting to address changing environmental conditions, failures in technocratic systems, or newly articulated social goals. Economies are uncertain based on shifting supply and demand, unpredictable prices, and interaction with other socio-political factors such as elections or income inequalities. Social values and behavior are often unpredictable, because they are pinned directly to people who are autonomous and act independently. Even more, these ecological and human systems interact interdependently, compounding the uncertainty in myriad ways. Nature, politics, economics, and society are all independently and interdependently uncertain (Langston 1995; McEvoy 1986).

And yet, amid the unpredictability of natural and human systems historical patterns often prevail, and so historians can help chart the changing contexts in which this uncertainty interacts and is constrained. Today, the complex interaction of nature and nations shapes everything significant that happens on the river, from headwaters to estuary. We like to think that the good things we derive from the Columbia River are a product of order: a benign nature and technological competence. On the other hand, problems such as floods, fluctuating salmon runs, and pollution we often blame on serendipity and chaos: freak storms, human error, unanticipated consequences. But order and chaos are simultaneously present at all times, just as natural systems and social systems are inextricably intertwined at all times in this modern age. We can take no more credit for full reservoirs than we can blame nature for drought. To credit ourselves for the water in Lake Roosevelt is hubris, while to blame nature when our water supplies run low is naïve. It distorts both the interconnectedness of nature and culture and the historical contingency of events like floods or droughts (Orsi 2004, 173). As historian Ted Steinberg vividly elucidates in *Acts of God*, most of what we call "natural disasters" are emphatically unnatural events; they are as much a product of human design as environmental irregularities. The disaster associated with hurricane Katrina is only one of myriad examples of this (Steinberg 2000). Just as the flooding of New Orleans was a combined result of a "natural" event, a design flaw, and a policy failure, so too the great challenges facing Americans and Canadians in joint management of the Columbia River reflect the interdependent relationship between nature's uncertain course and humankind's shifting intentions, technologies, and policies. Our effort to wrest order from disorder falls in the realm of governance.

The Columbia River Treaty (CRT) is an instrument of governance constructed at a particular place and time. It has weathered well, according to its administrators, but it is nevertheless also a dated document shaped by a generation now passing. The treaty culminated an era of profound technological optimism in humankind's ability to control nature. Corporate and government-sponsored schemes for comprehensive river development transformed the region's natural resources and political economy during the first half of the twentieth century setting the framework for the treaty. The CRT coordinated the two nations' efforts in a manner reflecting the prevailing political calculus of the times: to maximize hydropower and minimize floods on the Columbia River. Those goals, in hindsight, now appear surprisingly narrow—although few people were surprised or concerned about it in the 1950s and '60s. Historical context shaped the treaty half a century ago and contemporary historical context will shape its re-negotiation a few years hence.

The first part of this essay describes early twentieth-century social, political, and economic conditions and values in North America that led up to and influenced the CRT. We then profile critical changes since the treaty was ratified. Following that, we offer a brief critique of the narrowly framed aims of the treaty and suggest more realistic, sustainable, and equitable principles for managing transboundary rivers. Finally, we end with some reflections on the forces and concerns that will likely influence treaty renegotiation in the coming years.

Governing Institutions and Ideals

North Americans have largely been enamored with science and its offshoot technology as ways to understand and then control natural systems. In the nineteenth century, explorer-scientists traveled the continent seeking to describe scientifically the land and watercourses they encountered with an eye toward economic development (Sachs 2006). Agricultural reformers in antebellum America and after developed new technologies and farming methods and eventually applied chemicals to boost yields (Stoll 2002). Engineers sought to remodel harbors or river systems to control or conquer them and make them more efficient for agricultural and commercial purposes (Worster 1985). In each case, science became an instrument to advance economic agendas, and government's support ensured state power undergirded scientists' work (Robbins 2004; Nelles 2005).

To be sure, scientists believed their work to be in the public interest. A century ago, as concerns about the costs of modernization mounted, they offered palliatives to the often harsh environmental effects of emerging industrialization. In the Pacific Northwest, for instance, industrial fishing combined with drastic habitat changes caused by logging and agriculture decimated salmon runs throughout the Columbia River Basin. Scientific managers promised to ameliorate the salmon population declines through artificial propagation. Rather than approaching the problem holistically, managers identified one problem—too few fish—and moved to solve that by creating more fish through state-sponsored hatcheries without addressing the fundamental linkages between ecosystems or confronting the human systems that were damaging the fishery. The results failed to stem the decline of salmon, although hatcheries were a politically popular "solution," because they did not require fundamental reform of the causes of salmon decline: human behavior and "perverse" economic incentives (Taylor 1999; Myers and Kent 2001; Lichatowich 2001). This piecemeal technological approach to problem solving prevailed often but resolved little.

This management paradigm resulted from specific ways of viewing nature and its relationship to the public. Policymakers who funded scientific research and formulated resource development agendas, especially in the North American West, saw the region's forests, rangelands, and rivers as potential wealth. Public resources were to be exploited with government assistance to create private wealth. An exemplary statement of this approach toward water in the United States written at the height of the Progressive era is John L. Mathews' *The Conservation of Water* (1910), a sprawling, confident statement of the promise of water development. In chapter after chapter on water conservation topics—flood control, swamp drainage, irrigation, waterpower, and so on—Mathews offered sweeping surveys of current or potential projects nationwide. Never does Mathews express doubt. Seldom are people mentioned, other than as masses benefiting from the new plenty engineered for them from nature. Without greater efforts to reclaim water from waste—that is, disuse—the nation's wealth would suffer. Developing the Columbia River Basin, including tributary streams, would yield power, irrigation, and navigation. With the correct storage and irrigation system, managers could ensure water was "applied at *exactly* the *right* time in the *right* amount and under *perfect control*." Done right, power development would benefit "manufacturing that will support an enormous population in the picturesque valley of that [Columbia] river," a prospect that would enhance the nation's trade to Asia. In Mathews' final chapter, "The Results of the Conservation of Water," he takes an imaginative journey upstream: "We enter it at the sea gate, at the mouth of a great river, where we find a port bustling with commercial activity." Continuing up the engineered river, the reader finds booming cities without pollution since power comes from the dams. Further on, public buildings (paid for by water conservation) line the banks that are themselves protected by soil stabilization measures beside which are endless irrigation villages. In short, nature's "dilapidation has given way to perfection." Further upstream, a dam blocks the river but locks move ships and a power plant generates power. The engineers have perfected nature and society. We must, Mathews implores, bring this utopian vision to life by developing the Columbia River.[1]

This was the emerging professional conservation vision in the early twentieth century. In his concluding paragraph, Mathews acknowledged it would not be easy or quick: "We will not see it done, nor our children. But our grandchildren will come to it by birth and education and sweep it on rapidly to complete perfection. And the trip up the river which I have described, which you who read this may never make, your grandson will make while still a young man, and he will make it with that new pride and delight in his own land which he cannot have until he has taken care of it, saved and protected it and made it a beautiful and lovely place" (Mathews 1910, 281). To make this vision a reality, Mathews assumed it would take time, commitment, and, most of all, government leadership.[2]

The U.S. federal government obliged, because it, too, caught the fever of river development. From the 1902 Reclamation (or Newlands) Act that sponsored federal reclamation projects to the 1920 Federal Water Power Act that coordinated and regulated hydroelectric development on the nation's navigable rivers, the national government took on a key role in controlling and developing rivers. Meanwhile in Canada, national and provincial governments supported their own reclamation programs to develop water resources throughout the western provinces. For instance, the British Columbia government in 1897 enacted a consolidated Water Law confirming provincial authority over all rivers with the intent to rationalize and coordinate river development for maximum efficiency and social utility. B.C. expanded its authority and oversight with the B.C. Water Act of 1909, which established a water commission to grant licenses and a Board of Investigation to adjudicate conflicts among water claimants. Both at the national and provincial level, Canadian governing institutions sought control over rivers for multipurpose economic development (White and Vick 1919, 70–3; Denis 1911, 303–8). B.C. Premier Richard McBride in 1914 succinctly articulated this philosophy: "If it be for the purpose of irrigation, let every inch of water do its duty, and, if it be for the purpose of power, let us see that the works are so carried out as to get from the investment and from the water conservation, the very best and most profitable results" (White and Vick 1919, 15). On either side of the forty-ninth parallel, then, the dream of an ordered nature developed with government oversight for private wealth, and social progress dominated the vision of water and land resources in the early decades of the twentieth century (Worster 1985; Worster 2002; Evenden 2006; Lee 1966; Mitcher 1980; Murton 2008).

With a relatively unified vision favoring development and with legislative structures in place, governments and bureaucracies pursued increasingly large projects to transform rivers into engines of economic growth. As erstwhile engineer and then-current U.S. Secretary of Commerce Herbert Hoover put it in 1926, each river system should be developed to its "maximum utilization" (R. White 1995, 54). Eventually, the Great Depression of the 1930s provided a further impetus in the U.S. as federal and state governments used the exigency to put masses of unemployed on public works projects, such as building Bonneville and Grand Coulee dams on the Columbia River. To partisans of this vision, the Columbia River and the larger Columbia Basin stood as a vast and relatively untapped resource awaiting technological control. To be sure, various interests vied for competing development plans, including various local political maneuverings (Ficken 1995; Pitzer 1994). An Oregon journalist-turned-politician, Richard L. Neuberger, captured the enthusiasm of these days and this project of basin transformation in his classic book, *Our Promised Land* (1938), a celebration of the history and future of the Pacific Northwest. In his chapter "Hydroelectric!," Neuberger explained that the region contained forty-one percent of the nation's available hydropower passively awaiting human use. Neuberger could wax poetic: "The river rumbles and roars. It thrashes against granite walls and gnaws ominously at basalt cliffs. Around jagged rocks it booms like the surf. . . . Wild and defiant, the river tumbles unharnessed to the Pacific." But ultimately, the "hours of this freedom are numbered," because a greater good—electricity—could serve the public and reform society. Representing the confidence of the age and ideal of development, Neuberger knew this project would succeed, would be "an event of national significance," and would help rural residents and urban industries. There could be no doubt. The *efficient* development of the Columbia Basin would produce great and wide-ranging benefits (Neuberger 1989, 86–122; Robbins 2004, 215–47; R. White 1995, 64–74).

Not all issues could be met so confidently, though. Following "Hydroelectric!" came "No More Salmon," a decidedly more ambivalent (and significantly shorter) chapter. Here Neuberger targeted salmon and Indians. He explained the fish ladder system being constructed at the Bonneville Dam at a cost of seven million dollars, but noted that Grand Coulee Dam would be the end of the anadromous line, a thousand miles of upstream spawning habitat in Canada forever closed to salmon. Lurking in the background of Neuberger's prose were tribal members, gazing mournfully at the construction and declaring, "White man's dams . . . mean no more Indians' salmon." Even Neuberger's technological fervor could not assuage such a diminishment of cultural and economic tradition. Despite the elegiac tone and Neuberger's recognition of the concern of government scientists over the declining salmon runs, he saw the loss as a necessary, if regrettable, sacrifice. In other words, the efficient development of the Columbia River excused the inequitable results; in this respect, Neuberger represented the broad consensus of policy makers, developers and conservationists (Neuberger 1989, 123–39; also Barber 2005).

Developing the Basin

Neuberger wrote at a time of growing technological optimism. Indeed, the mid-twentieth century, from the 1930s to the 1960s, was perhaps the zenith of this scientific-technological faith. Besides the vision and desire, the Depression of the 1930s furnished the opportunity for unprecedented development with the rise of Franklin D. Roosevelt's New Deal policies. To rescue capitalism, the New Deal initiated broad-scale planning for economic and natural resource development because of perceived needs and a temporary willingness on the part of the American people to accept massive government intervention in the economy. One result was comprehensive river basin planning, the most famous example being the Tennessee Valley Authority. Armed with a faith that applied science could control nature for maximum profit and welfare for regional residents, along with the idea that public power would provide greater benefits than private power had achieved, federal planners and engineers sought to put the Columbia River to work (Blumm 1999; R. White 1995).

It was an ambitious and evolving project that took many forms and several decades. One centerpiece of the river planning effort was the Bonneville Power Administration (BPA), which marketed power generated from Bonneville Dam and other subsequent federal hydroelectric dams. From the late 1930s to the early 1950s, electricity for residential use alone quadrupled and could be met only through the important integration of these federal hydroelectric projects. A second centerpiece was the Columbia Basin Project (CBP), which would provide supplemental irrigation water and attract new basin farmers. Designed to support family farms, the CBP was often dominated by absentee corporate owners benefiting from irrigation water at highly subsidized rates. Such results demonstrate that federal investment in the Columbia River did generate economic growth although not always for the intended target, suggesting that efficiency tended to be easier to realize than equity. Underpinning these efforts was an instrumental view and use of nature. Both the BPA and the CBP extracted from the Columbia River its economic benefits for a greater Pacific Northwest (Blumm 1999, 268–72; Pitzer 1994; R. White 1995, 74).

Engineers sought to create a systematized and controlled Columbia River, maximally beneficial for regional residents. For instance, the growth of manufacturing in the region played a significant role during World War II, and rural electrification transformed much of the Northwest in powerful ways. River basin planning was producing a controlled environment for the greater good. Then, in 1948, the Columbia River flooded, hitting the former wartime boomtown of Vanport, Oregon. After more than $100 million in property damage and thirty-eight deaths, the flood showed that control over the river was elusive or at least incomplete (R. White 1995, 74; Brooks 2006, 60).

It also encouraged greater efforts at control. With the Vanport disaster fresh in their minds, policymakers and resource managers launched an aggressive campaign to do more. More dams would mean no more floods, they claimed. More dams would also forestall the looming threat of power shortages that BPA routinely predicted were just a few years away. Public power partisans proposed a Hells Canyon High Dam to help solve these problems. Meanwhile, fishery policy, supposedly "planning for permanent control," in historian Karl Boyd Brooks' terms, sacrificed salmon in the Columbia Basin. Grand Coulee Dam had ended anadromous fish runs upstream of the dam. In a nod toward compensation, federal planners simply sought to remove the fish from one part of the river system (upriver) to another (downriver) through investments in hatcheries, a process that rearranged salmon's geography, failed to conserve fish as promised, and entailed significant social inequities for Columbia Basin tribes. In short, salmon were sacrificed. And tribal members who depended on those fish for cultural and economic purposes were largely shut out of the decision process as they lost their access and rights (Barber 2005; Brooks 2006, 81–92 and passim; Taylor 1999, 235–36 and passim; R. White 1995, 89–113).

By the 1960s, dozens of large dams held back the waters of the Columbia and its tributaries. Despite treaty assurances, tribal fishers had been removed from much of the fishery. Despite millions of dollars invested in conservation, salmon had declined. How had this occurred? Several interlocking reasons contributed. An instrumental view of nature certainly set the intellectual stage. A faith that technology could mitigate any environmental problems similarly guided managers. Asymmetries in power among government agencies played a significant role, as well. Federal agencies with an interest in economic development, such as the dam-building Bureau of Reclamation and the U.S. Army Corps of Engineers, always wielded more power than fisheries agencies (e.g., Brooks 2006; Goble 1999, 247–49; Taylor 1999; R. White 1995). Bureaucratic self-interest led to "institutionalized conquest," in the words of water scholar Ellen E. Wohl, and "seeing like a state," in the memorable phrase from political anthropologist James C. Scott (Wohl 2004, 176–221; Scott 1998). That is, bureaucratic initiatives to generate wealth guided by narrowly framed principles of utility trumped both ecosystem values and equity concerns.

Some government biologists, tribal fishers, and other observers recognized the ecological problems this controlled river system was creating and exacerbating. But when they protested, they were excluded from meetings and decisions, such as frequently occurred in the 1940s as the Columbia Basin Inter-Agency Committee (CBIAC) helped decide the future of the Basin's development (Brooks 2006, 93–117). At other times, agencies simply did not care. For instance, the head of the Army Corps of Engineers reportedly stated, "We do not intend to play nursemaid to the fish" (Goble 1999, 248). Functioning ecosystems and healthy fisheries were not the goals of mid-twentieth-century basin development.

Ecological and economic uncertainty had been unsettling to people in the Columbia Basin, especially policymakers and resource managers. Governments' efforts over these decades had been dedicated to minimizing uncertainty in natural and human systems by asserting greater control and rationalizing the river system into what historian Richard White (1995) referred to as an "organic machine." Bureaucrats sought to produce more salmon in mass-production hatcheries to account for shortfalls based on habitat destruction, climatic changes, and over harvesting. They sought to regulate access to water and related resources through treaties with American Indian tribes and foreign nations, as well as state laws and regulations. Most visibly and powerfully, they dammed the Columbia River and its many tributaries to minimize the unpredictable damage of floods, to furnish water for irrigation to minimize the uncertainty of rainfall, and to generate hydroelectric benefits by storing the spring floods for controlled release in the summer and fall. These collective efforts largely were designed to control nature for maximum human benefit, usually defined strictly in economic terms.

A third of the Columbia River, though, rested in British Columbia, and that portion of the river remained largely undeveloped in the 1940s. It became clear to technological utopians and advocates of maximum control for greater wealth in the U.S. that removing uncertainty and developing the river resource fully depended on coordinating with Canada.

The Binational Relationship
The Columbia River Treaty structured a relationship between two nations that continues to evolve. The transnational trends are interesting. Formal cooperation

in the management of shared waters along the U.S.-Canada border is, historically speaking, a rather modern phenomenon. In the nineteenth century, relations between the two nations were competitive rather than cooperative. Each nation remained largely insular and somewhat distrustful following the parting of ways during the American Revolution and due to contentious boundary disputes (e.g., the division of the Oregon Country in 1846). In the nineteenth century, Canadians jealously guarded their lands and resources, and viewed their southern neighbors as aggressively expansionist—with good reason (Reimer 2002; Coates 2002).

Yet even in the midst of the United States' rush of imperial adventures in Latin America and the Pacific in the 1890s, that distrustful relationship began to change (LaFeber 1998). Americans and Canadians increasingly recognized their common interests and sought opportunities for enhanced trade and cooperation. The ratification of the U.S.-Canada Boundary Waters Treaty of 1909 was one important step in this direction. It established the International Joint Commission (IJC), which just celebrated its centennial in 2009. The treaty and the IJC marked a passage of sorts, reflecting the contentious past while ushering in a new era of cooperation. Established by the two nations primarily to adjudicate conflicts over shared boundary waters, the commission evolved during the twentieth century to also promote cooperative development, as in the case of the Columbia River. At the request of the two nations, the IJC investigated opportunities for U.S.-Canadian cooperation in developing the Columbia River from 1944 to 1959, leading up to the drafting of the Columbia River Treaty (Bloomfield and FitzGerald 1958; Mouat this volume).

Old habits died hard, though. Many business and opinion leaders in the U.S. continued to covetously eye Canada's water, power, timber, and fish during the twentieth century, while Canada continued to zealously guard them. International relations remained an uneasy mix of cooperation, complaint, and dispute resolution. The International Pacific Salmon Fisheries Commission (IPSFC), established in 1937 by the U.S. and Canada, exemplified this. While the IPSFC sought to resolve conflicts over the Fraser River sockeye fishery and equitably divide the catch between U.S. and Canadian fishers, it also sought opportunities for cooperation in regulating the fishery, restoring habitat and depleted runs on the Fraser, and conducting biological research (Evenden 2004; Taylor 2002).

Relations over the Columbia River fit this pattern of emerging cooperation from the 1930s to the 1950s. Prior to this period, Canadians developed their electric power grids in an east-west orientation rather than north-south, keeping Canadian resources in Canadian hands (Stadfeld 2003). They passed laws discouraging or prohibiting both water and power exports to the U.S.[3] To the south, Americans developed the Columbia River Basin paying little heed to how it might affect Canadians, as, for example, when Grand Coulee Dam blocked all salmon migration upstream into Canada. Still, there were notable efforts to get beyond mere conflict resolution and cooperate for mutual benefit in the management of natural resources. The United States was especially enthusiastic about comprehensive river basin planning and even produced a series of weighty government tomes under the series title "Our Rivers: Total Use for Greater Wealth." The volume on the Columbia River, prepared by the U.S. Bureau of Reclamation, appeared in 1947 and acknowledged "a broad mutuality of interest" between Canada and the U.S. and the desire for "appropriate coordination of plans for resource development through international agreement" (U.S. Department of Interior, Bureau of Reclamation 1947, 47).[4]

In the 1940s, through the IJC, the United States began pressuring Canada to cooperate on the joint development of the Columbia River. The U.S. was less interested in joint benefits than in enhancing flood control and hydropower system performance on its reach of the Columbia, but to achieve its goals required cooperation and the negotiation of shared costs and benefits. So the two set out to cut a deal that would benefit both nations. Treaty negotiations set a precedent for a much freer and more coordinated north-south flow of water and power and for shared responsibilities and joint development, but it was all framed in what Tina Loo, following James Scott, calls a "high modernist" intellectual milieu (Loo 2004; Scott 1998). Considerations of equity and efficiency profoundly influenced the negotiations, but the U.S. and Canadian calculus for measuring efficiency and equity remained very narrowly defined, reflecting the culture of "total use for greater wealth" in an age of technological enthusiasm.

In this period, the IJC produced several reports associated with the treaty negotiations that reveal these perspectives well. Consider a representative report prepared by the International Columbia River Engineering Board in 1959 for the IJC. This report catalogued the Basin's ecology and resources, and it discussed various development options. Great details explained flood damages, hydroelectric development, and economic resources. A few pages, at most, discussed issues related to tribal fishing rights or local communities threatened with removal because of the damage to the fishery and proposed dams. Indicative of this approach was the comment: "No specific fish and wildlife studies were made for this report. However, responsible agencies have the problems under study. Indications are that ways and means may be devised to satisfy the requirements of the water resource development programs, and those of the fish and wildlife resource as well. Research programs to be carried on by these agencies should lead the way towards solving conflicts of interest." Such faith that some "responsible" parties would solve problems necessarily led to disconnected development programs that neglected one or more central aspects of the river and social systems. The conclusion to the *Water Resources of the Columbia River Basin* report included eighteen specific recommendations.

They all boiled down to cooperative development of the river for greater power generation. They identified no obstacles to successful implementation, no problem with engineering the river into a better power generator, no equity concerns, no significant environmental constraints (International Columbia River Engineering Board 1959, 60, 109–10).

The treaty itself, signed in the very last days of President Dwight D. Eisenhower's administration, also incorporated a narrow range of values. Indeed, it is difficult to even recognize a river in the treaty language. A pulsing, vibrant, life-giving force became reduced to acre-feet of storage, kilowatt hours, and dollars. Phrases in the treaty's preamble reinforced its ultimate—and arguably only—purpose:

Recognizing that the Columbia River basin . . . contains water resources that are capable of contributing greatly to the economic growth and strength and to the general welfare of the two nations, and

Being desirous of achieving the development of those resources in a manner that will make the largest contribution to the economic progress of both countries and to the welfare of their peoples of which those resources are capable, and

Recognizing that the greatest benefit to each country can be secured by cooperative measures for hydroelectric power generation and flood control, which will make possible other benefits as well (U.S. Department of the Interior 1971, 2).

Instrumentalist language reflected the narrowly conceived utilitarian purposes of the Columbia River Treaty. There was no disguising its primary purposes and the consensus between nations.

When negotiating the treaty, the fundamental goal for both sides was to build dams to enhance river control and hydropower output to generate economic benefits for the binational region. The mid-century consensus regarding comprehensive river basin development, the primary focus of the treaty negotiators on costs and benefits, and the view of the Columbia River as a controllable natural resource greatly simplified the issues in ways that environmental managers today can only regard wistfully. The Columbia River Treaty seems a natural outcome of its historical milieu in which B.C.'s Social Credit government, the U.S. government, and the Northwest states all dreamed the same dream of total use for greater wealth (Fisher and Mitchell 1996; Belshaw and Mitchell 1996).

The Columbia River Treaty in a Transitional Context

Just as negotiators were no doubt congratulating themselves on a job well done, larger contexts shifted. The Columbia River Treaty was indeed a product of its time. Had it been negotiated a decade earlier or later the outcome would likely have been different. Much changed in the legislative landscape after the United States and Canada ratified the treaty. In the United States changes were afoot that challenged the total use for greater wealth consensus. In the same year the treaty was ratified, the U.S. Congress approved the Wilderness Act (1964), establishing a national land preservation policy endorsed by a growing number of Americans skeptical that development was always and everywhere a good thing. In the next dozen years, a remarkable reformation of environmental law permanently changed the landscape of natural resource policy and development. The U.S. Congress enacted the Wild and Scenic Rivers Act (1968), the National Environmental Policy Act (1969), the Clean Air Act (1970), Clean Water Act (1972), the Endangered Species Act (1973), the Renewable Resources Planning Act (1974), the Eastern Wilderness Areas Act (1975), and a veritable flood of laws in 1976, including the National Forest Management Act, Federal Land Policy and Management Act, Resource Conservation and Recovery Act, and Toxic Substances Control Act. Environmental management was transformed (Andrews 1999, ch. 14).

Collectively, these laws reflected changing environmental and political values. Further, they helped drive sentiment in favor of a more cautious and complex view of natural resource management. The laws also significantly democratized the decision-making arena, requiring greater federal agency openness, assessments and disclosure of social and environmental impacts of proposed developments, and extensive public participation through hearings and public comment periods. More information flowed, more people got involved, and more voices came to the table. The new environmental governance placed limits on agencies responsible for resource development and required multidisciplinary planning that significantly weakened the hegemony of economists and engineers. In addition, it empowered environmentalists, biologists, and affected communities to object to plans and demand accommodation and mitigation. In essence, decision making slowed down and became more complex but also better reflected the pluralist values of society and ecological complexities (Hoberg 1992). A similar reformation of environmental and natural resources policy occurred in Canada though not so quickly as in the U.S. (Hessing et al. 2005, ch. 2, 3, and 5; Boyd 2003, ch. 2–5).

This precedent-setting reformation of laws and values between 1964 and 1976 impacted river development all over North America, including the Pacific Northwest. While voices have long risen in opposition to damming rivers, those who advocated structural river control enjoyed essentially free rein in management regimes from the Progressive era until the 1960s.[5] But the tide was already turning during the Columbia River Treaty negotiations. In fact, the treaty might be seen as the culmination of the era of mega-river development schemes—at least south of the forty-ninth parallel. In the decade before the treaty was ratified, engineers completed six mega-dams on the Columbia River main stem: Rock Island Dam (1953), McNary Dam (1954), Chief Joseph Dam (1955), The Dalles Dam (1960),

Rocky Reach Dam (1961), and Priest Rapids Dam (1961). Only two dams had been built on the main stem earlier (Bonneville in 1937 and Grand Coulee in 1942) and only two dams have been built on the main stem since (Wells in 1967 and John Day in 1971). The river development elite in the U.S. turned to Canada once they had built out all the good dam sites in the U.S. portion of the river. British Columbia's Big Dam era (e.g., the Columbia and Peace River projects) ended later than in the U.S., in part because it started later. All of the main stem Columbia River dams in B.C. were built after and resulted from the 1964 Columbia River Treaty.

South of the border, where the impetus to develop the Canadian portion of the Columbia River originated, the big dam era was winding down as opposition to dams ramped up. The first shot across the bow came in 1955 when environmentalists successfully blocked a federal government proposal to build a large dam and storage reservoir at Echo Park in northern Utah at the confluence of the Green and Yampa rivers. More ominously, the Colorado River Storage Project Act of 1956 forbade the construction of any Colorado River project dams in any national park or monument (Harvey 1994). Closer to home in the Columbia Basin, the federal government's Hells Canyon High Dam proposed for Idaho's Snake River Canyon went down to defeat, mostly due to opposition from Idaho Power Company, which wanted to build three smaller dams in Hells Canyon, and a later U.S. Supreme Court decision that asserted the public interest of free-flowing rivers might be greater than any potential hydroelectricity generated from another dam (Brooks 2006, Sowards 2009, 120–22). Northwest environmentalists successfully resisted proposals to dam Idaho's Salmon River, today a designated Wild and Scenic River. The lesson of these defeats was not lost on the federal dam-building agencies. Instead of pushing for more American dams, they pushed for Canadian development from which they could import the benefits while exporting the social and environmental impacts. Consistent with their governing priorities, the Social Credit government in B.C. willingly obliged after extracting fairly generous payments and terms (Belshaw and Mitchell 1996, 328).

However, British Columbia was not without its dam opponents. They were just more concentrated on the west side of the Cascades and focused primarily on salmon-bearing streams, especially the Fraser River. In fact, historian Matthew Evenden makes a convincing case that fish advocates in B.C. were better organized, more politically influential, and more effective in fighting dams than fish advocates in the U.S. Northwest. Commercial, tribal, and provincial fishing organizations in B.C., supported by U.S. commercial fishers and state and federal fish agencies south of the border, succeeded in taking all proposals for main-stem dams on the Fraser River off the table by 1958. The B.C. government's growing interest in Columbia and Peace river development coupled with pressure and incentives from the U.S. for cooperative development of the Columbia Basin, convinced B.C. Premier W.A.C. Bennett to abandon the Fraser and focus on the Columbia (Evenden 2004). That maneuver provided substantial impetus to the Columbia River Treaty.

B.C. eventually built three dams on the Canadian portion of the Columbia River: Keenleyside Dam (1968), Mica Dam (1973), and Revelstoke Dam (1984). At the same time, the province built two large dams on the Peace River in northeast B.C.: W.A.C. Bennett Dam (1967), one of the largest earthen dams in the world, and the nearby Peace Canyon Dam (1980). By the 1980s, the Big Dam era in both B.C. and the United States seemed over. No new large dams have been built in the U.S. since 1980. British Columbia took a hiatus from mega-projects for a couple of decades but is now resurrecting plans for a third dam on the Peace River known as "Site C."[6] In tune with the contemporary milieu of environmental values, equity concerns, and public participation in decision making, opposition to the Site C proposal is building in parallel with the efforts of the proponents to generate support.[7] Its future at this point remains uncertain.

The controversy over Site C is a microcosm of the complex forces at play in river development schemes in this new century—the same forces that will shape Columbia River Treaty renegotiations a few years hence. Renegotiating the Columbia River Treaty will be much messier than the initial negotiations. Governing agencies must consult more broadly than in 1960, involve the public in analysis and planning, assess and disclose environmental and social impacts, subject any claims regarding project costs and benefits to public scrutiny, address more than economic efficiency in negotiations, accommodate values pluralism, and see rivers as ecosystems rather than as plumbing. In short, decision makers will have to embrace complexity and uncertainty, and settle for more humble objectives in their desire to "control" nature for human benefit. These changes are products of decades of historical developments since 1964.

Efficiency and Equity

Most of our institutions are designed to achieve social objectives—to establish rights and responsibilities, to allocate costs and benefits, and to identify and solve problems. The Columbia River Treaty exemplifies this, as multiple authors in this volume demonstrate. The treaty confirmed rights to water and power, established bi-national responsibilities for developing the river, sought to equitably distribute the flow of dollars and electrons from joint development, and hoped to alleviate the costs and hardships of damaging floods on the lower Columbia.

If we are optimists, we might rightly be proud that the treaty achieved most of its purposes. The dams were built, storage achieved, downstream power and flood control enhanced, and benefits shared between the U.S. and Canada. But this rationalist view of the treaty and of our laws and institutions in general can be easily challenged. By tilting our view a little we can see the other side of the coin: inefficiencies, inequities, and worsening environmental problems. How we evaluate the treaty and the institutions it generated depends on our point of view—specifically, what we consider important and what factors go into our evaluation calculus. Not surprisingly, these criteria change over time and place according to historical context.

For example, in 1967 Resources for the Future economist and Columbia River Treaty analyst John Krutilla, an American, argued that the treaty negotiations resulted in compromises that made the choice and design of projects inefficient. From a purely economic calculus, he argued, the two nations would have been better off developing their own reaches of the river separately (Krutilla 1967, 10–11, and chs. 4 and 5). In contrast, Neil Swainson, a Canadian writing about a decade later, argued that the treaty drafters eschewed narrow national interests and strict efficiency considerations in order to optimize a *joint and equitable* distribution of development costs and benefits on this transboundary river (Swainson 1979, 5–11).[8] Using economic efficiency as the prime consideration, Krutilla judged the treaty unfavorably based on how much bang the U.S. got for its buck. With international cooperation in river development as the goal, Swainson judged the treaty favorably based on how equitably the countries shared costs and benefits.

The above example introduces the tension between managing for efficiency and managing for equity. Krutilla complained that the treaty was inefficient because it obligated the U.S. to build a dam on an inferior site (Libby Dam) while it did not obligate Canada to build the much more efficient High Arrow Lakes Dam. Instead, the treaty endorsed a lower dam at Arrow Lakes in part due to public opposition to the high dam in Canada. J. W. Wilson, in his 1973 book *People in the Way*, painted a sympathetic portrait of the communities displaced by the Arrow Lakes dam, lamenting how the "progress" promised by the efficiency experts really meant gains for some and losses for others. BC Hydro had hired Wilson to coordinate the relocation efforts at Arrow Lakes, but his work there left him ambivalent about Columbia Basin development. Extending Wilson's critique, historian Tina Loo has recently argued that the narrow "modernist" lens of the treaty negotiators led to regional and class-based inequities in which rural communities in B.C.'s Columbia Basin were sacrificed to benefit urban power consumers and U.S. economic interests, a story increasingly common in post–World War II North America. Loo not only questioned the value and need for *any* dam at Arrow Lakes, but she challenged the whole comprehensive river development scheme envisioned in the Columbia River Treaty (Wilson 1973; Loo 2004, 161–96).

When viewed from an efficiency perspective, sacrificing the lives and livelihoods of a few for the benefit of the many appears rational. When viewed from an equity perspective, consistently advantaging one group over another appears unfair.

Drawing from the work of political scientist Helen Ingram, we suggest that good public policy seeks both efficiency and equity, and when the two are at odds a balance must be struck between them (Whitely et al. 2008). An efficient allocation of resources is of little use if it so violates principles of fairness that it generates social unrest and resistance. On the other hand, resource allocations that pay too little attention to economic efficiency are often fiscally unsustainable. We need a good measure of both.

The other principle suggested by these examples is that the calculus for assessing efficiency and equity in the management of the Columbia River ought to be as broad as possible—and it has not been. Maximizing upstream storage for the least cost is too narrow a criterion for "efficiency." Likewise, sharing the downstream power benefits of upstream storage is too narrow a principle for equity. There are more interests at stake than just the two nations, and the river is more than just a giant power stream. Without aligning diverse economic, political, and social agendas within ecological limits, any planning is doomed in the long run.

Admittedly, this can muddy the waters. Presumptions are questioned, conflicts highlighted, the decision process slowed. But caution and tempered ambitions can be an antidote to the hubris that has marred many of the river development schemes of the twentieth-century world. Although nature and society have always been complex and uncertain, our natural resource managing policies and institutions have been routinely and naively reductionist. Simplifying the factors we evaluate has made decisions easier, but it has not usually made them better. Just as we must accept complexity and uncertainty in nature, we must also embrace those characteristics in our societies.

To the Future from the Past

The past produces the present and influences the future. Historians cannot predict the future, and so we cannot predict with certainty how renegotiations will proceed. But we can be certain that historical context will inevitably shape those renegotiations. The century of changes we have analyzed allows us some perspective to contextualize the upcoming renegotiations, recognizing continuity and change, evolution and disjunction.

What likely will be the issues and pressures that drive renegotiation of the treaty in the next few years? How will our contemporary historical context shape U.S.-Canada relations in the next decade? Just like fifty years ago, energy supply is likely to be front and center in the negotiations. In 1960, the U.S. pushed for the Columbia River Treaty mainly to enhance power generation at American dams along the Columbia River. It needed more electricity and was running out of good dam sites. B.C. had plenty of good, large, undeveloped dam sites left, so the two nations cut a deal. Today, increases in energy consumption, especially electricity, are very likely to continue unabated for the foreseeable future, particularly with the trend toward plug-in hybrid and electric cars (replacing petroleum-based energy with electrical energy) and the increasing role of the Internet in modern life.[9] (Internet server farms are among the region's biggest consumers of electricity.)[10] Concern for global climate change and the imposition of carbon emission caps will discourage coal- and natural gas-fired power plants in favor of hydropower, nuclear, wind, and solar. The combination of those two trends will likely renew pressure for the development of more hydropower dams in the basin. However, unlike the early 1960s, there are few sensible places left for major dams in the basin, and public and professional opposition to new dams is much stronger on both sides of the border than fifty years ago. New dams are highly likely to be on the agenda, but much less likely to be an approved centerpiece of any future treaty negotiation. Not just because of organized opposition, environmental impacts, and poor cost-benefit ratios, but also because alternative clean energy technologies (wind, geothermal, and solar) are for the first time economically competitive with conventional energy sources and growing worldwide at a record pace, faster than other competing forms of electrical production.[11]

Joint management of the river and "downstream benefits" are likely to play a key role again in future negotiations. But what are the U.S. and Canada jointly managing? As in 1960, hydropower system performance certainly will be front and center, with Canada again bargaining for a fair share of the downstream benefits to U.S. hydropower production attributable to its storage reservoirs. Flood control played a critical role in the original treaty negotiations, but with few options for large new storage dams in the Columbia Basin, it is doubtful that flood control will play a significant role in treaty renegotiation unless it involves payments to Canada for downstream flood control benefits from its *existing* storage reservoirs.

Unlike 1960, it is very likely that salmon will play a bigger role in renegotiation. While salmon were hardly even mentioned in the context of the original treaty (Grand Coulee Dam had already blocked all upstream salmon migrations), salmon restoration in the lower river is now a billion dollar business prompted by the Endangered Species Act, and water releases for fish passage are likely to be on the agenda. Although in B.C. concerns for fisheries have been fairly well-integrated with power planning since at least the 1940s, it was not until 1980 that the Northwest Power Act married power planning with fish and wildlife planning through the aegis of the Northwest Power Planning Council (Evenden 2004, ch. 6; Lichatowich 2001, ch. 8). The close coupling of fish and power will remain with us for the foreseeable future.

Similarly, when the original treaty was forged, Native claims to fish and to water were scarcely mentioned, and Native nations were not involved in the negotiations. Since then, indigenous peoples have experienced cultural and political revitalization in both the U.S. and Canada, rooted significantly in issues related to natural resource management and sovereignty (Glavin 1998). In the U.S., long ignored nineteenth-century treaty rights are now being enforced, new laws have strengthened tribal authority and autonomy, and tribal governing institutions play significant roles in natural resource management in the region. Meanwhile in British Columbia, the B.C. Treaty Commission is negotiating treaties with Canadian First Nations since they never relinquished their land claims (www.bctreaty.net). As a result, issues of sovereignty and rights to the resources that are central to the Columbia River remain unresolved and fluid. Undoubtedly, any future negotiations will involve tribes as full partners.

While the past's painful inequities will echo through indigenous peoples' roles in the renegotiation process, it remains to be seen whether the historically suspicious and adversarial relationship that periodically recurred over the past century between Canada and the United States will hamper the process. National self-interest, of course, will play a large role, but for the past two decades, the bi-national Pacific Northwest has worked hard to integrate their economies for future development. The North American Free Trade Agreement (NAFTA) is but the most obvious example. In the 1990s, regional planners and other political and economic theorists envisioned a future in which British Columbia and the American Northwest would be well-integrated into a larger region to be known as Cascadia, the hallmarks of which would be a clean, vibrant, globally oriented and environmentally friendly economy (Coates 2002; Kaplan 1999, pt.VI). Such an economy will depend in part on power from the Columbia.

Whether such a joint vision can carry the day against global economic and ecological uncertainty remains to be seen. A successful treaty renegotiation will join equity to efficiency in planning for the Columbia Basin's future, and craft new institutions that are flexible, adaptive, and responsive to changing and uncertain economic needs, social values, and environmental priorities.[12] Whether the biologists, engineers, economists, lawyers, and policymakers can approach the Columbia River with humility—instead of hubris—and a sense of its past complexity—instead of its future certainty—and think like a river—instead of a state or a market—are perhaps the most critical issues for the renegotiation agenda.[13]

Notes

1. Mathews 1910, 63–64, 139, 267–81, and passim, quotations from 64, 124, 267, 270. Emphasis added.

2. Terms like "saved," "beautiful," or "lovely" have transformed in the last century, revealing the ubiquity of changing values. A century ago, reformers and visionaries who cared about nature and society sought to control rivers to make them regularized, dependable, and peaceful components of a bucolic pastoral human landscape. Then, like the proverbial machine in the garden, the engineered river presented a contradictory reality:

a large measure of control but also a large measure of loss and disappointment. Society wanted technology to furnish security, modernization, and wealth, but it was unprepared for tradeoffs. In other words, the realities of river development failed to match Mathews' and others' dreams. The utopian dreams of the technocrats eventually lost luster and fell out of favor. Those who benefited from the redistributed wealth embodied in these development projects continued to drive them along despite objection. But by the 1960s and 1970s, the tide had turned. All the sensible projects had been built, along with many unsensible ones. Those who cared about nature and society defined "save," "protect," and "beautiful" quite differently.

3. Prior to 1963, Canadian law forbade the export of electric power to the U.S. (Krutilla 1967, 12).

4. The report with additional comments from the affected states and federal agencies was reprinted in House Document 473, 81st Congress, 2nd Session (February 1950).

5. Protest against dams has a long history in America. U.S. state and federal laws throughout the nineteenth century sought to regulate activities that harmed the public interest in rivers, including harm to fisheries, water quality, and navigation. Indicatively, the 1849 state constitution of Oregon forbade any obstruction blocking any stream or river in which salmon spawned. Opposition to damming rivers was widespread and common in the nineteenth century, though increasingly drowned out by the prophets of progress by the early twentieth century. For a two-century view of this subject profiling Maine and Washington State, see Jeffrey Crane, *Finding the River,* 2011).

6. The main proponent of the Site C Peace River Dam is the provincially owned power company BC Hydro. Its public information on the Site C proposal is available at: http://www.bchydro.com/planning_regulatory/site_c.html

7. For one view of the organized opposition, see: http://www.ecobc.org/peace_valley_no_site/

8. Swainson asserts that "Canadian decision makers were genuinely concerned with the best interests of all the parties involved" and that there was a "genuine perception of mutual responsibility" (10–11).

9. The U.S. Energy Information Agency forecasts that energy consumption will continue to rise through 2030, even though it now predicts that per capita consumption will decrease due to higher prices and efficiency upgrades promoted by statutes and incentives. See: http://www.eia.doe.gov/oiaf/aeo/overview.html section titled "Energy Intensity."

10. See http://seattletimes.nwsource.com/html/localnews/2003114987_microsoft09.html and the National Public Radio story available at: http://www.npr.org/templates/story/story.php?storyId=5545145

11. The U.S. Energy Information Agency projections for energy production/consumption by source to 2030 indicate no increase in hydropower and large increases in non-hydro renewable energy. In fact, the EIA predicts that renewables will increase faster than any other source of electricity and will supply almost the same amount of power as nuclear energy by 2030.

12. Pragmatic adaptive management is one perspective that incorporates change and uncertainty into natural resource management regimes. See Langston 2003, ch. 6.

13. Thinking like a river plays off Aldo Leopold's famous essay, "Thinking Like a Mountain," from *A Sand County Almanac* and has been used in Worster 1993, and Scott 1998.

Works Cited

Andrews, Richard N. L. *Managing the Environment, Managing Ourselves: A History of American Environmental Policy.* New Haven: Yale University Press, 1999.

Barber, Katrine. *Death of Celilo Falls.* Seattle: University of Washington Press, 2005.

Belshaw, John Douglas, and David J. Mitchell. "The Economy since the Great War." In *The Pacific Province: A History of British Columbia*, edited by Hugh J. M. Johnston, 313–42. Vancouver: Douglas & McIntyre, 1996.

Bloomfield, Louis M., and Gerald Francis FitzGerald. *Boundary Waters Problems of Canada and the United States: The International Joint Commission 1912–1958.* Toronto: Carswell, 1958.

Blumm, Michael C. "The Northwest's Hydroelectric Heritage." In *Northwest Peoples: Reading in Environmental History*, edited by Dale E. Goble and Paul W. Hirt, 264–94. Seattle: University of Washington Press, 1999.

Botkin, Daniel. *Discordant Harmonies: A New Ecology for the Twenty-First Century.* New York: Oxford University Press, 1990.

Boyd, David Richard. *Unnatural Law: Rethinking Canadian Environmental Law and Policy.* Seattle: University of Washington Press, 2003.

Brooks, Karl Boyd. *Public Power, Private Dams: The Hells Canyon High Dam Controversy.* Seattle: University of Washington Press, 2006.

Coates, Kenneth S. "Border Crossings: Patterns and Processes along the Canada–United States Boundary West of the Rockies." In *Parallel Destinies: Canadian-American Relations West of the Rockies*, edited by John M. Findlay and Kenneth S. Coates, 3–27. Seattle: University of Washington Press, 2002.

Crane, Jeff. *Finding the River: An Environmental History of the Elwha.* Corvallis: Oregon State University Press, 2011.

Denis, Leo. *Water Powers of Canada.* Ottawa: Commission of Conservation of Canada, 1911.

Evenden, Mathew. *Fish Versus Power: An Environmental History of the Fraser River.* Cambridge: Cambridge University Press, 2004.

———. "Precarious Foundations: Irrigation, Environment, and Social Change in the Canadian Pacific Railway's Eastern Section, 1900–1930." *Journal of Historical Geography*, no. 32 (January 2006): 74–95.

Ficken, Robert E. *Rufus Woods, The Columbia River, and the Building of Modern Washington.* Pullman: Washington State University Press, 1995.

Fisher, Robin, and David J. Mitchell. "Patterns of Provincial Politics since 1916." In *The Pacific Province: A History of British Columbia*, edited by J. M. Jonston, 254_72. Vancouver: Douglas & McIntyre, 1996.

Glavin, James. *A Death Feast at Dimlahamid.* Vancouver, B.C.: New Star Books, 1998.

Goble, Dale D. "Salmon in the Columbia Basin: From Abundance to Extinction." In *Northwest Lands, Northwest Peoples: Readings in Environmental History*, edited by Dale D. Goble and Paul W. Hirt, 229–63. Seattle: University of Washington Press, 1999.

Harvey, Mark W. T. *A Symbol of Wilderness: Echo Park and the American Conservation Movement.* Albuquerque: University of New Mexico Press, 1994.

Hessing, Melody, Michael Howlett, and Tracy Summerville. *Canadian Natural Resource and Environmental Policy: Political Econonmy and Public Policy, 2nd Edition.* Seattle: University of Washington Press, 2005.

Hoberg, George. *Pluralism by Design: Environmental Policy and the American Regulatory State.* New York: Praeger, 1992.

International Columbia River Engineering Board. "Water Resources of the Columbia River Basin." Report to the International Joint Commission of the United States and Canada , Ottawa and Washington D.C., 1959.

Kaplan, Robert D. *An Empire Wilderness: Travels into America's Future.* New York: Vintage, 1999.

Krutilla, John V. *The Columbia River Treaty: The Economics of an International River Basin Development.* Baltimore: Johns Hopkins Press, 1967.

LaFeber, Walter. *The New Empire: An Interpretation of American Expansion.* Ithaca, NY: Cornell University Press, 1998.

Langston, Nancy. *Forest Dreams, Forest Nightmares: The Paradox of Old Growth in the Inland West.* Seattle: University of Washington Press, 1995.

———. *Where Land and Water Meet: A Western Landscape Transformed.* Seattle: University of Washington Press, 2003.

Lee, Lawrence B. "The Canadian-American Irrigation Frontier, 1884–1914." *Agricultural History*, no. 40 (October 1966): 271–84.

Lichatowich, Jim. *Salmon Without Rivers: A History of the Pacific Salmon Crisis.* Washington, D.C.: Island Press, 2001.

Loo, Tina. "People in the Way: Modernity, Environment, and Society in the Arrow Lakes." *BC Studies*, no. 142/143 (Summer/Autumn 2004): 161–96.

Mathews, John L. *The Conservation of Water.* Boston: Small, Maynard & Company, 1910.

McEvoy, Arthur F. *The Fisherman's Problem: Ecology and Law in the California fisheries 1850–1980.* New York: Cambridge University Press, 1986.

Mitcher, E. Alyn. "Western Waters and the Federal Government." *Alberta History*, no. 28 (1980): 1–5.

Murton, James. "Creating Order: The Liberals, the Landowners, and the Draining of Sumas Lake, British Columbia." *Environmental History*, no. 13 (January 2008): 92–124.

Myers, Norman, and Jennifer Kent. *Perverse Subsidies: How Tax Dollars Can Undercut the Environment and the Economy.* Washington, D.C.: Island Press, 2001.

Nelles, H.V. *The Politics of Development: Forests, Mines, and Hydro-Electric Power in Ontario, 1849–1941,* 2nd edition. Montreal: McGill–Queens University Press, 2005.

Neuberger, Richard L. *Our Promised Land.* Moscow: University of Idaho Press, 1989.

Orsi, Jared. *Hazardous Metropolis: Flooding and Urban Ecology in Los Angeles.* Berkeley: University of California Press, 2004.

Pitzer, Paul C. *Grand Coulee: Harnessing a Dream.* Pullman: Washington State Unviersity Press, 1994.

Reimer, Chad. "Borders of the Past: The Oregon Boundary Dispute and the Beginnings of Northwest Historiography." In *Parallel Destinies: Canadian-American Relations West of the Rockies,* edited by John M. Findlay and Kenneth S. Coates, 221–45. Seattle: University of Washington Press, 2002.

Robbins, William G. *Landscapes of Conflict: The Oregon Story, 1940 2000.* Seattle: University of Washington Press, 2004.

Sachs, Aaron. *The Humboldt Current: Nineteenth-Century Exploration and the Roots of American Environmentalism.* New York: Viking, 2006.

Scott, James C. *Seeing Like a State: How Certain Schemes to Improve the Human Condition Have Failed.* New Haven: Yale University Press, 1998.

Sowards, Adam M. *The Environmental Justice: William O. Douglas and American Conservation.* Corvallis: Oregon State University Press, 2009.

Stadfeld, Bruce. *Electric Space: Social and Natural Transformations in British Columbia's Hydroelectric Industry to World War II.* Ph.D. dissertation, University of Manitoba, 2003.

Steinberg, Ted. *Acts of God: The Unnatural History of Natural Disaster in America.* New York: Oxford University Press. 2000.

Stoll, Steven. *Larding the Lean Earth: Soil and Society in Nineteenth-Century America.* New York: Hill and Wang, 2002.

Swainson, Neil A. *Conflict Over the Columbia: The Canadian Background to an Historic Treaty.* Montreal: McGill–Queens University Press, 1979.

Taylor, Joseph E., III. "The Historical Roots of the Canadian-American Salmon Wars." In *Parallel Destinies: Canadian-American Relations West of the Rockies,* edited by John M. Findlay and Kenneth S. Coates, 155–80. Seattle: University of Washington Press, 2002.

———. *Making Salmon: An Environmental History of the Northwest Fisheries in Crisis.* Seattle: University of Washington Press, 1999.

U.S. Department of the Interior. *Treaty between the United States of America and Canada Relating to Cooperative Development of the Water Resources of the Columbia River Basin, and the Documents Associated Therewith.* Portland, Ore., 1971.

U.S. Department of the Interior, Bureau of Reclamation. *The Columbia River.* Washington, D.C.: Government Printing Office, 1947.

White, Arthur V., and Charles Vick. *Water Powers of British Columbia.* Ottawa: Commission of Conservation of Canada, 1919.

White, Richard. *The Organic Machine: The Remaking of the Columbia River.* New York: Hill and Wang, 1995.

Whitely, John M., Helen M. Ingram, and Richard Warren Perry, eds. *Water, Place, and Equity.* Cambridge, Mass.: MIT Press, 2008.

Wilson, J. W. *People in the Way: The Human Aspects of the Columbia River Project.* Toronto: University of Toronto Press, 1973.

Wohl, Ellen E. *Disconnected Rivers: Linking Rivers to Landscapes.* New Haven: Yale University Press, 2004.

Worster, Donald. *Rivers of Empire: Water, Aridity, and the Growth of the American West.* New York: Oxford University Press, 1985.

———. "Thinking Like a River." In *The Wealth of Nature: Environmental History and the Ecological Imagination,* 123–34. New York: Oxford University Press, 1993.

———. "Wild, Tame, and Free: Comparing Canadian and U.S. Views of Nature." In *Parallel Destinies: Canadian-American Relations West of the Rockies,* edited by John M. Findlay and Kenneth S. Coates, 246–73. Seattle: University of Washington Press, 2002.

Part II: Changes Informed by Science

The Effects of Dams and Flow Management on Columbia River Ecosystem Processes

Chris Peery

Introduction

The Columbia River is a dominating ecological feature of the Pacific Northwest. This river, second largest in the U.S., encompasses a basin containing much of the states of Oregon, Washington, and Idaho, and parts of Montana, Utah, Wyoming, Nevada, and British Columbia, Canada: about 660,500 square kilometers. It has physically shaped the regional landscape for millennia. Relatively speaking, the influence of the Columbia River on human culture and economy has spanned much less time, both pre- and post-European settlement, but the effects have been no less dramatic. Iconic of the Columbia River have been the remarkable runs of anadromous Chinook salmon (*Oncorhynchus tshawytscha*), believed to be the largest in the world historically, and steelhead (*O. mykiss*) that returned each year. But the system also supported sizable populations of coho (*O. kisutch*), sockeye (*O. nerka*), chum (*O. keta*) and pink (*O. gorbuscha*) salmon, anadromous Pacific lamprey (*Lampetra tridentata*), white and green sturgeon (*Acipenser transmontanus; A. medirostris*), and burbot (*Lota lota*) among many other aquatic species. Many of the existing salmon populations within the Pacific Northwest are also relative newcomers, likely developing since the retreat of the last ice event, approximately 16,000 years ago (Waples et al. 2009).

There is no real way to know, but estimates range from six to as many as sixteen million salmon returning to the Columbia River basin historically (Northwest Power Planning Council 1986). Most salmonid populations have been reduced significantly, with only a handful of populations considered healthy or even self-sustaining without significant management intervention. Return of all salmon to the Columbia River now is closer to one million annually and 80 to 90 percent of these are believed to be the result of hatchery production (National Research Council 1996). Decline of our salmon populations has been an obvious manifestation of the changes humans have wrought to the Columbia Basin. Causes for the declines have been much debated, as have methods to conserve and restore salmon populations. Significant resources have been devoted to salmon-related mitigation, research, and recovery efforts. Much of this work revolves around the use of technology to raise fish at hatcheries and to improve survival of juvenile and adult life stages as they migrate between freshwater spawning and juvenile rearing habitats (includes hatcheries) and marine areas where the bulk of growth occurs (Ruckleshaus et al. 2002). Unfortunately, after several decades of this effort, costing into the billions of U.S. dollars (General Accounting Office 2002), little improvement in salmon production within the basin has been realized. As a result, there is a renewed interest by the governments of the U.S. and Canada as well as scientists and scholars to understand the underlying causes of salmon decline and the factors that limit salmon production. One response has been a movement to use an ecosystem view of the Columbia River and the processes that maintain system functions (Williams 2006; Bottom et al. 2009).

The System

The ecosystem encompasses all the separate components that make up our environment. Ecosystem components are physical (landscape and climate) and biological (microbes, plants, animals). Although there may be a tendency to think of rivers primarily as water flowing through a channel, the river ecosystem comprises everything within the watershed. From the perspective of the salmon, the watershed consists of (at least), (1) the riverine section—wetted channel ranging from small headwater streams to the estuary and associated microhabitat features (pools, riffles, runs, large woody debris (LWD), etc.,) (2) riparian zone and flood plain—the cross sectional characteristics of stream valley with associated plants and animals, and (3) the hyporheic zone—areas where water flows through substrate of the flood plain. The latter influences base flows, nutrient dynamics, and temperature (Ward 1989). Although not physically included within the Columbia River Basin, we can add the ocean as a fourth component to the ecosystem when dealing with anadromous salmon.

Aquatic systems are inherently dynamic, driven by cyclical and chaotic forces that work at a range of spatial and temporal scales. These drivers are the ecosystem processes that guide interactions between the physical and biological components and maintain functioning systems. In the Columbia River, the primary *physical processes* are discharge and temperature (i.e., Climate) and their interaction with underlying geology and geography. These in turn affect sediment and nutrient supply and riparian and floodplain characteristics (Waples et al. 2009). For example, large flow events dictate the major geomorphological characteristics of a river— scour and depositional features (i.e., pools and gravel bars), lateral meanders, and recruitment and movement of LWD. Interaction of sustained discharge and substrate through surface runoff and groundwater exchange determines nutrient availability and the baseline productivity of a watershed.

Biological processes typically act as modifiers to the physical components. Some examples include riparian vegetation, which supplies shading, influences uptake and release of nutrients, and is the source for large woody debris. Terrestrial grazers, such as elk, reduce forage biomass in riparian areas and in turn are controlled by top level predators. Beaver dams absorb flood energy, may decrease connectivity for small fish and thin riparian vegetation, but the resulting ponds increase shallow water rearing habitat and can raise ground-water levels in valley bottoms (Moore 2008). Spawning salmon clean gravels of fine sediments and provide marine-derived nutrients to inland areas, which may be a significant source of system productivity (Gende et al. 2002). Humans, however, are by far the most powerful modifiers with an extensive list of ways we can directly and indirectly modify river systems. The most significant physical change we have engineered in the Columbia River system has been the network of dams and reservoirs.

Dams

Within the Columbia River there are now hundreds (two hundred to four hundred, depending on source) of "major" dams, including over two dozen main-stem and major storage projects (Figure 1), with a combined generating capacity of over 30,000 megawatts and total storage capacity of over 55.8 Maf (US Army Corps of Engineers n.d.). Dams have provided significant benefits to our society. Primarily they generate electricity, promote agriculture and commerce, and provide flood control. However, dams and impoundments, and associated flow management, have had multiple direct and indirect effects on aquatic systems. Some projects, built without fish passage, block access to spawning and rearing areas for anadromous salmonids. Cumulatively, it is estimated that 55 percent of original spawning and rearing habitat is currently inaccessible to salmon (Northwest Power Planning Council 1986). Even projects with salmon passage inhibit movements for other species, such as Pacific lamprey and sturgeon (U.S. Army Corps of Engineer 2009; Parsley et al. 2007). Dams flood valleys and eliminate complex river habitat such as meanders, braided and side channels, and other shallow-water areas and associated wetlands and riparian zones and replace them with slower moving, relatively uniform reservoirs (Williams 2006). Indirect effects of dams on river processes may be just as harmful. In general, dams and flow regulation disrupt the biological and physical connectivity in river systems (Ward and Stanford 1979, 1995; Friedl and Wuest 2002; Waples et al. 2009).

Flow

Flow manipulations in regulated rivers generally store water during spring runoff for flood control and release it later to generate electricity during summer and winter. This results in a substantial change in annual hydrograph as depicted in Figure 2.

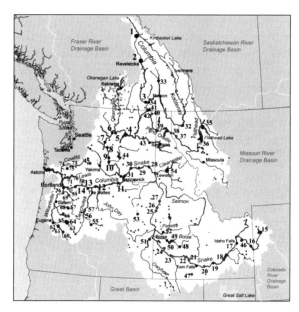

Figure 1. Dams in the Columbia River basin. Source: Karl Musser (http://en.wikipedia.org/wiki/File: Columbia_dams_map.png).

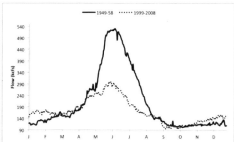

Figure 2. Ten-year average hydrograph of Columbia River at Bonneville Dam during 1949-58 (solid) and 1999-2008 (dashed). Source: University of Washington, School of Aquatic and Fisheries Science, Columbia Basin Data Access in Real Time, www.cbr.washington.edu/dart/.

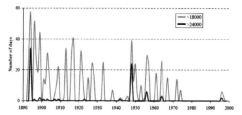

Figure 3. Peak flood events showing decline following development of dams.

One effect of this flattening of the hydrograph is a significant decline in flood events (Figure 3), which are known to be vital to the normal function of river systems, over time (US Department of Commerce 2005). Bottom et al. (2009) estimated that Columbia River peak spring time runoff at The Dalles, Oregon, has been reduced by about 43 percent: 17 percent by irrigation withdrawals and climate change, and the remaining 26 percent by upstream flood storage. Reducing over-bank flood events limits interactions with the floodplain, including channel migration, which maintains habitat complexity and enhances nutrient exchange (Figure 4) (Hauer and Lorang 2004). Often coupled with the buildup of levees and sediment retention within reservoirs, the separation of river from floodplain can lower groundwater levels and reduce hyporheic flows downstream from dams (Stanford and Ward 1993). For salmon, truncated spring discharge and lower river velocities through reservoirs influence the timing of ocean entry for smolt migrants and potentially affect their early ocean survival (US Department of Commerce 2005).

Temperature

Impounding water behind dams can alter natural temperature regimes in regulated rivers. Increased volume, lower velocities and mixing, and increased surface area have caused earlier warming, higher peak temperatures, and later cooling in the main stem Columbia River. Between 1938 and 1993, Quinn and Adams (1996) reported the date water temperatures reached 15.5°C in spring advanced thirty days, while the date temperatures cooled to 15.5°C in the fall was delayed seventeen days, and maximum summertime water temperatures increased 1.8°C on average. Summertime temperatures commonly reach levels stressful to salmonids in the main stem Columbia River now (Berman and Quinn 1991; U.S. Army Corps of Engineers 2003), resulting in increased straying of adult upstream migrants to tributary streams seeking cooler refuge areas (Goniea et al. 2006; High et al. 2006). In tributary streams, withdrawing water for agriculture results in lethal temperatures and even stream dewatering. Although not as well documented, winter flow augmentation may produce warmer water temperatures than occurred historically, with potential consequences for aquatic productivity, such as changes in the timing of incubation and hatching of salmonid larvae.

Sediment and Nutrients

Impoundments trap sediments and nutrients carried by rivers, disrupting habitat quality and productivity in downstream areas. One evaluation of the global effects of large reservoirs estimated that "more than 40% of global river discharge is intercepted locally by the large reservoirs analyzed here, and a significant proportion (~70%) of this discharge maintains a sediment trapping efficiency in excess of 50%" and that "[s]everal large basins such as the Colorado and Nile show nearly complete

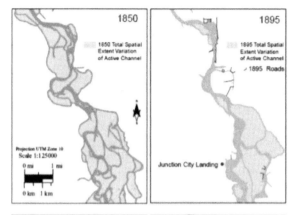

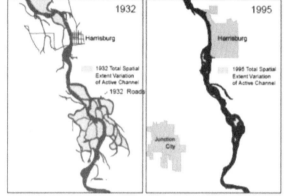

Figure 4. Representation of the simplification of channel complexity over time in segment of the Willamette River near Harrisburg, OR. Source: Hulse et al. 2002.

trapping due to large reservoir construction and flow diversion" (Vorosmarty et al. 2003). As a result, areas downstream in and near regulated rivers, including coastal systems, have experienced accelerated erosion and loss of freshwater and marine productivity worldwide (World Commission on Dams 2000). In the Columbia River, Bottom et al. (2009) estimated there was a 50 to 60 percent reduction in annual average sediment transport to the estuary between 1945 and 1999. Nutrient dynamics mirror those of sedimentation in regulated rivers (Bunn and Arthington 2002). For example, following construction of Duncan Dam on the Duncan River (1967) and Libby Dam on the Kootenay River (1973) there was a measurable decline in productivity of fish populations in Kootenay Lake, which flows into the Columbia River in British Columbia. It was later determined that phosphorous loading to the lake had declined significantly following closure of the dams and tighter pollution controls in the basin. This was followed by reduced phytoplankton, zooplankton, and then kokanee (land-locked sockeye salmon) production in Kootenay Lake (Friedl and Wuest 2002). A fertilization program was initiated in the lake starting in 1992 to compensate for the reduced nutrients and kokanee production began trending upward two years later (Ashley et al. 1997).

As important as the downstream nutrient flux, if not more so, are the marine-derived nutrients (MDN) that adult salmon migrants provide to salmon streams. Historically MDN likely supported a significant portion of the primary productivity in stream systems, especially in the interior areas of the Columbia Basin, which are naturally oligotrophic. Currently MDN have been eliminated from as much as 55 percent of the historical salmon range, including thousands of kilometers of Canadian rivers upstream from Grand Coulee Dam (1942). The reduction of MDN to the remainder of the basin is equal to the salmon population declines. One reason why salmon restoration efforts have not been more successful in the Columbia River may be that lower MDN loading from diminished salmon runs has created negative feedback through reduced juvenile rearing capacity in streams that otherwise appear to contain adequate rearing habitat (Williams 2006). One estimate suggests recent salmon escapement levels may only provide 6–7 percent of historical MDN inputs to salmon rivers in the Pacific Northwest (Gresh et al. 2000). Another analysis suggested less than 2 percent of historical marine-derived phosphorus currently returns to the Snake River (Scheuerell et al. 2005), or that, under some circumstances, there could even be a net export of nutrients when adult escapement is extremely low (Moore and Schindler 2004).

Conclusion

The Columbia River is integral to the ecology of the Pacific Northwest particularly to anadromous salmonid production, but it is also one of the most developed and regulated rivers in the world. The Columbia River Treaty has become central to the management of the Columbia River because of the role it plays in flood control, hydropower production and, more recently, fisheries enhancement. The costs and benefits of dams and reservoirs in the system can be both dramatic and subtle, but in most cases are difficult to quantify. This chapter provides a brief outline of some of the broad-scale ecological changes that have resulted from impoundment and regulation of the Columbia River. Primarily what we can see is that regulation and extraction of flow from the river, along with channelization for flood control, has interrupted the natural river-flood plain interaction and diminished or eliminated the up- and downstream flux of nutrients and substrate required to maintain habitat complexity and primary/secondary productivity in natural systems. Effects of dams and flow regulation extend downstream to the estuary and near-shore coastal areas that have been deprived of flood events, sediment, and large woody debris. The riverine environment has been further simplified by addition of uninterrupted series of impoundments that also affect thermal and velocity conditions, making them less compatible with native species but encouraging proliferation of exotic fish, invertebrate, and plant species throughout the basin.

Can these trends be reversed? I believe it is possible to improve the ecology of the Columbia River through use of new management tools such as implementing natural flow regimes that allow controlled low level flood events but limit catastrophic flood damage, and levee set-backs that allow more natural channel and riparian development. Adoption of these kinds of management scenarios will require quality science-based information on the ecology of the Columbia River, together with a comprehensive education program and proactive organization.

Much more detailed information is available on the ecology of the Columbia River than I have discussed here and more is revealed annually as we increase our ability to study and interpret watershed processes and ecological interactions. Also not addressed here but discussed elsewhere in this volume is the influence of regional climate change and how this may affect our capacity to balance resource use and maintenance of the natural system.

Regardless of these patterns, we must recognize that humans are part of the Columbia River ecosystem and our use of resources within the watershed significantly influences every biological component and process across the basin. Being mindful of this at critical junctures, such as the potential renegotiation of the Columbia River Treaty, will help managers and engineers make informed decision that will better balance ecological risks and societal desires. Vital to this will be educating decision makers and the general public of the ramifications of our resource use, which requires that we are proactive in disseminating research results and in identifying data gaps to be addressed by new research and monitoring efforts.

While it is inevitable that natural resource conservation and resource use will at some level be in opposition, the level of compatibility between these two efforts can be increased substantially through an enhanced understanding of ecosystem functions. Better models will translate into an improved ability to manage flood control, power generation, fisheries harvest, etc., so as to minimize impacts to the watershed and associated wildlife populations. Ultimately though, we must decide what we will be managing for. With Pacific Northwest anadromous salmon and steelhead, for example, are we content to have populations maintained primarily though intensive aquaculture production with only a remnant natural production in areas set aside as wildlife refuges (the current condition in other areas with larger population bases such as Japan and portions of California) or is it worthwhile to further limit human development and resource extraction in order to promote a healthy, self-sustaining wild salmon population across a broad landscape? With better knowledge of the workings of the Columbia River ecosystem, we will at least have the ability to make informed decisions.

Works Cited

Ashley, K I, L C Thompson, D C Lasenby, L McEachern, K E Smokorowski, and D Sebastian. "Restoration of an Interior Lake Ecosystem: The Kootenay Lake Experiment." *Water Quality Research Journal of Canada*, no. 32 (1997): 295–323.

Berman, C H, and T P Quinn. "Behavioral Thermoregulation and Homing by Spring Chinook Salmon, *Oncorynchus tshawytscha* (Walbaum), in the Yakima River." *Journal of Fish Biology*, no. 39 (1991): 301–12.

Bottom, D L, K K Jones, C A Simenstad, and C L Smith. "Reconnecting Social and Ecological Resilience in Salmon Ecosystems." *Ecology and Society* 14, no. 1 (2009): art 5.

Bunn, S E, and A H Arthington. "Basic Principles and Ecological Consequences of Altered Flow Regimes for Aquatic Diversity." *Environmental Management*, no. 30 (2002): 492-507.

Friedl, G, and A Wuest. "Disrupting Biogeochemical Cycles—Consequences of Damming." *Aquatic Sciences*, no. 64 (2002): 55–65.

Gende, S M, R T Edwards, M F Willson, and M S Wipfli. "Pacific Salmon in Aquatic and Terrestrial Ecosystems." *Bioscience*, no. 52 (2002): 917–28.

Goniea, T, C Peery, D Bennett, M Keefer, L Stuehrenberg, and T Bjornn. "Behavioral Thermoregulation and Slowed Migration by Adult Fall Chinook Salmon in Response to High Columbia River Water Temperatures." *Transactions of the American Fish Society*, no. 135 (2006): 408–19.

Gresh, T, J Lichatowich, and P Schoonmaker. "An Estimation of Historic and Current Levels of Salmon Production in the Northwest Pacific Ecosystem." *Fisheries* 25, no. 1 (2000): 15–21.

Hauer, F R, and M S Lorang. "River Regulation, Decline of Ecological Resources, and Potential for Restoration in a Semi-Arid River in the Western USA." *Aquatic Sciences*, no. 66 (2004): 388–401.

High, B, C Peery, Bjornn, and D Bennett. "Temporary Staging of Columbia River Summer Steelhead in Coolwater Areas and its Effect on Migration Rates." *Transactions of the American Fish Society*, no. 135 (2006): 519–28.

Hulse, D, S Gregory, and J Baker. *Willamette River Basin Atlas.* Corvallis: Oregon State University Press, 2002.

Moore, J W. "Animals as Ecosystem Engineers in Streams." *Bioscience*, no. 56 (2008): 237–46.

Moore, J W, and D E Schindler. "Nutrient Export from Freshwater Ecosystems by Anadromous Sockeye Salmon (*Oncorhynchus nerka*)." *Canadian Journal of Fisheries and Aquatic Sciences*, no. 61 (2004): 1582–89.

National Research Council. *Upstream: Salmon and Society in the Pacific Northwest.* Report on the Committee on Protection and Management of Pacific Northwest Anadromous Salmonids for the National Research Council of the National Academy of Sciences, National Research Council, DC: National Acadamy Press, 1996.

Northwest Power Planning Council. "Council Staff Compilation of Information on Salmon and Steelhead Losses in the Columbia River Basin." Northwest Power Planning Council, Portland, 1986.

Parsley, M, C Wright, B Van Der Leeuw, E Kofoot, C Peery, and M Moser. "White Sturgeon Passage at the Dalles Dam, Columbia River, USA." *Journal of Applied Ichthyology*, no. 23 (2007): 627–35.

Quinn, T P, and D J Adams. "Environmental Changes Affecting the Migratory Timing of American Shad and Sockeye Salmon." *Ecology* 77, no. 4 (1996): 1151–62.

Ruckleshaus, M H, P Levin, J B Johnson, and P M Kareiva. "The Pacific Salmon Wars:

What Science Brings to the Challenge of Recovering Species." *Annual Review of Ecology and Systematics*, no. 33 (2002): 665–706.

Scheuerell, M D, P S Levin, R W Zabel, J G Williams, and B L Sanderson. "A New Perspective on the Importance of Marine-Derived Nutrients to Threatened Stocks of Pacific Salmon." *Canadian Journal of Fisheries and Aquatic Sciences*, no. 62 (2005): 961–64.

Stanford, J A, and J V Ward. "An Ecosystem Perspective of Alluvial Rivers: Connectivity and the Hypoheic Corridor." *Journal of North American Benthological Society* 12, no. 1 (1993): 49–60.

U.S. Army Corps of Engineers. Columbia Basin Water Management Division. Web site at http://www.nwd-wc.usace.army.mil/

U.S. Army Corps of Engineers. "Adult Pacific Lamprey in the Lower Columbia River: 2007 Radiotelemetry and Half Duplex PIT Tag Studies." Report for Portland District, 2009.

U.S. Army Corps of Engineers. "Water Temperatures and Passage of Adult Salmon and Steelhead in the Lower Snake River." Technical Report, Walla Walla District, 2003.

U.S. Department of Commerce. "Salmon at River's End: The Role of the Estuary in the Decline and Recovery of Columbia River Salmon." NOAA Tech Memo, 2005, 246.

U.S. Department of Commerce. "Effects of the Federal Columbia River Power System on Salmonid Populations." NOAA Tech Memo, 2005, 150.

U.S. General Accounting Office. "Columbia River Basin Salmon: Federal Agencies Recovery Responsibilities." Expenditures and Actions, 2002.

Vorosmarty, C J, M Meybeck, B Fekete, K Sharma, P Green, and J P M Syvitski. "Anthropogenic Sediment Retention: Major Global Impacts from Registered River Impoundments." *Global and Planetary Change*, no. 39 (2003): 169–90.

Waples, R S, T Beechie, and G R Pess. "Evolutionary History, Habitat Disturbance Regimes, and Anthropogenic Changes: What Do these Mean for Resilience of Pacific Salmon Populations?" *Ecology and Society* 14, no. 1 (2009): art. 3.

Waples, R S, G R Pess, and T Beechie. "Evolutionary History of Pacific Salmon in Dynamic Environments." *Evolutionary Applications*, no. 1 (2008): 189–206.

Ward, J V. "The Four-Dimensional Nature of Lotic Ecosystems." *Journal of the North American Benthological Society*, no. 8 (1989): 2–8.

Ward, J V, and J A Stanford. "Ecological Connectivity in Alluvial River Ecosystems and its Disruption by Flow Regulation." *Regulated Rivers: Research and Management*, no. 11 (1995): 105–19.

Ward, J V, and J A Stanford. *The Ecology of Regulated Streams.* New York: Plenum Press, 1979.

Williams, R N. *Return to the River.* London: Academic Press, 2006.

World Commission on Dams. *Dams and Development: A New Framework for Decisionmaking.* London: Earthscan Publications Ltd., 2000.

When Courts Run Regulated Rivers:
The Effects of Scientific Uncertainty

Carmen Thomas Morse

Introduction

"Dams and Salmon Don't Mix," or so the story goes (Ellensburg *Daily Record* 1968). In the U.S., hydropower and anadromous salmon—fish that live in the ocean and migrate to freshwater rivers to spawn—have been in conflict for well over a century. In the Pacific Northwest, this issue has been continuously litigated since the early 1990s (see e.g., *Idaho Department of Fish and Game v. National Marine Fisheries Service* 1994; *American Rivers v. National Marine Fisheries Service* 1997; *National Wildlife Federation v. National Marine Fisheries Service* 2003, 2005, 9th Cir. 2005, 2008). This litigation is especially contentious because the region is home to the world's largest hydroelectric system (Northwest Power Planning Council 1986), comprising more than one hundred fifty dams. This system—which includes the Federal Columbia River Power System (FCRPS) – currently generates approximately 55 percent of the region's electricity (Northwest Power Planning Council 2005).

The FCRPS consists of fourteen dams and associated facilities and is located on the Columbia River, which flows from the west side of the Canadian and Rocky Mountains to the Pacific Ocean and drains areas of seven states and one Canadian Province (U.S. Bureau of Reclamation n.d.). The U.S. Army Corps of Engineers (the Corps) operates all but two of the FCRPS dams; the U.S. Bureau of Reclamation operates Grand Coulee and Hungry Horse dams (*National Wildlife Federation v. National Marine Fisheries Service* 9th Cir. 2005). Although many studies of the interactions between dams and anadromous salmonids have been conducted, numerous uncertainties remain. These uncertainties gave rise to the dispute between the National Wildlife Federation (NWF) and the National Oceanic and Atmospheric Administration (NOAA) over the impacts of the FCRPS (*National Wildlife Federation v. National Marine Fisheries Service* 2003, 2005).

In the FCRPS litigation, multiple parties (hereafter referred to as NWF) contested NOAA's approach to determining whether federal operation of dams jeopardizes listed runs of anadromous salmonids within the Columbia Basin. On one level, the litigation is a dispute over the trade-offs of maintaining or restoring native fish populations and the resultant reductions in the Basin's capacity to generate electricity through hydropower production. At a deeper level, this litigation is a dispute over the proper use of science and the proper approach to dealing with uncertainties. The deeper controversy centers on the difficulties inherent in natural resource management, in which choices must be made, often without clear answers or distinction between science-based and policy-based decisions.

This chapter uses the federal district court's grant of NWF's 2004 and 2005 requests for injunctive relief from FCRPS' proposed operations to examine the science, policy, and legal issues involved in managing the FCRPS and listed salmonids within the Columbia River Basin. Because the court's decision in issuing the injunction was clearly informed by prior agency actions and litigation, portions of that history are incorporated into the discussion to provide the necessary foundation.

The chapter begins with a brief description of the legal and scientific framework within which NWF requested injunctive relief. While the Endangered Species Act provided the legal basis for the FCRPS litigation, the Northwest Power Act influenced its timing, and the available methods for facilitating downstream migration of juvenile anadromous salmonids (smolts) defined the universe of alternatives. This is followed by an examination of NOAA's prescription of downstream fish passage methods through its biological opinions, identification of scientific uncertainties, and an explanation of how these uncertainties gave rise to the FCRPS injunction litigation. The chapter then uses the litigation to identify both parties' use of "science-policy" to cloak uncertainties and analyzes the impact of this use before concluding with recommendations for avoiding the "science-policy" conflation and advancing the discussion on the three-way intersection of science, policy, and law.

LEGAL FRAMEWORK
The continuing operation of the FCRPS is subject to a number of statutes, including the Endangered Species Act (ESA) and the Pacific Northwest Electric Power Planning and Conservation Act (Northwest Power Act). Both the ESA and the Northwest Power Act contemplate protection and restoration of the Columbia Basin's salmonid runs (see e.g., 16 U.S.C. §§1532, 1536).

The Endangered Species Act
The ESA requires the Fisheries Division of the National Oceanic and Atmospheric Administration (NOAA, previously named the National Marine Fisheries Service) to evaluate the status of oceanic and anadromous fish species and list as threatened or endangered those that are at risk of extinction within the foreseeable future (16 U.S.C. §§ 1532(3), (6), (19), 1533). Listing triggers substantive provisions (16 U.S.C. §§ 1536, 1538, 1539) that require federal agencies to avoid actions that jeopardize the continued existence of the listed species (16 U.S.C. § 1536). Federal agencies must consult with NOAA to determine whether these actions are likely to jeopardize listed species (16 U.S.C. § 1536(a)(2)). If, during the consultation, the

agencies determine that a proposed action is likely to adversely affect listed species, NOAA must prepare a biological opinion (BiOp) (16 U.S.C. §1536(b); 50 CFR § 402.14(h)). In a BiOp, NOAA evaluates the effects of the proposed action against the current status of the species and all other known factors affecting it (16 U.S.C. §1536(b); 50 CFR § 402.14(h)). If NOAA determines that the proposed action is likely to jeopardize the species, NOAA may recommend a "reasonable and prudent alternative" (RPA) to modify the proposed action (*National Wildlife Federation v. National Marine Fisheries Service* 9th Cir. 2005; 16 U.S.C. §1536(b)(3)(A)). An RPA generally seeks to accomplish the proposed project's original goal while avoiding or minimizing adverse effects to listed species (50 CFR § 402.02).

Northwest Power Act

The ESA places primary importance on the survival and recovery of species at risk of extinction. In contrast, the Northwest Power Act (NPA) was passed with the promise to treat fish and hydropower equally (H.R. Rep. No. 96-967(I) 1980). The NPA was enacted in 1980 (16 U.S.C. § 839), shortly after NOAA began a status review of anadromous salmonids within the Columbia River Basin (43 Fed. Reg. 45:628 October 3, 1978). The NPA directed the interstate council established under its provisions to provide both power planning and a fish and wildlife program for the basin. Although the NPA did not direct the Council to make tradeoffs between power and fish, due to the legislative history promising equal treatment of fish and hydropower and the NPA's detailed provisions designed to restore Columbia Basin fish, NOAA suspended its salmonid status review (Blumm and Simrin 1991). Ten years later, with the promise not realized and salmonid populations still declining, NOAA reinitiated its salmonid status review and began listing (see e.g., 55 Federal Register 13, 181; 37,342). After the initial listings of the Snake River chinook and sockeye salmon in 1992 (U.S. Fish and Wildlife Service n.d), NOAA began consultation with the Corps on the FCRPS operations, issuing its first FCRPS BiOp in 1993 (*Idaho Department of Fish and Game v. National Marine Fisheries Service* 1994).

In the 1993 and each subsequent BiOp, NOAA analyzed the impacts of FCRPS operations on listed salmonid survival and recovery. Central to the analysis and FCRPS injunction litigation were the impacts of FCRPS operations on downstream smolt migration; mortality estimates for upriver migration are generally lower (White 1995). Because NOAA's analysis builds upon its view of the best available science, it is necessary to examine the state of the science at the time.

SCIENCE FRAMEWORK
Despite different purposes, the ESA and the NPA both require use of "best available science" in making natural resource management decisions (15 U.S.C. § 1536(a) (2); 16 U.S.C. § 839(h)(b)(6)(B)). The question then becomes what was the "best

available science" relative to fish passage methods at the time of the injunctions? Merely examining the state of the science in 2004 and 2005 will not yield the answer because the best available "science" was the result of a gradual accumulation and modification of models, assumptions, and risk management approaches that began in the early 1990s with NOAA's initial analyses of the effects of FCRPS operations on listed anadromous salmonids. Although the impacts of dams on migrating salmonids were well established at the time of the FCRPS injunction litigation (NOAA 1995), the best method of providing fish passage around dams remained uncertain. This section briefly describes the impacts of dams on migrating salmonids before reviewing the available means of fish passage.

Dams and Salmonids

While many factors contributed to the anadromous salmonid declines, construction and operation of hydroelectric dams were a major factor—resulting in mortality rates up to ninety percent for some species (NOAA 1995). Dams have numerous adverse effects, both direct and indirect, on anadromous salmonids (NOAA 2000). An estimated 15 to 30 percent of salmonid smolts perish annually at each dam within the FCRPS (Blumm and Simrin 1991). Of the indirect effects of dams, altered flow regimes that disrupt spring and summer flows are particularly harmful, because these flows are critical to downriver smolt migration (Blumm and Simrin 1991).

Despite more than forty years of investigations of fish passage around the FCRPS hydroelectric dams (*American Rivers v. National Marine Fisheries Service* 1997), uncertainties remain about which method, or combination of methods, is most effective.

Passage Methods

Listed salmonids may navigate the FCRPS hydroelectric dams during their migration from upriver spawning and rearing areas to the ocean in two ways: through dams or around dams. Because passage through dams causes the greatest smolt mortality (*National Wildlife Federation v. National Marine Fisheries Service* 9th Cir. 2005), it is not a preferred passage method. There are three main alternate methods of fish passage around dams: in-river bypass systems, transportation systems, and increased spill (*Idaho Department of Fish and Game v. National Marine Fisheries Service* 1994). Although in-river bypass systems (NOAA 2000) are used in locations within the FCRPS, their operation is not contested. In contrast, use of transportation and increased spill to facilitate downstream smolt migration is hotly contested and the key issue in the FCRPS injunction litigation.

Transportation. Trucking or barging fish around dams and thereby avoiding the hazards of turbines and slack water is currently the primary means of passing fish downstream through the FCRPS because it minimizes costs to the hydropower system (Blumm et al. 2006). The system generally consists of a series of pipes that allow fish to pass from a reservoir to a collection facility, from which they are transferred to a truck or barge, transported around the hydropower system, and released below the last dam (Blumm et al. 2006). In the FCRPS transportation system, smolts are collected at McNary Dam on the Columbia River, and Lower Granite, Little Goose, and Lower Monumental dams on the Snake River (Blumm et al. 2006). All smolts are released below Bonneville Dam (Blumm et al. 2006).

While the transportation system avoids direct mortality associated with passage through turbines, it exposes smolts to higher levels of predation during transportation and upon release and increases vulnerability to disease—both of which result in indirect mortality (NOAA 1995). Efficacy studies of the FCRPS transportation program have yielded equivocal results (NOAA 1995). Only ten of twenty-four studies found that transportation resulted in higher adult return rates of spring/ summer chinook compared to smolts captured, marked, and released to migrate in-river. One of the studies found higher adult returns in the control group of non-transported fish; the remaining thirteen studies obtained too few adult returns to come to any conclusion. The studies also found that transport efficacy varied depending on the facility, fish species, life stage, and water year (NOAA 1995).

Consequently, in 1994 an independent peer review team consisting of scientists from NOAA, the U.S. Fish and Wildlife Service (FWS), state fisheries agencies, and treaty tribes concluded not only that the transportation program alone could not eliminate serious threats facing declining salmon populations, but also that it should not be viewed as a primary or even supporting means of salmon recovery (Independent Scientific Group 1996). Reviewers cited a number of reasons for their conclusion, including (1) the transportation program's underlying premise— that handling and marking "control" fish resulted in no mortality—had not been sufficiently tested and could be incorrect; (2) transportation potentially interfered with homing behavior, which could result in adverse genetic effects; and (3) transportation stressed fish and increased their vulnerability and exposure to disease and predation (Blumm et al. 1998). The primary alternative to transportation around the FCRPS is to leave smolts in-river and increase the amount of water released through dams (spill) to aid downstream fish migration.

Increased Spill. In general terms, any dam-released flows (intentional or otherwise) that do not pass through turbines are "spill" (see NOAA 2000). Spill releases to benefit downstream migrating smolts results in conflict because it requires such releases when dam operation for power favors storing water and using it to generate hydropower sales (Blumm et al. 1998; Blumm et al. 2006). In addition, spill can result in a potentially fatal fish disease induced by gas supersaturation (NOAA 1995; Weitkamp and Katz 1980).

Despite these problems spill is used to pass fish because it is most effective in aiding downstream smolt migration. By avoiding contact with turbine blades and rapid pressure changes associated with turbine operation, spill passage increases smolt survival at least four times over turbine passage. Spill also reduces reservoir travel time, an important factor for smolts vulnerable to warm water. Consequently, NOAA concluded that spillway passage was the preferred passage method for non-transported smolts (NOAA 2000).

REGULATORY FRAMEWORK AND AGENCY ACTIONS
This section examines NOAA's analyses of proposed FCRPS operations and its smolt passage prescriptions in three BiOps. Review of all three BiOps is necessary because each plays an important role in the FCRPS injunction litigation: the 1995 BiOp introduces NOAA's "spread the risk" policy which is central to the litigation, and the 2000 and 2004 BiOps continued NOAA's risk-spreading approach and were the final agency actions from which injunctive relief was sought. The 1995 BiOp followed a district court rejection of a finding of no jeopardy in the 1993 BiOp (*Idaho Department of Fish and Game v. National Marine Fisheries Service* 1994).

1995
NOAA's 1995 BiOp determined that the 1994 to 1998 FCRPS operations, as proposed, would jeopardize listed salmonids due to both the significant reliance on transportation and the failure to improve in-river migratory conditions (*National Wildlife Federation v. National Marine Fisheries Service* 2005). NOAA's jeopardy determination relied on the use of models to evaluate risk of various alternatives to salmonids. Following the determination of jeopardy, NOAA proposed a Reasonable and Prudent Alternative (RPA).

   Jeopardy Finding. NOAA turned to both passage and life cycle models to evaluate whether the proposed FCRPS operations offered equivalent or greater risk reduction for listed salmonids than its Recovery Plan. It considered three fish passage models all of which analyzed downstream survival: the Columbia River Salmon Passage model (CRiSP), the Passage Analysis Model (PAM), and the Fish Leaving Under Several Hypotheses (FLUSH) model. Although each passage model produced survival estimates, each relied on different hypotheses and thus predicted very different outcomes. For example, FLUSH assumed higher delayed mortality for transported fish than CRiSP. Consequently, FLUSH output suggested that increased flows had "the best chance of ensuring [salmonid] survival, while a strategy that [relied] on transportation [was] likely to lead to extirpation." In contrast, CRiSP output suggested "that increased and improved transportation [had] the best chance of ensuring species survival," while a strategy that relied on increased flows was "likely to lead to extirpation" (*National Wildlife Federation v. National Marine Fisheries Service* 2005).

Model assumptions are made to address areas of uncertainty. Because of the conflicting predictions, NOAA concluded that management decisions made "solely and conclusively on any single set of assumptions" would result in great risk to listed species, and consequently combined all three model outputs. In doing so, NOAA weighed the CRiSP model most heavily because it believed that it most closely approximated actual conditions (*National Wildlife Federation v. National Marine Fisheries Service* 2005).

NOAA then used the combined passage model results as inputs to life-cycle models, which assessed adult salmonid survival and recovery under the proposed FCRPS operations. Because the life-cycle models yielded a low likelihood of survival and recovery of each species, NOAA concluded that "without major modifications to the Snake and Columbia River dams," it was likely that proposed FCRPS operations would "impede the survival and recovery of listed [salmonids]." Accordingly, NOAA provided an RPA to the proposed FCRPS operations which included "major modifications" to proposed flow, spill, and transportation (*National Wildlife Federation v. National Marine Fisheries Service* 2005).

Reasonable and Prudent Alternative and "Spread the Risk" Program. NOAA's RPA required both an "adaptive management" approach and immediate actions. Under the adaptive management approach, management actions would be iteratively adjusted as new information became available, thereby gradually decreasing uncertainties about effects of various actions (NOAA 2007). Among the immediate actions, NOAA's RPA required a program to aid downstream smolt migration.

Although the states and tribes preferred use of increased spill to aid migration, NOAA determined that transportation was still appropriately used as a "major means to mitigate the adverse impacts of the FCRPS" and required a combined transportation and spill program. NOAA expected this combination of transportation and spill to result in transportation of approximately 56 to 74 percent of spring/summer chinook smolts, with the residual 44 to 36 percent remaining as in-river migrants. The agency included both components because it believed "that there was sufficient uncertainty about the benefits of transportation to warrant an evaluation of whether improved in-river migration may result in adult returns that are higher than adult returns from the transportation program" (NOAA 1995). NOAA termed this combination of transport and spill a "spread the risk" approach, and intended to use it only on a short-term, interim basis until the results from the comparative study could direct the proper amount of reliance on either component in aiding downstream smolt migration (NOAA 1995; Blumm et al. 2006). Such "risk spreading," NOAA concluded, would "not result in an unacceptably low likelihood of survival" in the short term.

NOAA based its transportation determination both on its preference for the CRiSP model and on empirical data indicating that transportation benefitted two of the four listed salmonids: Snake River spring and summer chinook. Because these two salmonids benefitted from transportation, NOAA reasoned, the other two would also likely benefit (NOAA 1995).

The spill program was intended to create conditions under which adult returns from in-river migration and transportation could be compared, and required spill at all Corps dams during the juvenile spring/summer chinook migration season (April through June), and at three dams during the fall chinook migration season (June through August). To limit the predation risk to smolts migrating in-river, spill was only required in the years when Snake River runoff was predicted to average at least 85,000 cubic feet per second. NOAA believed that in lower water years, in-river migrating smolts were too exposed to predators to properly test spill benefits and to spill at such times would compromise any efficacy comparison between the transportation and spill components (NOAA 1995).

The District Court upheld this BiOp and RPA against challenges (*American Rivers v. National Marine Fisheries Service* 1997).

2000

In 2000, the Corps proposed to continue the FCRPS operations largely as mandated by the 1995 BiOp and RPA. Accordingly, the Corps would determine flow objectives each year using "a sliding scale based on forecasted runoff," and continue to provide spill levels that "spread the risk" between transported and in-river migrants. Summer spill, however, would be reduced relative to the 1995 RPA requirements to maximize the number of smolts collected and transported from all four transport facilities (NOAA 2000).

Jeopardy Finding. NOAA's 2000 BiOp determined that continuing FCRPS operations would jeopardize eight listed salmonid species. More specifically, the agency used a five-step approach to determine whether the proposed FCRPS operations would jeopardize any listed salmonids (NOAA 2000). Steps three and four are important here.

In step three of its effects analysis, NOAA used its Simulated Passage model (SIMPAS) to "[apportion] the run to various passage routes based on empirical data and input assumptions for fish passage parameters" (NOAA 2000). In brief, the SIMPAS model accounts for successful fish passage (survival) and losses (mortality) through each alternative passage route to estimate total smolt survival past each project. It also accounts for the proportions of smolts transported or left to migrate in-river, and provides survival estimates at each project and for the system as a whole. When available, average measured rates through each passage route serve as model inputs for fish survival. When actual rates are not available, either passage parameter estimates at other similar projects or professional judgment are substituted (NOAA 2000).

The fourth step in NOAA's analysis was central to the 2000 BiOp and its absence was critical to the 2004 BiOp. The fourth step consisted of two parts. First NOAA evaluated the effects of the action within the FCRPS area against the biological requirements of the listed species (NOAA 2000). The agency then looked beyond the FCRPS area to evaluate the impacts of the proposed action on the entire life cycles of listed salmonids, including the ocean-going life stages. This analytical approach is referred to as the "aggregate approach" (*National Wildlife Federation v. National Marine Fisheries Service* 2005).

NOAA concluded that while proposed FCRPS operations would improve survival of a few listed salmonids, operations would jeopardize most listed salmonids because they needed "additional survival improvements beyond those expected from the proposed action and all other reasonably foreseeable recovery activities" (NOAA 2000). As in 1995, NOAA proposed an RPA to avoid jeopardy that included both transport and spill measures to improve survival of downstream migrating salmonid smolts (*National Wildlife Federation v. National Marine Fisheries Service* 2003).

Reasonable and Prudent Alternative Development. The 2000 RPA required seasonally specific spill and transportation measures. During spring migration, NOAA required the Corps to collect and transport smolts from three lower Snake River dams. During summer migration, they required the Corps to collect and transport smolts from three lower Snake dams and McNary Dam on the Columbia River. In addition, NOAA required sufficient spill to ensure the "highest survival" for in-river migrants during the spring, and to provide for summer research. Contrary to the 1995 BiOp, NOAA made no attempt in the 2000 BiOp to mandate a specific ratio of smolts remaining in river to those transported. NOAA did, however, estimate that 43 to 91 percent of Snake River spring/summer chinook would be transported under this strategy. The purpose of the summer transport program was to minimize in-river smolt mortality rates due to exposure to adverse water quality (NOAA 2000).

This combination of transport and spill continued NOAA's "spread the risk" approach to managing FCRPS impacts to listed salmonids. NOAA explained that spill measures were necessary in part"[b]ecause listed Columbia River basin anadromous fish are in such fragile condition, an immediate focus on areas and measures that provide gains within 1 to 10 years is essential" and estimated that the 2000 RPA would increase Snake River fall chinook survival from 31 to 63 percent (NOAA 2000). Most of this benefit came from increased spill and flow augmentation (*National Wildlife Federation v. National Marine Fisheries Service* Plaintiff's Memo 2004). Because these measures alone were insufficient to avoid jeopardy to listed salmonids, NOAA added additional non-flow related measures to the 2000 RPA to avoid jeopardy (*National Wildlife Federation v. National Marine Fisheries Service* 9th Cir 2005).

Although the federal district court remanded the 2000 BiOp, it generally approved of NOAA's analysis (*National Wildlife Federation v. National Marine Fisheries Service* 2003). In May 2001, the Corps agreed to implement the 2000 BiOp and RPA, including the spring and summer spill measures.

2004

The Corps' Proposal to Curtail Summer Spill. After the Corps agreed to implement the 2000 BiOp and following a change in administration, agency heads of NOAA, the Corps, the Bonneville Power Administration, and the Bureau of Reclamation jointly pronounced that the costs of the summer spill program were "exceedingly high relative to the biological benefit," and all agencies shared a responsibility to devise an approach that was "less costly while maintaining the ability to achieve the biological objectives of the BiOp" (*National Wildlife Federation v. National Marine Fisheries Service* Defendant's Opposition 2004). In 2004, while the 2000 BiOp and RPA were still in effect, the Corps and BPA proposed to change FCRPS operations to realize this goal.

More specifically, the Corps and BPA proposed to curtail summer spill because fewer listed fish were in the river at that time—most were being collected and transported. The Corps used both the CRiSP and SIMPAS models to evaluate effects of the proposed action on juvenile salmonids; both models indicated that curtailing summer spill would have minimal impacts on listed salmonids. After two rounds of public comments and subsequent modifications to the original proposal, the revised action would eliminate 39 percent of the entire spill volume for the season as well as the last two weeks of spill in August (*National Wildlife Federation v. National Marine Fisheries Service* Defendant's Opposition 2004; NOAA 2000). Because the proposed action would ostensibly produce the survival rates necessary to avoid jeopardizing listed salmonids, however, thereby providing biological benefits equivalent to those of the 2000 BiOp and RPA, NOAA approved it (*National Wildlife Federation v. National Marine Fisheries Service* Defendant's Opposition 2004). Shortly thereafter, NWF moved the court for injunctive relief to enjoin the spill curtailment, which the federal district court granted on June 10, 2005 (*National Wildlife Federation v. National Marine Fisheries Service* 2005).

NOAA's No Jeopardy Determination. In 2004, the Corps proposed to continue the 2000 BiOp's basic spring and summer spill program and to implement a three-stage approach to spreading the risk (U.S. Army Corps of Engineers 2004). In early spring, approximately 48 percent of the smolts would be passed through spill, and 52 percent by transportation. After mid-April, approximately 95 percent of juvenile salmonids would be transported, and in late summer, approximately 50 percent of migrating smolts would be passed via each means. Thus, the Corps proposed to continue using NOAA's "spread the risk" approach nine years after its introduction as an interim policy. Transportation remained warranted, the Corps maintained, because analyses of the relative benefits of transport and spill remained inconclusive (*National Wildlife Federation v. National Marine Fisheries Service* Defendant's Opposition 2004). The Corps also pushed back the comparison of spill versus transportation survival for summer migrating fish, initially planned in 1995, to 2007 or 2008 (U.S. Army Corps of Engineers 2004).

Using a third analytical approach, NOAA's 2004 BiOp determined that the proposed FCRPS operations for the summer of 2005 would not jeopardize listed salmonids (*National Wildlife Federation v. National Marine Fisheries Service* 2005). Instead of using the aggregate analysis of the 2000 BiOp, NOAA used a "comparative approach," which deviated from the aggregate approach in two important respects.

First, NOAA included in the environmental baseline both the effects of past FCRPS operations, as well as effects of continuing FCRPS operations that were not subject to agency discretion. While agencies may choose not to proceed with the original construction of a dam, here, NOAA explained, the dams already exist, and that existence was "beyond the scope of the present discretion of the Corps ... to reverse." Hence, NOAA included continuing FCRPS operations in the environmental baseline (NOAA 2004). This inclusion was unorthodox because ESA regulations restrict the environmental baseline to two categories: past and present impacts of all human activities within the action area, and anticipated future impacts from all proposed federal projects in the action area that have already undergone section 7 consultation (50 CFR § 402.02). Because the Corps had not yet consulted on the continuing effects of the non-discretionary FCRPS operations, these effects could not be properly included in the environmental baseline. By including some continuing FCRPS operations in the baseline (NOAA 2004), NOAA both reduced the adverse impacts attributable to future FCRPS operations and avoided consulting on effects of ongoing FCRPS operations. Instead, NOAA only consulted on certain future FCRPS effects.

Second, NOAA's 2004 comparative analytical approach differed from its 2000 aggregate approach in its comparison of the environmental baseline (inclusive of continuing FCRPS operations) to estimated future conditions under the proposed FCRPS operations. NOAA conducted this comparison in three steps. NOAA first evaluated, at a project-specific level, the difference in fish survival between implementation of the proposed action and a "reference operation," or surrogate for the hydropower portion of the environmental baseline. NOAA next analyzed the effects of actions proposed to minimize or mitigate that difference. Finally, NOAA analyzed the net effect of the entire proposed FCRPS operation and determined that it did not jeopardize listed anadromous salmonids. As a result, NOAA's 2004 BiOp did not require summer spill (NOAA 2004).

The court held NOAA's comparative effects analyses flawed for two reasons. First, the court explained, NOAA erred by partitioning FCRPS impacts into continuing and future effects and only considering future effects on listed salmonids. Second, NOAA erred in using a fictional reference operation as a backdrop for comparing additional mitigation measures because it overestimated the possible beneficial impacts of the FCRPS. The court also noted NOAA's failure to explain its departure from the previously accepted 2000 analytical approach (*National Wildlife Federation v. National Marine Fisheries Service* 9th Cir. 2005).

Following the 2004 BiOp remand, NWF moved for a second, broader preliminary injunction which included an order requiring the Corps to provide spill as set forth in the 2000 RPA (*National Wildlife Federation v. National Marine Fisheries Service* 2005; NOAA 2000). On December 29, 2005, the court granted injunctive relief regarding late spring spill and summer spill but denied NWF's request regarding early spring flows (*National Wildlife Federation v. National Marine Fisheries Service* 2005).

Science and Policy Intersect in the FCRPS Injunction Litigation
When NWF initially moved the court for injunctive relief in 2004, the "best available science" was fraught with uncertainty. Between 1993 and 2004, NOAA had analyzed the impacts FCRPS operations on listed salmonids using three distinctly different analytical approaches in four BiOps, and the court had upheld only one of the four BiOps. Because the spill and transportation efficacy studies—originally slated to occur in 1995—had not yet been conducted, NOAA's *interim* "spread the risk" approach remained in place. The few available comparative data on the efficacy of the two passage methods were equivocal at best. As this section will show, this scientific uncertainty was the source of the spill versus transportation dispute.

It is well established that implementation of the ESA requires a combination of science and policy (Ruhl 2007; Ruhl and Salzman 2006; Doremus 2005; Wagner 1995). Such use of policy is at least anticipated, if not required by the ESA. This section first describes the interaction of science and policy in ESA interagency consultation before demonstrating how NOAA used policy to bridge scientific uncertainties in making the required determinations. The section ends by explaining that this use of policy was the source of the dispute that gave rise to both of NWF's requests for injunctive relief.

Science and Policy in Interagency Consultations
Science and policy intersect in the ESA because both are required for full implementation of the statute (Ruhl 2007; Ruhl and Salzman 2006; Doremus

2005; Wagner 1995). More specifically, legal scholars have identified sections 4 and 10 of the ESA as creating programs that require both science and policy to implement (Ruhl 2007). For example, while science provides information such as a species' population size, distribution, and life history requirements essential to determining whether a species should be listed under section 4, science will not reveal whether, due to economic impacts on human inhabitants, an occupied area should be excluded from critical habitat designation (Ruhl and Salzman 2006).

Similar to sections 4 and 10, section 7—interagency consultation—also created a program that requires both science and policy to implement. Although science may reveal the threats to a species and describe the environmental baseline, it may not provide sufficient information upon which to base a jeopardy decision. Given that most interagency consultations are conducted on listed species, and information on these species is often scant, NOAA frequently must use "some basis in addition to [science] for reaching a jeopardy decision" (Ruhl and Salzman 2006). Even if scientific data is not lacking on a species, determining whether an action is likely to cause a species to become extinct "ultimately requires some judgment" (Ruhl and Salzman 2006).

In addition, the section 7 time frame does contemplate agency research or information gathering to supplement scant science. Indeed, section 7 requires NOAA to complete a BiOp within 135 days of consultation initiation, regardless of the state of the available information (16 U.S.C. § 1536(a)(2)). Moreover, section 7 directs NOAA to use the "best scientific and commercial data available" in making its determination. Congress' mandate to use the "best available science" illustrates its awareness both that the science may be incomplete and that "science alone [may not] definitively provide the final answers" (Ruhl 2007). Thus, section 7 jeopardy determinations are likely to require alternative or additional nonscientific bases; professional judgment and policy are frequently employed (Ruhl 2007).

This combination of science and policy is frequently necessary to resolve natural resource issues. Problems arise, however, when parties fail to explicitly distinguish between science-based and policy-based decisions, and instead conflate science and policy into "science-policy" and present it as "science" (Wagner 1995). Over the course of the FCRPS litigation, both parties used "science-policy" to resolve scientific uncertainties.

"Science-Policy" Was Used to Address Scientific Uncertainties in the FCRPS Litigation
During the eleven years of FCRPS consultation leading to NWF's first motion for injunctive relief, NOAA made numerous decisions that were not purely science-based. Because uncertainties involved in each issue precluded science from providing "the answer," NOAA employed "science-policy" to fill the gaps and

make the decision required by the ESA. Two examples will illustrate this use of "science-policy."

First, throughout the FCRPS consultation, NOAA used "science-policy" each time it selected an approach to analyzing the potential impacts of the Corps' proposed actions. In 1993, NOAA decided to narrowly define its environmental baseline, thereby protecting FCRPS capabilities but failing to accurately portray existing conditions and risks to listed salmonids (1997). In 1995 and 2000, NOAA decided to use a complex aggregate effects analysis in which all existing impacts to listed salmonids were evaluated prior to analyzing the effects of proposed FCRPS operations (NOAA 2000). Finally, in 2004, NOAA decided to use a comparative effects analysis which included pre-existing FCRPS impacts in the environmental baseline and effectively avoided consultation on those effects (NOAA 2004).

Second, NOAA's decisions to combine outputs from the three passage models in 1995 and to adopt its spread the risk approach were not wholly science-based. While science provided information about potential outcomes of proposed FCRPS operations, it did not direct NOAA to combine model outputs to moderate the risk of exclusively relying on any one model, or to hedge its bets by using both passage methods in combination (Ruhl and Salzman 2006). Instead, these decisions were policy-based. Indeed, NOAA explained its decision to combine models as a "risk-management" decision, which by definition is not science-based (Garner 2004).

Although NWF did not quibble with either of these decisions initially, it later disputed NOAA's continued adherence to them, especially its combined transportation and spill program. It was largely NOAA's implementation of the spread-the-risk program that led to NWF's motions for injunctive relief. NWF requested injunctive relief in 2004 after NOAA accepted the Corps' proposal to curtail summer spill in 2005. Both proposal and acceptance were based on output generated by the SIMPAS model indicating that implementation of the proposal would not adversely affect listed salmonids. NWF criticized NOAA's reliance on the SIMPAS model both generally and with regard to specific results.

Generally, NWF complained that fisheries experts had "consistently pointed out to [NOAA] and the Corps that [SIMPAS] is ill-suited to predicting the effects of reducing spill on Snake River fall chinook," the primary species out-migrating during the proposed spill curtailment. NWF argued that the SIMPAS model was flawed because it ignored uncertainty and risk and consequently, model results failed to provide a "robust foundation for fish passage management decisions." In addition, the plaintiffs argued, elimination of summer spill increased risk to affected salmonids and the proposal's offset measures were insufficient mitigation for such risks (*National Wildlife Federation v. National Marine Fisheries Service* Plaintiff's Memo 2004).

NWF's fisheries experts also criticized SIMPAS at a project-specific level for parameters that were overly sensitive to change, and confidence intervals that exceeded the anticipated benefits of the proposed action. For example, increasing juvenile mortality at the Dalles dam by only 1 percent resulted in a 37 percent increase in impacts to Snake River fall chinook migrants above the impact level estimated by NOAA. NWF argued that the combination of uncertainties present both in individual parameters and overall survival estimates at each dam rendered "scientifically meaningless" the Corps' conclusion that curtailing spill in August would result in "minimal harm" (*National Wildlife Federation v. National Marine Fisheries Service* Plaintiff's Memo 2004).

In response, NOAA acknowledged the "considerable uncertainty" inherent in its proposed spill curtailment analysis, but declared that it used the "best available information." Even assuming the worst-case scenario, NOAA argued, only a small percentage of a single species would be impacted by the spill curtailment because the 2000 RPA required transportation of approximately 90 percent of smolts during August. In addition, potential mortality resulting from spill curtailment would be compensated by increased smolt survival below Bonneville Dam. Further, NOAA asserted, acceptance of the Corps' proposal was consistent with the adaptive management component of its "spread the risk" policy because SIMPAS output indicated higher survival of out-migrating smolts under the Corps' proposal than under the 2000 BiOp and RPA (*National Wildlife Federation v. National Marine Fisheries Service* Defendant's Opposition 2004). This difference in opinion about the proper amount of reliance to place on a single, potentially flawed, model led directly to NWF's second request for injunctive relief in 2005.

NWF requested additional injunctive relief in 2005 to reduce the prominence of the transportation program in passing out-migrating smolts. In its request, NWF cited the 2000 BiOp and state and tribal comments on the 2004 BiOp to support its position that spill "indisputably provides the safest passage past the FCRPS dams." Again NOAA argued that the benefits of spill had not been established, and again NWF argued that the benefits of transport for downstream migrating smolts had not been established and that NOAA's decision to maximize transportation was based on "limited information with considerable uncertainty" (*National Wildlife Federation v. National Marine Fisheries Service* Plaintiff's Memo 2004).

Thus, in the FCRPS injunction litigation, both NWF and NOAA exploited scientific uncertainties to advance their respective values and policies: both stated the certainty of their interpretation of the science while emphasizing the uncertainty in the opposing party's stance.

Positions of the Parties

The disputes between NWF and NOAA regarding modeling, curtailing summer spill, and the use of transportation or spill to pass smolts are essentially arguments about the proper policy approaches to addressing scientific uncertainty in natural resource management. Although NOAA acknowledged the "considerable uncertainty" inherent in analyses of the proposed FCRPS operations, it concluded that absent definitive data showing the need for summer spill or higher adult returns from spill-assisted in-river smolt migration relative to transportation, the benefits of spill and in-river migration are speculative and do not warrant reliance. Instead, NOAA argued for continued reliance on transportation to aid smolt outmigration because clear scientific evidence had not yet undermined the program's efficacy (Blumm et al. 2006). Moreover, NOAA emphasized, while the ESA requires the use of the "best available science," it does not require the "best available conclusion" from a range of options (*American Rivers v. National Marine Fisheries Service* 1997; *National Wildlife Federation v. National Marine Fisheries Service* Defendant's Opposition 2004). Hence, NOAA asserted, its conclusion and explanation of how risk should be managed were entitled to deference over NWF's opinions. Finally, NOAA explained, NWF's approach would preclude any adjustments in FCRPS operations under the 2000 BiOp for lack of "perfect information" or predictive models without some uncertainty (*National Wildlife Federation v. National Marine Fisheries Service* Defendant's Opposition 2004). Thus, NOAA used science "defensively" to delay difficult salmonid management decisions and to protect against requiring potentially unpopular changes to the FCRPS system (Doremus 2005).

On the other hand, NWF believes that since NOAA is permitting actions that may jeopardize the continued existence of multiple species, NOAA must affirmatively demonstrate that the proposed action will not harm those species before permitting the action. Merely showing that available data does not demonstrate adverse effects from the proposed action is not sufficient, according to NWF, because it forces the species, rather than the agencies proposing an action, to bear the risk of uncertainty (*National Wildlife Federation v. National Marine Fisheries Service* Plaintiff's Memo 2004). Thus, NWF has used science "offensively" to support a position by exaggerating the role of scientific data in their call for salmonid protection (Doremus 2005).

In sum, the parties' positions in the injunction litigation are based on opposing policy approaches to addressing scientific uncertainties. Both parties rationalized their position as science-based, thereby conflating science and policy into "science-policy." Such use of "science-policy" obscured the boundaries between science-based and policy-based choices and made it more difficult for the court to evaluate NOAA's determinations and NWF's requests for injunctive relief. The next section evaluates the impact of scientific uncertainties and parties' use of "science-policy" in the FCRPS injunction litigation and shows that the court properly granted both preliminary injunctions before concluding with recommendations for moving forward.

Impacts of Science and Policy on the FCRPS Injunction Litigation
The scientific uncertainties inherent in salmonid management and the effects of FCRPS operations are central to the FCRPS injunction litigation. At its core, the injunction litigation is about who or what should bear the risks of scientific uncertainties, and what level of risk is acceptable in management strategies for species listed under the ESA. The scientific uncertainties that permeate the FCRPS injunction litigation and the "science-policy" used to address those uncertainties affected the FCRPS litigation in a number of interrelated ways, including the amount of deference afforded to NOAA by the court.

Legal Standard of Review for Biological Opinions and Injunctions
Since the 2005 and 2006 injunctions arose from the court's rejection of NOAA's 2004 BiOp, an analysis of the injunctions must begin with a brief explanation of the proper standard of review for a BiOp and an injunction before the district court's analysis and holdings may be examined for conformity to these standards.

The BiOp that results from completion of formal consultation under section 7 of the ESA constitutes final agency action and is subject to judicial review (*Bennett v. Spear* 1997). The scope of judicial review is governed by the Administrative Procedures Act (APA) because section 7 of the ESA has no specific provision for judicial review. Under the APA, an agency action must be upheld unless it is found to be "arbitrary, capricious, an abuse of discretion, or otherwise not in accordance with law" (5 U.S.C. § 706(2)(A)). In evaluating an agency action, a court must consider whether the agency considered only and all relevant information and "articulated a rational connection between the facts found and the choices made" (*Pacific Coast Federation of Fishermen's Association, Inc., v. National Marine Fisheries Service* 9th Cir. 2001). Although courts are responsible for ensuring that an agency acts within its statutory limits, courts typically defer to an agency's "technical expertise and experience," particularly when gaps exist in a statutory framework and on complex scientific issues (*Chevron U.S.A., Inc. v. Natural Resources Defense Council, Inc.* 1984; *United States v. Alpine Land and Reservoir Co.* 9th Cir. 1989). The presumption of agency expertise is rebuttable, however, even on scientific matters, if the agency's decisions are not well reasoned or its interpretation of a regulation is inconsistent or varies over time (*Watt v. Alaska* 1981). Thus, agency determinations must either be consistent or thoroughly explain any deviations from past actions in a logical, reasoned manner to warrant judicial deference.

In determining whether to issue an injunction under the ESA, a court's discretion is limited (*National Wildlife Federation v. Burlington Northern Railroad, Inc.,* 9th Cir.

1994). Courts are not required to balance equities in determining whether to grant injunctive relief to alleviate a harm alleged under the ESA and instead are to afford listed species "the highest of priorities" (Tennessee Valley Authority 1978). Courts are not, however, required to mechanically grant injunctions for each violation of the ESA. Instead, plaintiffs must demonstrate a "threat of irreparable harm" to a listed species from the contested activity (*National Wildlife Federation v. Burlington Northern Railroad, Inc.* 9th Cir. 1994). In addition, where an agency has failed to fulfill its statutory obligations over a long time period, injunctions ordering the agency to take specific steps to remedy its failure may be proper (see e.g., *Alaska Center for the Environment v. Environmental Protection Agency* 9th Cir. 1994). Moreover, the district court "has broad latitude in fashioning equitable relief," and may direct an agency to take specific steps to ensure progress toward statutory objectives, provided that the court requires only those steps that are "undeniably necessary" (*Alaska Center for the Environment v. Environmental Protection Agency* 9th Cir. 1994).

In 2005 Oregon's district court analogized to the Alaska Center case while evaluating NOAA's BiOps. The court viewed the fact that the BiOps were repeatedly invalid as a failure to comply with the ESA's substantive provisions, and thereby a failure to enforce the ESA (*National Wildlife Federation v. National Marine Fisheries Service* 2005). Thus, because the court based its injunction against NOAA on the agency's continued inaction, the injunction was valid under the Ninth Circuit's Alaska Center holding. The question of whether the injunction was properly granted, however, turns on whether NOAA's determinations warranted deference.

Legal Outcomes: Reduced Deference to NOAA Was Appropriate
Despite a deferential standard of review for agency actions, the court's refusal to defer to NOAA regarding NWF's 2004 and 2005 requests for injunctive relief was appropriate both because the analytical approach NOAA used in its 2004 BiOp constituted a radical and unexplained departure from its previous analytical method, and because NOAA's analysis failed to consider all relevant data.

Courts look for consistency in determining whether agency regulatory decisions have a rational basis (see e.g., *Chevron U.S.A., Inc. v. Natural Resources Defense Council, Inc.* 1984). "When an agency's new interpretation of a regulation conflicts with its earlier interpretations, the agency is entitled to considerably less deference than [when the agency articulates] a consistently-held agency view" (*Immigration and Naturalization Service v. Cardoza-Fonseca* 1987). Lack of consistency, however, does not necessarily preclude deference. An agency may modify its analytical approach as more effective or more accurate methods are developed or uncertainties are resolved, provided that it explains each deviation (see e.g., *Industrial Union Department, AFL-CIO v. American Petroleum Institute* 1980). If the agency clearly explains the transition, and the change is logical and rational, courts will likely afford the agency deference. Such explanation is critical to the court's understanding, however, and a prerequisite for judicial deference. Absent the explanation, the court will lack any basis for finding the agency's decision rational, and will not likely defer.

In the NOAA/NWF case, NOAA used three different analytical approaches in four BiOps to evaluate the effects of very similar proposed actions between 1993 and 2004 and failed to explain the reasoning behind its apparently inconsistent approaches. In reviewing NOAA's BiOps, the court repeatedly noted inconsistent effects analyses and expressed frustration at the lack of explanation, particularly upon remanding the 2004 BiOp (*National Wildlife Federation v. National Marine Fisheries Service* 2005). These course changes were apparently policy-driven because the science had not substantially changed.

In addition to using new analytical methods without explaining the basis for the change, NOAA failed to consider relevant data in its 2004 effects analysis. More specifically, NOAA failed to consider the "nondiscretionary" effects of the operations (NOAA 2004). In both cases, NOAA selectively tailored the data informing its effects analysis rather than considering all existing conditions or consulting "on the entirety of the proposed action" (*National Wildlife Federation v. National Marine Fisheries Service* 2005). In both cases, NOAA made a policy-based decision to exclude certain scientific information; these exclusions reduced the apparent impacts of FCRPS operations on listed salmonids, and perhaps avoided jeopardy determinations.

Under the general standards of judicial review of agency determinations, either of NOAA's errors—its substantial departure from previously approved analytical methods or its failure to consider all relevant data—would be sufficient to rebut the presumption of agency expertise and accompanying judicial deference. In addition, as the Ninth Circuit concluded, after the district court rejected the 2004 BiOp there simply "was no formal agency finding to which deference might arguably be owed" (*National Wildlife Federation v. National Marine Fisheries Service* 9th Cir. 2005). Hence, NOAA would benefit from clearly explaining both the policy and science behind its FCRPS BiOps. NOAA remains an "expert" agency, both in fisheries science and in administration of the ESA. As such, NOAA is more likely to be afforded deference if its decisions are transparent, well reasoned, and logical (*see Citizens to Preserve Overton Park, Inc. v. Volpe* 1971).

Preliminary Injunctions Were Properly Granted
The district court properly enjoined the Corps' proposed curtailment of 2005 and 2006 spills because the court determined that such operations, in conjunction with maintenance of the 2004 BiOp conditions, would result in immediate and irreparable harm to listed species. In making its determination, the court compared

remedial measures contained in the 2000 and 2004 BiOps and found that only the 2000 BiOp contained an RPA that required summer spill at four lower Snake and Columbia river dams "at a level minimally necessary to allow for a meaningful in-river migration program." The court also reviewed expert witness testimony that increased spill was the best method of increasing downriver survival of juvenile salmonids because it resulted in the lowest smolt mortality rates. Because the record indicated that the spill measures were critical to listed fish, and summer spill had been part of the 2000 RPA but the 2004 BiOp failed to contain such measures, the court determined that there was sufficient evidence that FCRPS operations cause significant mortality to migrating salmonids, strongly contribute to their endangerment, and "irreparable injury will likely result if changes are not made" (*National Wildlife Federation v. National Marine Fisheries Service* 2005). Accordingly, the court properly struck a balance in favor of listed species by issuing injunctions requiring seasonal spill.

Thus, after more than a decade of NOAA's inconsistent analytical methods and exclusion of certain factors relevant to its effects analyses, the court refused to defer to the agency's 2004 determinations that proposed FCRPS operations would not jeopardize listed anadromous salmonids. Instead, in 2005, the court enjoined curtailment of both 2005 summer spill and 2006 late spring and summer spill. In addition, the court ordered NOAA to take specific steps to fulfill its statutory mandates under the ESA. In doing so, the court began to "run the river" (*National Wildlife Federation v. National Marine Fisheries Service* 2005).

Functional Outcome: The Court Runs the River
The federal district court effectively began running the river in June 2005 by individually evaluating and approving spill actions in its June 2005 opinion. NWF's 2004 motion for preliminary injunction consisted of five distinct requests; the court individually examined each one, but chose to grant only the motion requiring summer spill and denied NWF's two other flow-related requests (*National Wildlife Federation v. National Marine Fisheries Service* 2005). In doing so, the court selectively dictated the timing, amount, and location of spill releases, thereby taking control and "running the river."

The court continued running the river in October of 2005. In its second injunctive relief motion, NWF requested orders requiring early spring, late spring, and summer spill as well as orders mandating both the amount and timing of the summer spill. The court granted the motions to require late spring spill and the full period of summer spill, and denied the motions to require early spring spill and an increase in the amount of summer spill (*National Wildlife Federation v. National Marine Fisheries Service* 2005). By again selectively dictating the timing and amount of spill releases, the court continued "running the river."

The court still runs the river today, partially due to both parties' use of "science-policy" to address scientific uncertainties involved in the FCRPS 7 consultations. This chapter concludes with three recommendations for reducing the "science-policy" issues that commonly manifest in ESA section 7 consultations such as those leading to the FCRPS injunction litigation.

WHAT NEXT?

The "science-policy" problem in natural resource management has been investigated by a number of legal scholars (Ruhl 2007; Ruhl and Salzman 2006; Doremus 2005; Wagner 1997). Although each scholar suggests slightly different ways of addressing the problem, all approaches focus on distinguishing decisions based on science from those based on policy. Recommended approaches generally differ in the means by which the decisions are delineated. For example, Ruhl and Salzman (2006) recommend using regulatory peer review, while Doremus and Wagner would rely on the courts (Doremus 2005; Wagner 1997). Both Doremus and Wagner recommend the courts defer to ESA-implementing agencies if their administrative records clearly distinguish between policy and science-based decisions (Doremus 2005; Wagner 1997). Implementation of these recommendations may prevent the conflation of science and policy and therefore ease courts' analyses of natural resource management decisions.

Recommendation One: Separation of Science and Policy Is Not Enough

While separating science and policy decisions is beneficial to ensure courts apply appropriate standards of review, such separation is more or less a line drawing exercise of limited import because it fails to resolve the tension between law and science inherent in implementation of ESA and other natural resource statutes. The issue of how to address scientific uncertainty within the ESA section 7 timeframe remains. On one hand, science is both driven and guided by an iterative process of hypothesis formulation and testing. On the other hand, law, through *stare decisis* principles, is guided by previous actions and driven by twin needs of predictability and finality. Hence, a process that merely separates science from policy, despite clarifying boundaries, does little to resolve the tension. Instead, resolution requires integration of the practices of law and science.

Recommendation Two: Use Adaptive Management

Within the section 7 consultation context, the ESA currently provides two alternatives for such integration. First, section 7 of the ESA allows inclusion of "adaptive management" into BiOps. Scientists frequently recommend using "adaptive management" to resolve scientific uncertainties and accordingly adjust natural resource management strategies. Despite the short timeframes of section

7–BiOps must be completed within 135 days of consultation initiation–the ESA regulations allow for processes like "adaptive management" (16 U.S.C. §§ 1533(b), (c)). More specifically, section 402.16 provides four situations in which consultation must be reinitiated if requested by NOAA, including when "new information reveals effects of [an] action that may affect listed species … in a manner or to an extent not previously considered" (50 CFR § 402.16). A hypothetical will illustrate how this can operate.

NOAA engages in formal consultation with the Corps on a proposed action. In the BiOp, NOAA determines whether a proposed action jeopardizes a species. Very little scientific information is available about the species, and NOAA's determination is based on numerous uncertainties. Assume that the action does jeopardize the species, but NOAA is able to avoid making a jeopardy determination by prescribing RPA measures. NOAA can build adaptive management into its BiOp, completely within the strictures of section 7(b) and 7(c) of the ESA, in four steps. First, NOAA conservatively estimates the amount of take authorized in its incidental take statement. Second, NOAA restricts the BiOp by adding an "expiration date," or a date beyond which the BiOp is no longer valid. Third, NOAA requires the Corps to complete studies designed to resolve scientific uncertainties, and synchronizes the timing of study completion and BiOp expiration. Finally, NOAA reinitiates consultation upon expiration of the initial BiOp and incorporates the newly produced information into its second analysis. This process continues in an iterative fashion until the uncertainties are resolved to an extent mutually agreeable to both agencies. This approach should allow NOAA to complete a BiOp within the 135 day timeframe while providing a mechanism by which initial assumptions may be tested and management strategies adjusted as appropriate.

This hypothetical should apply equally to any formal section 7 consultation on species about which little scientific information exists. In a "no jeopardy" BiOp, NOAA can require studies either as a "Reasonable and Prudent Measure" to minimize take, or as a "Term and Condition" of implementing such Reasonable and Prudent Measures (40 CFR § 402.16). An obvious hitch to this hypothetical is the question of who must fund the studies. Appropriate resolution of this question is beyond the scope of this article, however, it bears noting some project proponents have funded needed studies (see www.salmonrecovery.gov/researchreportspublications.aspx).

Although NOAA's FCRPS BiOps have included adaptive management since 1995 and numerous studies have been conducted on juvenile salmonid passage and survival, these studies do not address the core issue underlying the litigation: how to avoid the science-policy conflation and clearly identify the boundaries of science and policy in agency determinations.

Recommendation Three: Avoid the God Squad

The ESA also provides a second avenue for resolving scientific uncertainty within interagency consultations (16 U.S.C. § 1536(e)(1)). Section 7(e) provides for formulation of an "Endangered Species Committee" (commonly known as a "God Squad") consisting of six cabinet-level officials and one individual, appointed by the president, from each affected state (16 U.S.C. §§ 1536(e)(1), 1532(e(1)-(2)). The God Squad reviews an agency's proposed project to determine whether to grant an exemption from section 7 consultation requirements (15 U.S.C. §1536(e)(2)). For example, if NOAA determines that an action proposed by the Corps would jeopardize a species and no RPA is possible, the Corps may apply to the Secretary of Commerce for an exemption (15 U.S.C. § 1536(g)). If the God Squad grants an exemption, the Corps may proceed with its proposed action without completing consultation under section 7(a)(2), even if the action may lead to the extinction of a species.

While the "God Squad" approach has been used in the past to resolve science-policy disputes (U.S. Fish and Wildlife Service; 57 Fed. Reg. 23,405, 1992), it is not preferable because it relies upon a group of appointed officials that typically possess less legal expertise than the judiciary, less scientific expertise than top level career agency officials, and are likely subject to political influence. In contrast, because the judiciary is comprised of legal experts largely independent of political influence and is able to incorporate scientific expertise in a variety of ways to assist the court in understanding complex scientific issues, the judiciary should resolve "science-policy" disputes that arise within the ESA context.

Although the courts typically lack scientific expertise, this deficiency may be readily addressed in a number of ways. First, some judges could be trained in scientific principles and then preside over the complex scientific cases. Second, courts can appoint independent scientific experts to assist the court in deciphering the issues (*see* Redden 2008). Third, the judiciary could develop specialized courts to hear scientific cases; such courts are already in use to a limited extent in Idaho to adjudicate water rights in the Snake River Basin. Once the courts' lack of scientific expertise is remedied, a quasi-legal science-policy dispute is best resolved by engaging the courts' considerable legal expertise. In addition, court decisions are less easily circumvented outside the public eye than God Squad decisions. For example, the Tellico dam was completed without section 7 consultation despite the convening of a God Squad and its unanimous denial of exemption from this requirement (Findley and Farber 2006). In contrast, to change a judicial ruling on the ESA Congress would have to pass legislation which, given the controversial nature of the ESA, would likely occur in a public forum.

Thus, although legal scholars recommend separating science and policy decisions to ensure proper scrutiny or deference, this approach alone is insufficient to resolve the circumstances leading to, and the problems resulting from, agency use of "science-policy" to address scientific uncertainties. Sections 7(b) and 7(c) of the ESA provide a mechanism to allow resolution of scientific uncertainties within ESA timeframes. By integrating science and law in this manner, uncertainties may be resolved and the need for "science-policy" reduced because there will be fewer gaps to bridge. Undoubtedly, a few "science-policy" disputes will remain; these disputes are best resolved by an impartial, well-informed judiciary to ensure adherence to existing statutory mandates.

Conclusion

This chapter has shown that the science supporting summer spill and alternate means of fish passage is fraught with uncertainties. Both parties to the FCRPS litigation have drawn on the same science but advocate opposing policy approaches to managing scientific uncertainties and use "science-policy" to justify their positions. The effect on the FCRPS injunction litigation of these uncertainties and the parties' blending of science and policy is an apparently irresolvable dispute. In reviewing the dispute and granting injunctive relief, the district court applied the proper standards of review. NOAA's inexplicable inconsistency and failure to consider all relevant factors virtually ensured a lack of judicial deference to its determinations. In addition, NOAA's failure to explain its 2004 BiOp determinations forced the court to review each component of NWF's requests for injunctive relief. Accordingly, the court individually approved or denied each part of NWF's motions, selectively dictating the timing and amount of spill releases, and thereby running the river.

As the FCRPS injunction litigation demonstrates, there is a fundamental tension between science and law that is likely to surface in many "science-policy" or scientific uncertainty disputes. Although legal scholars recommend as a solution distinguishing decisions based on science from those based on policy, this approach alone is insufficient. Instead, a mechanism for resolving scientific uncertainties within the legal framework is needed. Section 7 of the ESA allows for such integration, and if fully utilized, these provisions may facilitate progressive resolution of scientific uncertainties and reduce uncertainty-based disputes such as the FCRPS injunction litigation.

Works Cited

Alaska Center for the Environment v. Environmental Protection Agency, 20 F.3d 981 (9th Cir. 1994).

American Rivers v. National Marine Fisheries Service, No. Civ. 96-384-MA, 1997, WL 33797790. (D. OR. April 3, 1997).

Bennett v. Spear, 520 U.S. 154 (1997).

Garner, Bryan A., editor. *Black's Law Dictionary*, 8th edition. Thomson West, 2004. Blumm, Michael, Laird J. Lucas, Don B. Miller, Daniel J. Rohlf and Glen H. Spain, "Saving Snake River Water and Salmon Simultaneously: The Biological Economic, and Legal Case for Breaching the Lower Snake River Dams, Lowering John Day Reservoir, and Restoring Natural River Flows." *Environmental Law* 28: 997 (1998).

Blumm, Michael, and Andy Simrin. "The Unraveling of the Parity Promise: Hydropower, Salmon, and Endangered Species in the Columbia Basin." *Environmental Law* 21:657 (1991).

Blumm, Michael, Erica J. Thorson, and Joshua D. Smith. "Practiced at the Art of Deception: The Failure of Columbia Basin Salmon Recovery Under the Endangered Species Act." *Environmental Law* 36: 709 (2006).

Bureau of Reclamation. Web site at http://www.usbr.gov/pn/programs/fcrps/fcrps.html.

Chevron U.S.A., Inc. v. Natural Resources Defense Council, Inc. 467 U.S. 837 (1984).

Citizens to Preserve Overton Park, Inc. v. Volpe, 401 U.S. 402 (1971).

Doremus, Holly. "Science Plays Defense: Natural Resource Management in the Bush Administration," *Ecology Law Quarterly* 32: 249 (2005).

Ellensburg *Daily Record* (Aug. 22, 1968). Htt://news.google.com/newspapers?nid-860&dat =19680822&id=oXENAAAAIBAJ&sjid-4ksDAAAAIBAJ&pg=6585,2798846/.

Environment Court of New Zealand Web site at http://www.justice.govt.nz/ environment/home.asp (last visited April 11, 2009).

Federal Register, Volume 55, 13181-03 (1990).

Federal Register, Volume 55, 37342-02 (1990).

Findley, Roger W., and Daniel A. Farber. *Cases and Materials on Environmental Law.* West Publishing Co. 1995).

Idaho Department of Fish and Game v. National Marine Fisheries Service, 850 F.Supp. 886 (D. OR. 1994).

Immigration and Naturalization Service v. Cardoza-Fonseca, 480 U.S. 421 (1987).

Independent Scientific Group (ISG). Return to the River: Restoration of Salmonid Fishes in the Columbia River Ecosystem (September 10, 1996) (available at http://www. nwcouncil.org/library/1996/96-6/00_Summary.pdf).

Industrial Union Department., AFL-CIO v. American Petroleum Institute, 448 U.S. 607 (1980).

National Wildlife Federation v. Burlington Northern Railroad, Inc., 23 F.3d 1508 (9th Cir. 1994).

National Wildlife Federation v. National Marine Fisheries Service, 254 F.Supp 2d 1196 (D. OR. 2003).

National Wildlife Federation v. National Marine Fisheries Service, 254 F.Supp 2d 1196 (D. OR. 2003). Memo in Support of Plaintiff's Motion for Preliminary Injunction. (D. OR. July 16, 2004) 2004 WL 3336374.

National Wildlife Federation v. National Marine Fisheries Service, 254 F.Supp 2d 1196 (D. OR. 2003). Defendants' Opposition to Plaintiffs' motion for Preliminary Injunction (D. OR. July 22, 2004). 2004 WL 3336381.

National Wildlife Federation v. National Marine Fisheries Service, No. CV 01-640-RE, 2005 WL 1278878 (D. OR. 2003).

National Wildlife Federation v. National Marine Fisheries Service, No. CV 01-640-RE, 2005 WL 1278878 (D. OR. May 26, 2005).

National Wildlife Federation v. National Marine Fisheries Service, No. CV 01-640-RE, 2005 WL 1398223 (D. OR. June 10, 2005).

National Wildlife Federation v. National Marine Fisheries Service, No. CV 01-640-RE, 2005 WL 2488447 (D. OR. Oct. 7, 2005).

National Wildlife Federation v. National Marine Fisheries Service, No. CV 01-640-RE, 2005 WL 3576843 (D. OR. Dec. 29, 2005).

National Wildlife Federation v. National Marine Fisheries Service, 422 F.3d 782 (9th Cir. 2005).

National Wildlife Federation v. National Marine Fisheries Service, 524 F.3d 917 (9th Cir. 2008).

NOAA, National Marine Fisheries Service. [1995 BiOp] Reinitiation of Consultation on 1994–1998 Operation of the Federal Columbia River Power System and Juvenile Transportation Program in 1995 and Future Years (Mar. 2, 1995) (unpublished manuscript, available at https://.pcts.nmfs.noaa.gov/pls/pctspub/sxn7.biop_results_detail?reg_inclause_in=(%27NWR%27)&idin=14092).

NOAA, National Marine Fisheries Service. [2000 BiOp] Reinitiation of Consultation on 1994–1998 Operation of the Federal Columbia River Power System (FCRPS), Including Juvenile Transportation System and 19 Bureau of Reclamation Projects in the Columbia Basin (Dec. 21, 2000) (unpublished manuscript, available at https://.pcts.nmfs.noaa.gov/pls/pcts-pub/sxn7.pcts_upload.summary_list_biop?p_id=12342).

NOAA, National Marine Fisheries Service. [2004 BiOp] Reinitiation of Consultation on 1994–1998 Operation of the Federal Columbia River Power System (FCRPS), Including Juvenile Transportation System and 19 Bureau of Reclamation Projects in the Columbia Basin (Nov. 30, 2004) (unpublished manuscript, available at https://.pcts.nmfs.noaa.gov/pls/pcts-pub/pcts_upload.summary_list_biop?p_id=14756).

NOAA, National Marine Fisheries Service. Adaptive Management for ESA-Listed Salmon and Steelhead Recovery: Decision Framework and Monitoring Guidance 91 (May 1, 2007) (available at http://www.nwr.noaa.gov/Salmon-Recovery-Planning/ESA-Recovery-Plans/upload/Adaptive_Mngmnt.pdf).

Northwest Power Planning Council. 1986. Northwest Conservation and Electric Power Plan (available at http://www.nwcouncil.org/library/1986/PowerPlan.htm).

Northwest Power Planning Council. 2005. The Fifth Northwest Conservation and Electric Power Plan. (available at http://www.nwcouncil.org/energy/powerplan/5/Default.htm).

Pacific Coast Federation of Fishermen's Association, Inc., v. National Marine Fisheries Service, 265 F.3d 1028 (9th Cir. 2001).

Redden, James A. Letter from Hon. James A. Redden, District Court Judge for the United States District Court, District of Oregon, to counsel of record in *National Wildlife Federation v. National Marine Fisheries Service*, CV 01-640, RE and *American Rivers v. NOAA Fisheries*, CV 04-000061RE (Aug. 7, 2008) (on file with author).

Ruhl, J. B. "Reconstruction of the Wall of Virtue: Maxims for the Co-Evolution of Environmental Law and Environmental Science." *Environmental Law* 37: 1063 (2007).

———, and James Salzman. "In Defense of Regulatory Peer Review." *Washington University Law Review*. 84:1 (2006).

Snake River Basin Adjudication Web site at http://www.srba.state.id.us/.

Tennessee Valley Authority v. Hill, 437 U.S. 153 (1978).

United States v. Alpine Land and Reservoir Co., 887 F.2d 207 (9th Cir. 1989), *cert. denied*, 498 U.S. 817 (1990).

U.S. Army Corps of Engineers. Final Updated Proposed Action for the FCRPS Biological Opinion Remand (Nov. 24, 2004) (available at http://www.salmonrecovery.gov/Biological_Opinion/FCRPS/biop_remand_2004/docs/upa_final/FinalUPANov242004.pdf).

U.S. Fish and Wildlife Service. Threatened and Endangered Species System. Web site at http://ecos.fws.gov.tess_public/.

U.S. Fish and Wildlife Service. The Endangered Species Act at 35, *available at* http://www.fws.gov/Endangered/ESA35/ESA35Timeline.html.

Wagner, Wendy. "The Science Charade in Toxic Risk Regulation." *Columbia Law Review* 95:1613 (1995).

Watt v. Alaska, 451 U.S. 259 (1981).

Weitkamp, Don E., and Max Katz. "A Review of Dissolved Gas Supersaturation Literature." *Transactions of American Fisheries Society* 109: 659 (1980).

White, Richard. *The Organic Machine: The Remaking of the Columbia River* (New York: Hill and Wang, 1996).

The Effects of Climate Change on Snow and Water Resources in the Columbia, Willamette, and McKenzie River Basins, U.S.A: A Nested Watershed Study

Anne Nolin, Eric Sproles, and Aimee Brown

Introduction

Significance and Motivation

Water management relies on predictions of water availability through time, and these predictions are necessary to the U.S. Climate Change Science Program Strategic Plan (US Global Change Research Program 2001; US Climate Change Science Program 2003). The snowmelt from the mountains of the Columbia River Basin provides critical water supply for agriculture, ecosystems, hydropower, and municipalities. Throughout the western United States, current and projected analyses show rising temperatures resulting in diminished snowpacks, leading to declines in summertime in-stream flow (Service 2004; Barnett et al. 2005; Knowles et al. 2006). Annual average precipitation in the Cascade Range in the Pacific Northwest ranges from approximately 50 cm on the eastern side to about 3800 cm at high elevations on the western side of the range (Taylor and Hannan 1999). Even with high amounts of precipitation, watersheds on the wet, western side of the Cascade Range experience seasonal drought and summertime low stream flows on an annual basis. Winter precipitation (November–March) accounts for 70% of the annual total and roughly 40-60% of that falls as snow (Serreze, et al. 1999). Because snow in much of the Cascades accumulates close to the melting point, a 2°C warming would result in a projected 22% decrease in midwinter snow-covered area in the Oregon Cascades, as snowfall would convert to rainfall (Nolin and Daly 2006). The ensuing effects of such a change would be higher peak flows and greater flood risk in the winter and lower summer lows when the water is needed most. Such changes have already been documented in this region. Using measurements of April 1 snow water equivalent (SWE) dating back to 1950, Mote et al. (2005) noted that the Pacific Northwest has experienced the largest declines in snowpacks of any region in the western United States. This change can be primarily attributed to an increase in winter temperatures (Mote 2003, Mote et al. 2005). In addition, this increase in temperature appears to account for a nine- to eleven-day shift towards earlier snowmelt runoff in the Pacific Northwest (Stewart et al. 2005). While these studies suggest general trends across much of the mountainous western

U.S., a closer look reveals that not all areas will be affected to the same degree by these broad climatic shifts. For example, the lower elevations of the Cascade Range are predicted to exhibit the greatest differences in snow accumulation (Nolin and Daly 2006) and in the timing and magnitude of snowmelt (Payne et al. 2004). This western coastal mountain region is especially sensitive because its Mediterranean climate is characterized by cool, wet winters (when water demand is low) and warm, dry summers (when demand is high). The changing patterns of streamflow and the timing of peak flow are particularly relevant to the Columbia River Treaty because they present a challenge to water managers.

In this chapter, we describe the projected effects of rising temperatures on snowpack at three nested watershed scales (Figure 1): the Columbia River Basin (the U.S. portion), the Willamette River Basin, and the McKenzie River Basin. Within the U.S. portion of the Columbia and Willamette River Basins, we map the extent of "at-risk" snowpack—areas of present day snow accumulation that, in a warmer climate, would experience rainfall instead of snow. In a more detailed case study, we then focus on the McKenzie River Basin, a sub-basin of the Willamette River Basin in western Oregon. For the McKenzie River Basin, we model present-day and future snow distribution, which we examine in the context of the present-day snow-observation network, and then offer some perspectives on snowpack monitoring now and for the future. We then briefly discuss the implications of changing snowpack on the Columbia River Treaty.

Background

Approximately 50–60% of the annual precipitation in the Cascades Range falls as snow during the winter months (Serreze et al. 1999). Much of the snow cover in this maritime mountain region accumulates at temperatures close to the melting point (Sturm et al. 1995). This is especially true of snow at elevations ranging from 1000–1800 m. Such lower-elevation snow in this relatively warm climate regime is at risk of falling instead as rain.

The McKenzie River Basin (Figure 2) located in the central western Cascades of Oregon, is typical of many river systems in the Columbia River Basin. In this watershed, farmers, fish, hydropower, and municipal users compete for a limited water supply—especially in summer when in-stream flows reach a minimum. Future climate projections anticipate warmer but wetter winters and longer, drier summers (Christensen et al. 2007) but the impacts of these regional projections on the entire watershed are not well understood. For example, we do not have spatially distributed measurements of snow water equivalent for this or other watersheds in the Oregon Cascades. While snowpack has been measured at the local scale for decades, accurate basin-wide measurements of snowpack do not exist.

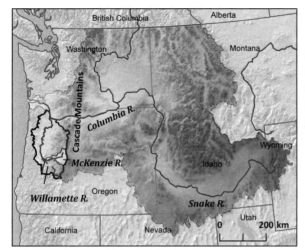

Figure 1. Locator map of the study area showing the three nested river basins: Columbia, Willamette, and McKenzie.

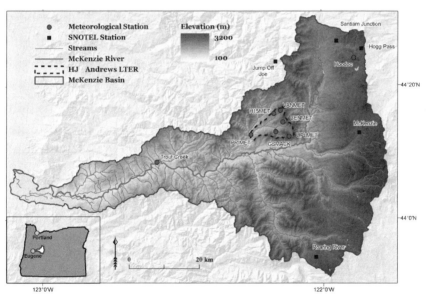

Figure 2. Map of the McKenzie River basin showing elevation and measurement sites.

Calculations for a surface water budget for the McKenzie River Basin have recently been completed (Lane Council of Governments and Eugene Water and Electric Board 2008), but do not incorporate the impacts of rising temperatures. Recent studies address the effects of climate on the hydrology of the upper reaches of the basin (Jefferson et al. 2008; Tague and Grant 2004), but do not incorporate a basin-wide prediction of snowpack, particularly transient snow at mid-elevations.

Regional Context: The Columbia River Basin and the Willamette River Basin
In order to map the area of "at-risk" snow in the U.S. portion of the Columbia River Basin, we followed the methodology described in Nolin and Daly (2006) using the 800-m gridded PRISM climate data (Daly et al. 1994; Daly et al. 2002), which span the period of 1971–2000 (PRISM Climate Group 2009). To represent the potential impacts of climate warming, we increased the monthly mean temperature values by 0.5°C increments for a projected warming of 1.0, 1.5, 2.0, 2.5, and 3.0°C and then showed the change in the snowfall area for each case. The accuracy of the present-day snow cover simulated with the 800-m PRISM data was validated using snowfall data from the NOAA/National Climatic Data Center Snow Climatology Project (NOAA National Climatic Data Center). Additional details of this study are provided in Brown (2009). It is important to note that this approach considers only the impacts of increasing temperature on snow accumulation; it does not address potential changes in the melt rate of snow during the ablation season nor does it address potential changes in the amount of precipitation. Moreover, this work focuses on changes in accumulation of midwinter snow (from December through February), the coldest months of winter. Thus, projected changes in snow accumulation are somewhat conservative since we are only simulating such changes during the core winter months when temperatures are at their coldest and precipitation has a higher likelihood of falling as snow than as rain compared with other months. Using the present-day temperature climatology from PRISM we find that, on average, snow accumulates over about 38% of the area of the Columbia River Basin for the December—February period. For a 2°C increase in monthly mean temperature during the study period, our results indicate that only 1—2% of this snow-covered area is at-risk of turning to rain. Figure 3 shows that, while this represents only a small fraction of the basin's snow covered area, it is highly concentrated in the lower elevations in the Oregon Cascades. The Willamette River Basin would experience the greatest impact with a projected 25% decrease in the area of snow cover under this 2°C warming scenario (Figure 4). Figure 4 also shows that seven of the fifteen Natural Resources Conservation Service (NRCS) Snowpack Telemetry (SNOTEL) sites in the Willamette River Basin are located in the "at-risk" snow zone. For the Willamette River Basin, the average SNOTEL site elevation is 1132 m (maximum = 1500 m, minimum = 609 m). In this basin, 13% of all the land area is above 1200 m. However, there are no SNOTEL sites in

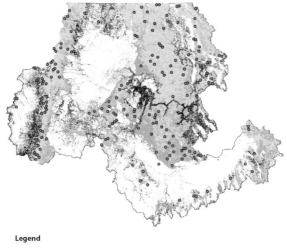

Figure 3. U.S. Columbia
River Basin snow covered
area and "at-risk" snow
(dark gray) for a 2°C
temperature increase.
The locations of NRCS
SNOTEL sites are shown
as dots.

Legend

- Snotel Site
- Warm temp., high precip., mixed winds
- Cold temp., high precip., low winds
- Cold temp., high precip., high winds

0   65   130   260 Kilometers

Figure 4. Willamette River Basin
snow-covered area and "at-risk" snow
for a 2°C temperature increase.

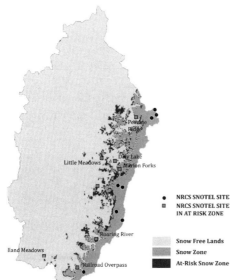

the upper 1700 m of elevation which makes up about 50% of the snow-covered area in the basin. Most of the regional climate projections for 2050 presented in Christensen et al. (2007) show increases in both winter temperatures and winter precipitation in the Pacific Northwest. If this were to be the case, then the lower elevations of the Oregon Cascades would transition to a rain-dominated winter regime, but the higher elevations might actually experience an increase in snowfall. Thus, these high-elevation snow areas may be considered stable in terms of the effects of increased temperature on snow accumulation and they may become more hydrologically important as water storage if indeed winter precipitation were to increase as well.

In the following section, we closely examine present-day and potential future snow accumulation and ablation patterns in the McKenzie River Basin, a sub-basin of the Willamette River Basin.

The Watershed Scale: The McKenzie River Basin

The McKenzie River Basin covers an area of 3,465 km^2 occupying less than 12% of the entire Willamette River Basin but supplying nearly 25% of the late summer discharge at the city of Portland (PNWERC 2002). This makes the McKenzie River Basin an important resource for ecological, urban, and agricultural interests. The basin extends from a maximum elevation of 3157 m at the summit of South Sister, to 114 m at its confluence with the Willamette River (Figure 2). Precipitation falls predominantly during the months of November through March, with precipitation increasing with increasing elevation. Snow cover typically accumulates at elevations above about 1000 m (Harr 1986). Monthly mean discharge for the McKenzie River follows the seasonal precipitation pattern with a maximum in February (283 m^3 per second, near Armitage Park) and a minimum of 57 m^3 per second in September (USGS). This relatively large percentage of late season flow is primarily due to the influence of groundwater via springs, providing both a muted and delayed stream response to snow melt (Jefferson et al. 2006). Land cover in the basin includes dense forests of western hemlock, Douglas fir, and western red cedar; clear cut timber harvests and alpine meadows; and unvegetated lava flows and snowy volcanic peaks.

Modeling Basin-Wide Distribution of Snow Water Equivalent (SWE)

The overall approach to modeling snow water equivalent is to use a physically based snow model, driven by meteorological observations and calibrated by observations of temperature and snow. For this we used SnowModel (Liston and Elder 2006a) and MicroMet (Liston and Elder 2006b), which compute the full winter season evolution of SWE (the amount of water stored in the snowpack) across a watershed including accumulation, redistribution, sublimation/evaporation, and melt. SnowModel is comprised of four sub-models: MicroMet spatially distributes meteorological values that are used to drive the model; SnowTran-3D computes wind redistribution and forest canopy; EnBal computes the internal energetics of the snowpack; and SnowPack calculates changes in snow depth and snow water equivalent. Meteorological forcing data from selected stations (Figure 2) drive MicroMet where they are spatially interpolated over the basin using elevation and topographic relationships to provide physically realistic distributions of air temperature, humidity, precipitation, temperature, wind speed and wind direction, and incoming solar and incoming long-wave radiation. At a minimum, MicroMet requires measurements of air temperature, relative humidity, wind speed and direction, and precipitation. Model grid resolution was set to 100 m in order to represent the physiographic variability of the study area. The model was run at a daily time interval. Basin topography is represented using a 10-m digital elevation model resampled to 100-m resolution. For Snow-Tran 3D, vegetation types (which influence both wind speed and canopy interception) are defined, for example, as coniferous forest, deciduous forest, scattered short-conifer, clear-cut conifer, subalpine meadow, etc. Vegetation types in the model are represented using data from the 30-m USGS National Land Cover Dataset; also resampled to 100 m. Our current case study simulates a single water year (2009) and will eventually be extended to years 1985–2010.

To calibrate the model we used a combination of point based measurements and remote sensing data. The point based measurements of SWE data were acquired from snow pillows located at SNOTEL sites and at three climate stations in the H. J. Andrews (HJA) Experimental Forest for 2009 (Figure 2). Because SNOTEL and HJA snow-measurement sites are located in relatively small, flat clearings within the forested landscape, it is unlikely that they are representative of SWE in the neighboring forested areas and steeper terrain (Brown 2009). However, SNOTEL sites serve as an index for intraseasonal and interannual variations in SWE and are valuable for gross comparisons with model output. Remote sensing data from NASA's Moderate Resolution Imaging Spectroradiometer (MODIS) were used to assess the spatial distribution of modeled snow cover. We computed fractional snow-covered area (Nolin et al. 1993) within each 500-m MODIS pixel and compared those values with modeled extent of snow cover for multiple dates during the snow season.

We find that SnowModel simulations are in good agreement with measured SWE (within ~10%) and the MODIS-derived snow covered area (7.1% on April 1). The largest differences between the satellite observations and SnowModel were in the valley bottoms, with slightly less snow detected by MODIS. However, the 500-m spatial resolution compared with the 100-m spatial resolution may be partly responsible for this difference.

Figure 5. Modeled April 1 SWE for present day (2009) climate conditions in the McKenzie River Basin.

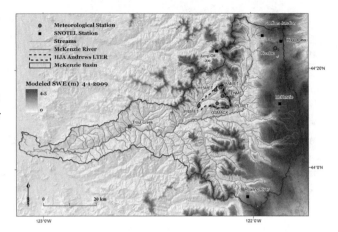

Our SnowModel results for the 2009 SWE are shown in Figure 5. Snow distributions follow a strong elevation gradient with SWE increasing with increasing elevation. As expected, the highest elevations develop the highest snow water equivalent though local differences are expressed due to slope, aspect, and vegetation cover.

To simulate potential impacts of climate change on SWE in the McKenzie River Basin, the model's temperature and precipitation forcing data were perturbed in alignment with projected climate changes. These perturbations are based on a composite of nineteen Intergovernmental Panel on Climate Change (IPCC) climate models downscaled to the Pacific Northwest region (Table 1). The composite represents the A1B emissions scenario that assumes economic growth, new technologies, and balanced energy production (Christensen et al. 2007). The perturbed temperature and precipitation values represent the monthly 20-year averages for each corresponding time period. The 2020 perturbation values were calculated from 2010–2039 mean model output values and the 2040 perturbations were calculated from 2030–2059 mean model output values; hereafter, these are referred to as SC2020 and SC2040, respectively.

With the changes in climate inputs, SnowModel simulates a loss of snow-covered area in the McKenzie Basin of 487 km² (SC2020) and 820 km² (SC2040) (Figure 7). Figures 8 and 9 show the spatial distribution of the declines in SWE for SC2020 and SC2040, respectively. There are significant declines of total basin-wide SWE with losses of 46% (SC2020) and 74% (SC2040) (Table 2). For SC2020, maximum SWE loss was 0.4–0.6 m and was located within an elevation band from 1300–1450 m (Figure 8). Losses on north-facing slopes were slightly less and the elevation band of maximum SWE loss was about 1350–1500 m. Maximum SWE losses for SC2040 are even higher (0.6–0.8 m) and are situated in an elevation

Table 1. Projected changes in monthly mean temperature and precipitation for 2020 and 2040.

	Jan	Feb	Mar	Apr	May	Jun	Jul	Aug	Sep	Oct	Nov	Dec
Temperature (°C)												
SC2020	1.22	0.99	1.11	0.99	1.01	1.28	1.59	1.60	1.37	1.00	0.83	1.17
SC2040	1.99	1.75	1.90	1.74	1.68	2.13	2.79	2.72	2.50	1.86	1.56	1.94
Precipitation (%)												
SC2020	0.01	0.16	2.04	1.30	-1.24	-5.87	-9.89	-9.78	-8.53	2.41	5.66	2.93
SC2040	4.38	0.77	6.28	5.75	-0.56	-9.97	-15.45	-12.17	-12.51	6.94	8.11	5.53

Table 2. Changes in snow water equivalent and snow covered area simulated by SnowModel.

	Volume of Snow Water			Snow-Covered Area		
	Total (km3)	Loss (km3)	% loss	Total (km2)	Loss (km2)	% loss
2009	0.78	-	-	1576	-	-
SC2020	0.42	0.36	46%	1089	487	31
SC2040	0.21	0.57	74%	756	820	52

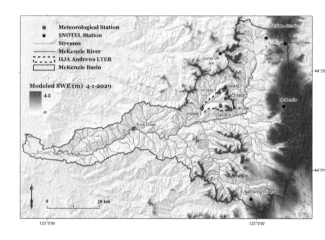

Figure 6. Modeled April 1 SWE for the McKenzie River Basin for SC2020.

Figure 7. Modeled April 1 SWE for the McKenzie River Basin for SC2040.

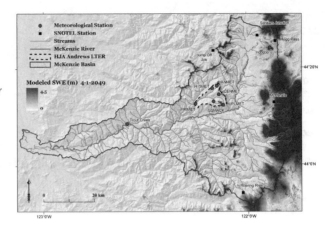

Figure 8. Difference between modeled 2009 SWE and SWE for SC2020.

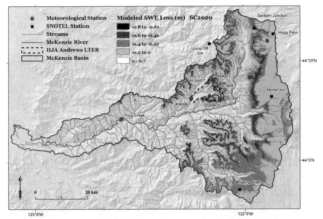

Figure 9. Difference between modeled 2009 SWE and SWE for SC2040.

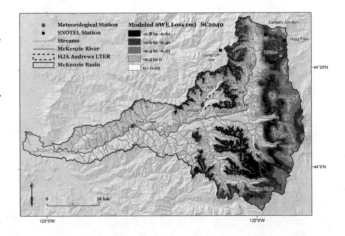

band from 1350–1480 m, primarily on south-facing slopes. As with SC2020, the elevation band of maximum SWE loss is about 50 m higher on north-facing slopes.

Examining simulated SWE for the full season at a high elevation site and a low elevation site, we see that the McKenzie SNOTEL station (1454 m, Figure 2) would show a decline in maximum SWE of about 0.25 m for SC2020 but the date of maximum SWE does not shift significantly for that scenario (Figure 10). The model also projects snow disappearance approximately 12 days earlier for SC2020 compared with 2009. For the SC2040 case, the model shows a 0.9 m decrease in maximum SWE, occurring in mid-January. The model also shows a significantly earlier date of snow disappearance of about April 20.

The HI15 meteorological station (923 m, Figure 2) in the H. J. Andrews Experimental Forest would experience a decline in maximum SWE of 0.14 m. For the SC2020 case, the snowpack would become transient with most snow disappearing by late January except for storm events that could deposit small amounts later in winter (Figure 10). For SC2040, maximum SWE would decline by 0.21 m and would occur 19 days earlier compared with 2009. As with SC2020, SC2040 shows a transient snowpack where snow is present only intermittently during the winter.

The Representativeness of Snow-Monitoring Locations at the Watershed Scale
In the previous section we demonstrated that accurate simulations of snow water equivalent could be developed at the basin scale and that we could use such simulations to explore potential impacts of increasing winter temperatures on snowpacks. Adequate meteorological forcing data are needed but when available, they can be spatially distributed within SnowModel. We can further use such simulations to identify areas within the basin that one might consider representative for the purposes of monitoring intraseasonal and interannual changes in snow water equivalent. Our current system for monitoring snowpacks is limited by economic considerations (i.e., the cost of constructing and maintaining each SNOTEL site) as

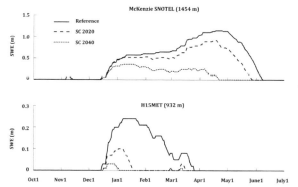

Figure 10. Site-specific modeled changes in SWE throughout the snow season for 2009, SC2020 and SC2040.

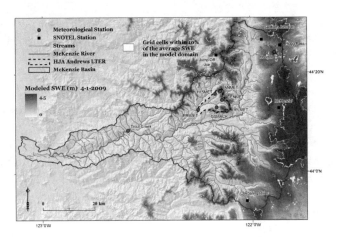

Figure 11. Modeled SWE for present day (2009) climate conditions (as for Figure 5) with grid cells whose SWE values are within 10% of the April 1, 2009 basin-average SWE shown in white.

well as by land ownership and wilderness management issues. By way of example, SNOTEL sites in the McKenzie River Basin are located over only a small range of elevations (1200–1500 m), but snow typically ranges from 1000 to over 3100 m, so about 25% of the basin area is excluded from SNOTEL data.

Using the basin-wide SWE from SnowModel, we can compute the basin-wide average SWE by calculating the mean SWE over all snow-covered grid cells within the watershed for any specified date. Thus we identify the portions of the McKenzie River Basin that are within 10% of the average April 1 SWE for 2005 (Figure 11). We can also assess the physiographic characteristics of the basin that are associated with "representative" SWE. Using an approach similar to that of Molotch and Bales (2006) we developed a binary regression tree classification using modeled SWE as the predictand and various physiographic variables as the predictors. This non-linear technique assesses the role of each variable on SWE distributions in the basin. We used elevation, slope, aspect, distance to a wind barrier, upwind fetch distance and solar radiation as the predictor variables. Our training sites were SWE values from the four SNOTEL sites. The binary regression tree classification results showed that only elevation was consistently significant in determining changes in SWE over the course of a season and particularly for determining differences in peak SWE. This is an expected result since elevation is a proxy for temperature, which, for these "warm" snowpacks, is a critical determinant in both accumulation (rain vs. snow) and ablation (melt rate). Based on this information, an appropriate design strategy for situating snow monitoring sites would be to place them along an elevation gradient that spans the full range of the snow zone. Thus, the modeling approach described above allows us to (a) identify those areas in a watershed that are representative of average snow water equivalent and (b) identify the physiographic variables that may lead to a more optimal sampling strategy for snowpack monitoring.

Conclusions

Mountain snowpacks play an important role in the hydrology and water resources of the Columbia River Basin. Projected increases in temperature are likely to lead to an increased fraction of winter precipitation that falls as rain rather than as snow. Our limited abilities to monitor snow at the watershed scale means that we must rely on additional tools including physically based models that can simulate snowpack distributions in space and time (assuming that forcing data are available). Our case study in the McKenzie River Basin shows that we can successfully model snow water equivalent for present day conditions. Incorporating projected future changes in temperature and precipitation, we see substantial declines in snow water equivalent for the McKenzie, particularly at the highest elevations. We also show an earlier shift in the date of snow disappearance in the timing of maximum SWE (Figure 10). In the context of the Columbia River Basin, we see that most of the basin's snowpack is not at risk for a 2°C warming. However, the "at-risk" snow is concentrated in the Willamette River Basin, which is home to about 70% of Oregon's residents and which faces the often-competing interests of endangered salmonids, irrigation, hydroelectric power generation, municipal drinking water, and recreation. These modeling results suggest that in the McKenzie River Basin and other similar watersheds, future winter precipitation will shift from snow to rain, especially at lower elevations leading to higher discharge in mid-winter, lower peak accumulation of snow and earlier disappearance of snowpacks. In the coming decades, this system will likely experience an increased probability of winter flooding and summer drought. While the Willamette River Basin will experience these impacts before some of the other sub-basins of the Columbia, increased winter temperatures will affect low elevation snowpacks throughout the Columbia River Basin. Dam operations will likely need to be modified to accommodate these shifting patterns of snow accumulation and loss of natural storage in the snowpack and may ultimately need to be considered in potential modifications to Columbia River Treaty provisions.

Our current system for measuring snowpack was not designed for climate monitoring purposes. It has gaps in coverage at low elevations where snowpacks are declining and at high elevations where future climate impacts on snow are highly uncertain. We need a comprehensive approach that includes observations and modeling so that we can provide critical information to water managers and decision makers in this time of change.

Works Cited

Barnett, T P, J C Adam, and D P Lettenmaier. "Potential Impacts of a Warming Climate on Water Availability in Snow Dominated Regions." *Nature* 438 (2005): 303–9.

Brown, A. *Understanding the Impact of Climate Change on Snowpack Extent and Measurement in the Columbia River Basin and Nested Sub Basins.* Master's thesis, Oregon State University, 2009.

Christensen, J H, B Hewitson, et al. "Regional Climate Projections." In *Climate Change 2007: The Physical Science Basis. Contributions of Working Group I to the Fourth Assessment Report of the Intergovernmental Panel on Climate Change,* edited by S Solomon et al. Cambridge: Cambridge University Press.

Daly, C, R P Neilson, and D L Phillips. "A Statistical-Topographic Model for Mapping Climatological Precipitation Over Mountain Terrain." *Journal of Applied Metereology* 33 (1994): 140–58.

Daly, C, W P Gibson, G H Taylor, P J Pasteris, and P Johnson. "A Knowledge-Based Approach to the Statistical Mapping of Climate." *Climate Research* 22 (2002): 99–113.

Harr, R D. "Effects of Clearcutting on Rain-On-Snow Runoff in Western Oregon: A New Look at Old Studies." *Water Resources Research,* no. 22 (1986).

Jefferson, A, A Nolin, S Lewis, and C Tague. "Hydrogeologic Controls on Streamflow Sensitivity to Climate Variability." *Hydrological Processes,* no. 22 (2008).

Jefferson, A, G Grant, and T Rose. "Influence of Volcanic History on Groundwater Patterns on the West Slope of the Oregon High Cascades." *Water Resources Research,* no. 42 (2006).

Knowles, N, M D Dettinger, and D R Cayan. "Snowfall Versus Rainfall in the Western United States." *Journal of Climate,* no. 19 (2006): 45–59.

Lane Council of Governments and Eugene Water and Electric Board. "McKenzie River MIKE BASIN Model." Eugene, OR, 2008.

Liston, G E, and K Elder. "A Distributed Snow-Evolution Modeling System (SnowModel)." *Journal of Hydrometeorology,* no. 7 (2006a): 1259–76.

Liston, G E, and K Elder. "A Micrometeorological Distribution System for High-Terrestrial Modeling Applications (MicroMet)." *Journal of Hydrometeorology,* no. 7 (2006b): 217–34.

Molotch, N P, and R C Bales. "SNOTEL Representativeness in the Rio Grande Headwaters on the Basis of Physiographics and Remotely Sensed Snow Cover Persistence." *Hydrological Processes* 20, no. 4 (2006): 723–39.

Mote, P. "Trends in Snow Water Equivalent in the Pacific Northwest and their Climatic Causes." *Geophysical Research Letters,* no. 30 (2003).

Mote, P W, A F Hamlet, M P Clark, and D P Lettenmaier. "Declining Mountain Snowpack in Western North America." *Bulletin of the American Meteorological Society,* no. 86 (2005): 9–49.

NOAA National Climatic Data Center. *Snow Climatology Project.* Web site at http://lwf. ncdc.noaa.gov/ussc/.

Nolin, A W, and C Daly. "Mapping at-Risk Snow in the Pacific Northwest." *Journal of Hydrometeorology,* no. 7 (2006): 1164–71.

Nolin, A W, J Dozier, and L A K Mertes. "Mapping Alpine Snow Using a Spectral Mixture Modeling Technique." *Annals of Glaciology,* no. 17 (1993): 121–24.

Payne, J T, A W Wood, A F Hamlet, R N Palmer, and D P Lettenmaier. "Mitigating the Effects of Climate Change on the Water Resources of the Columbia River Basin." *Climatic Change* 62, no. 1–3 (2004): 233–56.

Pacific Northwest Ecosystem Research Consortium. *Willamette River Basin Planning Atlas: Trajectories of Environmental and Ecological Change.* Corvallis: Oregon State University Press, 2002.

PRISM Climate Group. 2009. Web site at http://prism.oregonstate.edu/products/.

Serreze, M C, M P Clark, R L Armstrong, D A McGinnis, and R S Pulwarty. "Characteristics of the Western United States Snowpack from Snowpack Telemetry (SNOTEL)." *Water Resources Research* 35, no. 7 (1999): 2145–60.

Service, R. "As the West Goes Dry." *Science*, no. 303 (2004): 1124–27.

Stewart, I, D R Cayan, and M D Dettinger. "Changes Toward Earlier Streamflow Timing Across Western North America." *Journal of Climate*, no. 18 (2005): 1136–55.

Sturm, M, J Holmgren, and Liston G E. "A Seasonal Snow Cover Classification System for Local to Global Applications." *Journal of Climate* 8 (1995): 1261–83.

Tague, C, and G E Grant. "A Geological Framework for Interpreting the Low-Flow Regimes of Cascade Streams, Willamette River Basin, Oregon." *Water Resources Research*, 2004: 3031–39.

Taylor, G H, and G E Grant. *The Climate of Oregon: From Rain Forest to Desert.* Corvallis: Oregon State University Press, 1999.

Taylor, G H, and C. Hannan. *The Climate of Oregon.* Corvallis: Oregon State University Press. 1999.

U.S. Climate Change Science Program (USCCSP). "Strategic Plan for the Climate Change Science Program, Chapter 5: Water Cycle." Final Report, 2003.

U.S. Geological Survey (USGS), Water Watch. Web site at http://water.usgs.gov/waterwatch.

U.S. Global Change Research Program (USGCRP). "A Plan for a New Science Initiative on the Global Water Cycle, Report to the USGCRP from the Water Cycle Study Group." 2001. http://www.usgcrp.gov/usgcrp/Library/watercycle/wcsgreport2001/default.htm

Part III: Rethinking the Columbia River Treaty

Rethinking the Columbia River Treaty

John Shurts

Earlier chapters in this volume have described the Columbia River Basin, the treaty, various perspectives of stakeholders, and some implications of science. This chapter, in contrast, is a beginning, a work in progress intended simply to organize information, frame issues, and speculate on these matters. It explores three ways of looking at the treaty and its benefits, costs, and underlying assumptions: First, at the time of adoption; second, as it currently operates; and third, at a hypothetical termination. The purpose of this exercise is to clarify what the people of the two nations want out of this river, or think they want, including biological qualities but also economic and other qualities, and whether they can have those qualities and at what cost—for what we are dealing with here is a biological river that is also a key component in a western North American electrical energy system.

Why and the Negotiations

The treaty is not a water allocation agreement. What prompted it instead was the promise of storage in the upstream nation to optimize certain uses of the river in the downstream nation. It was all about changing the natural hydrograph. High river flows ran at the wrong time of year to be truly useful, from the frame of reference of the dominant culture in the developing nation south of the border in the mid-twentieth century.

The Columbia is one of the great rivers of North America. It is the fourth largest river in North America, behind only the Mississippi, St. Lawrence and Mackenzie rivers. Most of the water storage for the Columbia is natural—snowpack in the headwaters mountains. River flows are affected most by snow runoff, with a late spring/early summer big-peak hydrograph. The peak flows are obvious: Average unregulated fall–winter (September–February) flows measured at The Dalles in the lower Columbia are around 100,000 cubic feet per second (or 94 kcfs), while May–June unregulated flows at the same point average more than 440 kcfs. The highest peak flood flow at The Dalles of 1.24 million cfs occurred on June 6, 1894, and about half of that flow came from Canada. The average annual total runoff of the Columbia is nearly 200 million acre feet at the river mouth, and 134 million acre feet measured at The Dalles, but the year-to-year variation is large—unregulated peak flows have run as high as 1,240 kcfs (1894) and as low as 36 kcfs (1937), a 1:34 ratio (compared to the St. Lawrence's ratio of 1:2 or the Mississippi's ratio of 1:25). (*See* the map in figure 1 in the Introduction to Parts I, II, and III).

People living in the United States portion of the basin saw two particular problems with the way the river ran. First, peak spring and early summer flows in higher runoff years brought the river out of its banks at various points, presenting an obvious obstacle to efficient development of cities and farm lands. This was especially troublesome in the more developed areas of the lower river. The problem came to a particular head with the huge May 31, 1948, flood flow of over 1 million cfs that damaged homes, farms, and levees from British Columbia to Astoria, caused greater than $100 million in damage in 1940s dollars, and in particular destroyed Vanport, Oregon, a city of 35,000 on the river next to Portland, killing more than fifty people. The USGS estimated that the 1948 flood flow was over 1 million cfs, and the river crested *25 feet* above normal at Vancouver, directly across from Vanport. Flows at Grand Coulee Dam, 600 miles inland, peaked at 633 kcfs that spring, more than three times the average at that location at that time of the year.

The second problem concerned power generation. Hydroelectric dams on the part of the river in the United States began powering the developing economy of the region in the 1930s and '40s, with plans (that came to fruition) to add far more hydropower generation after the Second World War. Hydroelectric power became the dominant form of electricity generation in the Pacific Northwest as nowhere else in North America. But the relatively cool Northwest developed a winter peak in electricity use, while the biggest slug of the water in the Columbia ran in the late spring/early summer, fueled by the melting snowpack—nearly three-quarters of the annual average natural flow is in the late spring and summer months. With the big exception of Grand Coulee, the Columbia hydroprojects built in the United States were largely run-of-the-river type, meaning they stored little water. By mid-century, the United States had built two large storage projects (Grand Coulee in the upper Columbia main stem and the smaller Hungry Horse on the South Fork Flathead, a tributary of the Columbia in Montana). Still, a lot of water ran wasted to the sea, as the saying went, and definitely ran at the wrong time to match generation to peak power demands.

With these two problems in mind, it was not lost on people that while Canada had 15 percent of the basin's drainage area, it contributed greater than 35 percent of the average annual flow at The Dalles *and* had several plausible storage reservoir sites. Also, flows at the U.S.-Canada border varied as much as from 14 kcfs to 555 kcfs, a 1:40 variation or even higher than for the river as a whole. People began to inquire about the feasibility of constructing storage projects in Canada to hold a big part of the snowpack runoff. This would help control spring/summer floods by knocking down the big peak flows, especially protecting U.S. lands from floods. It would also help optimize the generation of electricity, again especially at the U.S. projects. The stored water could be released in the winter, augmenting the lower winter base flows and thus increasing the amount of electricity that could be generated in winter at the hydroprojects in the lower river, better matching the Northwest winter peak demand.

These realities, however, created an odd dynamic for any negotiations between the two countries. The storage projects, and thus the costs (both monetary and in terms of inundated lands and dislocated land uses and communities), would be in Canada. Almost all of the expected benefits would occur in the United States— significant flood protection; increased generation from the U.S. hydroprojects; *and* the bulk of electricity demand and use, at that time and as projected into the future. So, how might Canada be induced to build the projects? How might the United States be induced to share the costs and benefits?

The United States and Canada began discussing the idea of storage dams on the Columbia in British Columbia for flood control and power generation purposes at least by the 1940s. In 1944, the two governments formally asked the International Joint Commission (IJC) to investigate. The IJC is an institutional byproduct of the two nations' Boundary Waters Treaty of 1911. The IJC formed what is known as the International Columbia River Engineering Board to conduct water management studies in the British Columbia part of the basin. The Engineering Board was particularly asked to assess whether "a greater use than is now being made of the waters of the Columbia River system would be advantageous and feasible."

Fifteen years later, in a 1959 report, the Engineering Board confirmed the practical potential of developing storage sites in the Canadian arm of the Columbia to regulate the river for flood control and power generation benefits. The IJC itself soon followed suit. Negotiations began in earnest in February 1960 and proceeded rapidly. The key issues were to identify dam sites, determine costs, and determine the amount and method of payments/benefits to Canada.

Canadian Prime Minister Diefenbaker and President Eisenhower signed the Columbia River Treaty on January 17, 1961. The U.S. Senate soon ratified. Canada did not. The Province of British Columbia had an interest in building dams on both the Columbia and Peace rivers, but needed money to do so. To get that money, the province requested that Canada sell what was going to be its share of the downstream power benefits that would result from the operation of the Columbia storage dams under the Treaty. These benefits, otherwise known as the Canadian Entitlement are explained in more detail below.

The Canadian government initially opposed such a sale. British Columbia objected to Canadian ratification without first working out this and other implementation details. The result was additional negotiations from 1961 to 1964 between the United States and Canada and, especially, between the national and provincial governments of Canada and British Columbia. These negotiations led to, among other things, two agreements between Canada and British Columbia, one in July 1963 and the other in January 1964, whereby Canada transferred its rights and obligations under the treaty to British Columbia and allowed for a sale of the Canadian treaty benefits by British Columbia.

The U.S. and Canada then executed a Treaty Protocol, also in January 1964, which allowed the sale of the Canadian Entitlement to the United States and clarified several issues raised by the treaty, one of which increased the amount of the Canadian Entitlement from earlier estimates. Additional negotiations among Canada, British Columbia, the United States, and a consortium of American utilities then resulted in a 1964 agreement on a thirty-year sale by Canada of the Canadian Entitlement into the United States at a price of $254 million. Canada then ratified the treaty, and it came into effect on September 16, 1964 with an exchange of diplomatic notes and a proclamation implementing both the treaty and the Entitlement sale.

Details of the Columbia River Treaty, with commentary

This section summarizes the most important provisions of the Treaty and related developments. It also offers analysis and commentary on certain points, such as (for example) on the nature of the flood control responsibility under the Treaty and how that might relate to the future of the Treaty.

Parties, Entities and Provisions

Parties and Provisions: The United States and Canada negotiated the treaty. Signed in 1961; ratified and effective in 1964. The treaty has 21 articles, two annexes, one statistical table and a subsequent Protocol containing additional provisions.

Canada immediately transferred its rights and obligations to the Province of British Columbia in a separate agreement. This is an important point—British Columbia is really the United States' treaty partner, not Canada as a whole, although it would again be Canada (through its Foreign Ministry) that would send a formal notice of termination and/or formally negotiate with the United States (through the State Department) in any effort to modify the treaty. One other point important to remember is that the main population, economic and political center of British Columbia is in the western portion of the province, especially the great city of Vancouver and the capital Victoria on Vancouver Island. That is where control over treaty negotiations, operations, and benefits rests, not in the eastern portion of the province where runs the Columbia and people realize the impacts from the dams.

Entities: The Treaty required each country to designate an Entity to implement it. The Canadian Entity is really a British Columbia Entity: B.C. Hydro, a provincial corporation that generates and sells most of the electricity in the province, from a host of province-owned hydro generation units on the Peace, Columbia, and other rivers. The U.S. Entity is *shared* between the Administrator of the Bonneville Power Administration (marketer of power from U.S. dams on the Columbia) and the

Division Engineer of the Northwest Division of the U.S. Army Corps of Engineers (operator of most of the federal Columbia dams and responsible for implementing systemwide flood control). The Entities have appointed Treaty "Coordinators" and a "Secretary" each, and established a joint Treaty Operating Committee from agency personnel to oversee annual planning and operations under the treaty.

Structures and physical effects

Treaty projects: The treaty obligated Canada to construct three dams and storage reservoirs to hold 15.5 million acre feet (maf) of treaty-dedicated storage (*see* map):

Mica Dam (7.0 maf of Treaty storage): Located just downstream of the big bend in the river at its most northern point, B.C. Hydro overbuilt Mica Dam to hold 20 maf in the reservoir known as Kinbasket Lake. Besides the 7 maf of Treaty storage, Mica has 8 maf of dead storage, and 5 maf of active storage called "non-Treaty storage."

Keenleyside Dam (7.1 maf): Also often referred to as Arrow, in that Keenleyside Dam created one long Arrow "Reservoir" out of what were two natural lakes or "widenings" of the river, Upper Arrow Lake and Lower Arrow Lake.

Duncan Dam (1.4 maf): Duncan Dam is on the Kootenay River tributary (spelled Kootenai in the U.S.), at one arm of Kootenay Lake.

Note on Libby Dam: Built and operated by the U.S. Army Corps of Engineers to provide additional flood control and power generation benefits. Libby is in northwestern Montana, and provides 5 maf storage on the Kootenai (spelled Kootenay in Canada), at a point where the U.S. is both downstream and *upstream* of Canada (the Kootenai/y begins in Canada, flows into the U.S., then back into Canada, and eventually joins the Columbia in Canada). The Columbia River Treaty authorized the United States to build Libby, necessary because building the dam required Canadian agreement, as the Libby impoundment (known as Lake Koocanusa) backs into Canada. But Libby storage is not considered part of "Treaty storage," and Libby operations are not officially part of treaty operations, although the United States cannot operate Libby in such a way as to interfere with treaty operations. Downstream power and flood control benefits resulting from Libby operations belong to the country where the benefits occur. British Columbia realizes some downstream power generation benefits from Libby flow regulation at power generation dams along Kootenay; the U.S. benefits from additional generation at Libby and in the lower Columbia.

Since the 1990s Libby Dam has been subject to operational requirements imposed in the United States to benefit salmon and steelhead in the lower Columbia and Kootenai River White Sturgeon and bull trout in the upper Columbia, all species listed as threatened or endangered under the federal Endangered Species Act. These operations led to concerns and complaints in Canada about effects in the Kootenay, especially power losses at Kootenay River hydroprojects. The dispute threatened the ability of the Entities to develop the required annual plans under the treaty (annual planning is described below in this section), even though Libby operations are not part of the treaty. The Entities ultimately resolved the dispute through the execution of a "Libby Coordination Agreement" in 2000 coordinating the operation of Libby with the operation of the hydroelectric plants on the Kootenay River and other considerations. In recent years, the coordination of Libby operations with treaty project operations has also been the subject of year-by-year supplemental operating agreements between the Entities, described below particularly in Section VI.

Note on Revelstoke Dam: Built and operated by B.C. Hydro, Revelstoke sits on the Columbia between Mica and Arrow. Flows out of Mica essentially go right into Revelstoke Lake. Revelstoke is also outside of the treaty, and again must be operated in a way so as not to interfere with treaty operations. Revelstoke adds storage and significant Canadian generation, and benefits greatly from Mica storage. In some senses, Revelstoke can work as a re-regulation dam for flows out of Mica.

Significance of Treaty Project Storage: Total active Columbia storage is approximately 55 maf, or about 33 percent of Columbia average annual runoff. (Compare to the Colorado River, for example, with storage capacity four or more times the average annual runoff. *Columbia River Treaty* storage is 15.5 maf, or over 25 percent of total basinwide storage. Non-Treaty storage in the Treaty projects, almost all at Mica, is another 5 maf. Libby then provides another 5 maf of Columbia storage. These three together (Treaty storage, non-Treaty storage at Treaty projects, and Libby storage) total more than 25 maf, or nearly 45 percent of total Columbia storage.

Effects on Flows from Storage: The effect of total Columbia storage (not just Treaty storage) has been to: reduce average peak Columbia runoff flows by greater than 40 percent; reduce the total amount of flow in the runoff season by approximately 30 percent; stretch the runoff season longer, with an earlier peak to the flow; significantly increase fall and winter minimum flows; almost eliminate over-bank flood flows; and greatly increase short-term flow variations. While the development of storage has significantly reduced annual and year-to-year flow variations, use of the projects for power production, load following, and power peaking has greatly *increased* the number and size of short-term (daily and hourly) flow variations; and eliminated the highest runoff flows, which has also greatly reduced the river's sediment transport.

Benefits

United States: Receives significant flood control and power generation benefits. Described in greater detail elsewhere.

Canada: Canada (or more precisely British Columbia through the agreement with the national government of Canada) is entitled to half of what are called the "downstream power benefits." The downstream power benefits are defined by determining the difference in hydroelectric generation in the United States with and without the use of Canadian storage. The point is that with the storage and thus controlled flows, more of the flow can be run through the generators (big runoff peaks must be spilled, and even if not, there was not always a need to generate at the time of spring/summer runoff) and at a time of greater value (the winter use peak). The parties agreed to treat Treaty storage on a "first-added" basis—that is, as adding the first increment of value on top of the base run-of-the-river generation in the United States at the time of the treaty. This means adding value ahead of generation benefits derived from subsequent storage in the United States such as at Libby Dam and Dorsal Dam (in the Snake River), from new thermal plants, and from power imported into the northwest on the Pacific Northwest–California transmission Intertie (more on that below). The Canadian half share of the downstream power benefits is known as the Canadian Entitlement.

British Columbia did not need the additional power represented by its share of the downstream power benefits in the 1960s. It did need money to build the dams, so as noted above the province promptly sold its share of the downstream power benefits in 1964 for thirty years at what was determined to be a net present value of $254 million. Sale was to a consortium of utilities in the United States organized as the Columbia Storage Power Exchange (CSPE), who then largely sold the rights to this power south to California. (More precisely, the CSPE was made up of eight public and three private utilities in the Northwest, with the dominant players being the three Mid-Columbia Public Utility Districts, who market power from their own dams in the mid-Columbia region. The CSPE then sold the rights to this power to 41 utilities, both public and private, many of them in California and the desert Southwest.)

This sale meant British Columbia turned the Canadian Entitlement into a direct monetary return right off the bat, but then did not receive on-going power benefits for the next thirty years. British Columbia largely used the money from the sale to build the storage projects. The nature of the sale also prompted the United States to make real a set of developments already contemplated, including the North-South Intertie and the Pacific Northwest Coordination Agreement discussed below.

Payments from the United States for Flood Protection: As compensation from the United States for the flood control benefits provided by the treaty dams, British Columbia received approximately $65 million, representing one-half of the estimated value of the reduction in flood damage in the United States through 2024. The United States paid this amount in three increments as each of the three dams became operational. The money went to the province as, at least in part, compensation for inundated lands, but that money did not go directly to or necessarily benefit the part of province adversely affected by the inundation, that is, in the Columbia basin. British Columbia is also entitled to payments to offset the actual costs of what is known as "on-call" additional flood control. However, the United States has never exercised its rights to "on-call" flood control, and so the province has never received these payments. System flood control issues are discussed in more detail below.

Flood control in British Columbia: British Columbia receives some flood control benefits, particularly in the Trail and Castlegar area.

Incidental generation: As it has turned out, perhaps the greatest value the treaty brought to British Columbia has been the incidental but substantial power generation opportunities *in* British Columbia. B.C. Hydro realized these incidental generation opportunities especially by adding generators at Mica (1740 mw) and by the construction and operation of the Revelstoke (almost 2000 mw) and Kootenay Canal projects (580 mw). Incidental generation must operate consistent with and subordinate to the required operation of treaty projects to serve the treaty purposes of optimizing flood control in the U.S. and power generation at dams in the United States and Canada. By one estimate, generation on the Columbia and its tributaries in Canada accounts for nearly 50 percent of the hydroelectric power produced in British Columbia.

Implementation: Principles, Procedures, and Responsibilities for Planning and Operations
Treaty Provisions
The treaty itself has operating or implementation principles and a basic plan for flood control and power operations and for determining the downstream power benefits. *See* Treaty Articles IV, V, VII, and XIV, Annexes A and B and the 1964 Protocol (Appendix A).

Planning: Based on these principles, the Entities develop "Assured Operating Plans" and "Detailed Operating Plans" to guide treaty project operations.

Assured Operating Plans: The treaty requires the Entities to develop each year an Assured Operating Plan (or AOP) for the treaty projects for the sixth succeeding operating year. The Entities also determine the downstream power benefits that would result from those operations. For example, in November 2004, the Entities agreed to the Assured Operating Plan for the 2009-10 operating year. (The annual planning gets delayed at times; the Entities did not complete the AOP for Operating Year 2012–13 until January 2008, and the AOP for Operating Year 2013–14 in February 2009). The AOPs are based on system regulation studies that take into account expected loads, resources, and generation under a historical range of water conditions. The AOPs provide the Entities with information and guidance necessary to plan for the actual operation of the power systems that are dependent on or coordinated with the operation of the treaty projects. Non-power and non-flood control considerations cannot be included in an AOP.

The AOP is also the vehicle for determining what downstream power benefits are to accrue from treaty operations in that future operating year, and thus what the Canadian Entitlement or Canadian share of those benefits will be. Actual operations in that future year do *not* affect the determination of the downstream power benefits—those benefits are set in the AOP and are not to be changed. The treaty allows the Entities to base the AOP and the determination of downstream power benefits on projected operations that allow for joint optimum power generation at both the United States and Canadian projects, rather than just optimum generation at the United States projects alone, so long as any resulting reduction in the downstream power benefits (when compared to operations that would optimize generation in the U.S. alone) is kept within an allowable limit.

Detailed Operating Plans: The Entities then also agree to a Detailed Operating Plan (or DOP) each year just in advance of that operating year. For example, the Entities agreed in June of 2004 to the Detailed Operating Plan for the operating year that ran from August 1, 2004 through July 31, 2005; in June 2008 the Entities adopted the current Detailed Operating Plan, for the operating year from August 1, 2008 through July 31, 2009; and the DOP for 2009–10 in July 2009. The Detailed Operating Plan is, as its name implies, a more detailed plan of operations based on more extensive and up-to-date studies than in the AOP for that year from five years before. The Detailed Operating Plan may produce treaty operations and results that are "more advantageous" to the parties than would result from operations under the AOP. But as noted above, the treaty does *not* allow the entities to recalculate the downstream power benefits/Canadian Entitlement at the time of the Detailed Operating Plan (or after the fact of actual operations). Those benefits remain as determined in the AOP.

Supplemental operating agreements: The treaty and its associated documents say nothing about supplemental operating agreements or any operating arrangements other than the Assured Operating Plan and the Detailed Operating Plan. But the Entities also understand the treaty to allow them to agree to mutually beneficial arrangements to store above or draft below the storage project levels that would otherwise result from strict adherence to the AOP and DOP, and that this may be done for both power and non-power objectives. This is the source of what are called the supplemental operating agreements. Much more on this subject below.

Project Operations, including Obligations as to Operations

B.C. Hydro operates the treaty projects to the operating plan specifications, as potentially modified by agreement through the year (when to store, how much to store, when to release, how deep to draft, etc.) The Entities understand the treaty to allow B.C. Hydro flexibility in how it operates the individual projects—especially in how it varies and balances outflows from Arrow and Duncan—so long as the net weekly flow at the United States/Canada border is as planned.

In turn, the United States Entity is required to operate the hydropower facilities in the United States in a manner that optimizes or maximizes the downstream power benefits from the flows that will result from treaty project operations. That is, the treaty requires the United States to operate its projects "in a manner that makes the most effective use of the improvement in stream flow resulting from operation of the Canadian storage for hydroelectric power generation in the United States." And just to make sure, the determination of the downstream power benefits to which Canada is entitled must *assume* the United States projects are operated in this manner, whether they are in fact or not—the United States must deliver power benefits for the Canadian Entitlement on a basis that assumes the optimum generation assumed in the AOP, however the United States actually operates the projects. The downstream power benefits owed to Canada are determined from the Assured Operating Plan, and not from the Detailed Operating Plan or actual operations.

Columbia River Treaty Operating Committee and Day-to-Day Implementation
The Entities have established a Columbia River Treaty Operating Committee, with members from each of the Entities, and use the committee and additional staff drawn from the Entity agencies to draft the annual plans and supplemental operating agreements and for monthly, weekly, and occasionally daily coordination to implement treaty operations. The Entities keep in close contact through the year through the committee and agency staff, and vary aspects of operations based on current runoff, flow, load, and other conditions in order to optimize results. Again, this in-season operation does not alter or affect the determination of the downstream power benefits.

Permanent Engineering Board
The treaty created a Permanent Engineering Board (or PEB) to assist in implementation. It has these characteristics: Two members from each country, with alternates. The responsibilities of the Permanent Engineering Board include reporting on any substantial deviations from operating plans and assisting the Entities in reconciling technical or operational differences. As far as I understand, the Board has only rarely performed these functions, as the Entities almost always work issues out themselves. In the most notable exception, the PEB had to render an opinion in the 1980s denying a request from the United States Entity to include "water budget" flows for salmon migration in the AOP. The Permanent Engineering Board has also had to, on occasion (although not recently), gently

remind the Entities to produce the required operating plans in a timely fashion. The Permanent Engineering Board also monitors treaty operations, keeps records and prepares reports for the two nations on treaty compliance. This is largely what the Board has done over the life of the treaty. Permanent Engineering Board meetings and records are not public, with the exception that it is possible to see the Board's annual report to the governments of the U.S. and Canada.

The Treaty Process Is Not Public: Treaty implementation, annual planning, and operations have no public aspect—no public process, input, meetings, review, comment, or accountability. The public can get access to copies of plans and other documents, but until recently only if you knew who to ask and what to ask for; availability is not publicized, and nothing is available to see until final or decided. Access to adopted operating plans, annual reports, and other relevant documents greatly improved in 2008 with the establishment of a Columbia River Treaty website jointly sponsored by the Entities, maintained by the Engineering Committee of the Permanent Engineering Board, and hosted by the Corps of Engineers (Columbia River Treaty website).

The Entities fully control implementation of the treaty and the operation of treaty projects, with little review by anyone else, public or private. As Nigel Bankes has written, *the culture of the Entities is the culture of the Treaty.*

Termination/Modification

Are there provisions for expiration or amendment? No: As noted above, the treaty has no expiration date (with the caveat that certain flood control responsibilities expire in 2024; see below). The treaty also does not have a provision or process for amendment or modification. Of course, the two nations could *mutually* decide at any time to terminate or modify the treaty, following the same process for negotiating the treaty in the first place.

Is there a provision for unilateral treaty termination? Yes: As also noted above, the treaty does have a unilateral termination provision. Article XIX provides that either nation may terminate certain provisions after the treaty has been in force for sixty years. The treaty came into force upon ratification, in 1964, so unilateral termination may first occur in 2024. But as also noted above, termination requires at least a ten-year notice to the other country. Thus a country that desires to terminate treaty provisions at the earliest opportunity will have to send the termination notice no later than 2014. Notice can be given more than ten years ahead; either nation could send the notice tomorrow.

But What Gets Terminated? And What Does Termination Mean?

Power provisions terminate: The treaty allows for complete termination of the power generation provisions. This means that upon termination Canada would no longer need to store or release water from the projects for any purpose related to generation in the United States. On the other hand, the United States would no longer owe Canada any share of downstream power benefits that might accrue in the United States from operation of the Canadian projects.

Flood control is more complicated: The treaty's termination provisions allow for the end of what could be called system flood control. In fact, British Columbia's system flood control responsibilities come to an end under the treaty in 2024 *whether or not* either country terminates the treaty. On the other hand, 2024 begins what is known as the "called-upon" flood control responsibility, and "called upon" flood control *may not* be unilaterally terminated. British Columbia's responsibility to provide "called upon" flood control will remain *even if* the rest of the treaty is terminated.

To explain briefly, Article IV(2) of the treaty prescribes a particular type of systematic flood control responsibility at the treaty projects that BC Hydro must follow for sixty years. More precisely, the treaty calls for 8.45 maf of flood control storage operation at Mica, Arrow and Duncan to provide flood control benefits in the United States and Canada. That provision ends after sixty years (or in 2024) even if the treaty is otherwise not unilaterally terminated. The understood reason that this responsibility comes to an end is that the United States, in its flood control payments at the time of dam completion, essentially "pre-paid" for system flood control for sixty years but no more.

The treaty also authorizes the U.S. to call for greater flood control than provided by the annual, prescribed system flood control operation, up to an additional 7.05 maf of Treaty storage (and another 5 maf of non-Treaty storage), if it ever needs to in an emergency. The United States would have to pay extra to British Columbia if it were to call upon this additional flood control, set at $1.875 million for each of the first four events. The United States has never exercised its rights to additional on-call flood control, although it came close in 1996 and 1997.

Finally, Article IV(3) provides that even after the expiration of sixty years, and "for so long as the flows . . . continue to contribute to potential flood hazard in the United States," Canada must, "when called upon by" the United State Entity (the Corps of Engineers in this case), operate "within the limits of existing facilities any storage in the Columbia River basin in Canada as the [United States] entity requires to meet flood control needs for the duration of the flood period for which the call is made." This is the "called upon" flood control, and it is what will be operable after 2024 if nothing else happens. "Called upon" flood control may provide up to (but not more than) the 8.45 maf of flood control now guaranteed. A request by the United States for called upon flood control is limited to potential floods that cannot be adequately controlled by the use of all effective United States storage. Also important, the United States will have to compensate Canada for any operating costs and economic losses due to requested called upon flood control operations. The treaty termination article provides that this provision for called upon flood control may *not* be unilaterally terminated until either "the end of the useful life" of the storage facilities or the end of the flood hazard conditions.

Given these provisions, it is not entirely clear what would be the practical effect of treaty termination, given British Columbia's continuing obligation to provide called upon flood control. The automatic end of the automatic flood control provisions *will* mean a new and greater allocation of flood control costs to the United States, which could be quite significant in its own right, but that will happen with or without treaty termination. On the other hand, the need to be able to provide the "called upon" flood control would seem to seriously constrain how free British Columbia would be to change the structures or significantly alter their operation, even upon termination of the other treaty provisions. The retention of the called upon flood control provision by itself would seem to forbid Canada from making radical changes in infrastructure, such as removing the dams, even if it were so inclined, as to do so would not allow for the same degree of flood control when called upon as currently available. The provision would also seem to limit Canada from making significant operational changes that would compromise the current ability to draft the reservoirs in the winter/spring before the runoff, in order to assure flood control space is available if called upon. My guess is that the flood control provisions more than any other aspect of the treaty will drive the parties to the negotiation table for a modified agreement.

Power System Developments in the United States as a Byproduct of the Treaty
There were a number of important developments in the United States contemporaneous with and largely in response to the Columbia River Treaty. These include: the transmission intertie with California; passage of the Pacific Northwest Preference Act of 1964; development of the Pacific Northwest Coordination Agreement; and development of additional generating units in the United States.

Transmission Intertie with California: The idea of a north–south transmission connection did not begin with the treaty, but the treaty made it happen. Neither Canada nor the Pacific Northwest portion of the United States needed the Canadian share of the downstream power benefits that were going to result from treaty operations. California utilities could make use of the power, and were eager to buy, but only if they had a guarantee that they could in fact obtain firm power from the north. So, Congress authorized the construction of the Intertie in 1964 with the ratification of the treaty. Expanded greatly since then, this was the beginning of what has become the significant linkage of the Northwest and the California/Southwest systems and, largely through the southern half of that tie, the link of the Pacific Northwest power system to a West-wide power grid. The

Intertie functionally allows a winter peak system in the Northwest and a summer peak system in the Southwest to benefit each other through reciprocal surplus power sales. For one thing, this has reduced the amount of peak generating capacity that each region would have to build if isolated.

Pacific Northwest Preference Act of 1964: This was passed by Congress at the urging of the Pacific Northwest delegation at the same time Congress authorized the Intertie, precisely because of concerns about the impact of the Intertie. The regional preference act assures that utilities in the Northwest states ultimately have first call to buy power generated at the United States dams on the Columbia. Power generated at United States-owned projects in the Columbia cannot be sold permanently out of the region (that is, to California); sales can occur only through limited-year contracts with recall provisions of "surplus" power, that is, power first offered for sale in the region with purchase declined. The obvious idea was for the region to benefit financially from the possibilities of the Intertie and the additional treaty generation, without having to worry about losing the power permanently to the more powerful economy to the south.

Pacific Northwest Coordination Agreement (PNCA): First developed in 1964 and subsequently renewed, this is an agreement among utilities and others that generate and/or take power from the Columbia. The signatories include BC Hydro, Bonneville, the Corps of Engineers, and a number of utilities in the United States. The purpose of the PNCA is to coordinate and, as far as possible, optimize power generation and delivery from the Columbia. This is necessary due to the patchwork of generating and distributing entities along the river and in the region. The generation and storage projects in the United States portion of the Columbia and its major tributaries include a collection of federal dams operated by two different agencies in two different departments (the Corps of Engineers and the Bureau of Reclamation), with a power marketer in a third (Bonneville), and a less numerous but quite significant set of nonfederal dams. The five dams in the mid-Columbia region below Grand Coulee in the state of Washington that are owned and operated by three public utility districts are the most important of these nonfederal projects in regard to coordinated operations. Then there are a host of utilities in the Northwest (and in the Southwest) that receive some or all of their power from generators on the river, and need to coordinate that power delivery with their loads and other generation. As this patchwork was coming into being in the post-war Northwest, the need for some coordination of operations was becoming obvious. The treaty simply made it essential, and added the Canadian projects to the mix, and the PNCA came into being just after the treaty. The PNCA yields an annual planning and operations exercise much like treaty implementation itself. The PNCA is subordinate to and works with the product of treaty operations.

Additional generating units in the United States: With the flow control promised by the Treaty and other storage, it made economic sense for the United States to develop additional generation at its dams. This included the Second Powerhouse at Bonneville Dam and the Third Power plant at Grand Coulee Dam.

Assumptions

It is fair to say that the people involved in the treaty negotiations and early implementation shared certain assumptions about the future. The treaty reflects these assumptions. The future did not turn out precisely as expected, a source of some of the present concerns and opportunities. Four of the key assumptions at the time of the treaty, at least from my vantage point follow.

Assumption no. 1. Treaty negotiators believed that the Northwest power system would change significantly over the life of the treaty in terms of loads and generating resources, significantly reducing the value of the hydropower system and thus of the treaty projects. A key assumption permeating the original treaty negotiations and documents was that changes in the Northwest power system that had nothing to do with the treaty would begin fairly quickly after the parties began implementing the treaty. One result would be a steady decline in the value of treaty operations.

People expected the regional demand for electricity to continue its dramatic post-war rise. The region would need additional generation, but there were few sites left for significant additions of low-cost hydroelectric generation. Increasing demand for electricity in the region would require the development of other generating resources, especially large-scale, higher-cost thermal plants, likely to be coal and nuclear. These types of plants would be cheaper and easier to run *if* built large (economies of scale) and then turned on and left on to meet firm power baseloads. Because big thermal plants of this type would be harder or more expensive to turn on and off, they would be less useful for peaking and load-following purposes. Hydropower would be more efficient for those peaking and load-following uses, assuming no particular constraints on flow regulation. So, the assumption was that the amount of the regional baseload to be served by new thermal plants was expected to increase dramatically over the life of the treaty. The corresponding proportion, significance, and value of hydropower generation (and thus the value of the downstream power benefits) in this overall developing system would diminish significantly over time, even given the obvious value to any system of hydropower peaking capability. These assumptions were reflected, for example, in the net present value assigned to the first thirty years of the downstream power benefits when British Columbia made the initial block sale of those benefits— British Columbia received far less than what turned out to be the actual value of those benefits.

Assumption no. 2. Another assumption about the Northwest power system (if less explicitly stated) was that it would continue to be largely isolated and self-contained. Loads in the region would be served by generation in the region, largely owned by regulated utilities in addition to the wholesale sales from the hydropower system by Bonneville and others. Generation in the region would be needed and dedicated to serving regional loads. Having sufficient firm-power capacity and energy in the region, and an adequate and reliable system in the region overall, would largely depend on what could be built in the region. Northwest generation would not play a role in the capacity, adequacy, and reliability calculations of other regions, and vice versa. To be sure, one reality underlying the Treaty itself—the fact that British Columbia was entitled to a share of the downstream power benefits, but had no need for the power, and neither did the Northwest part of the United States— promoted the original development of the north-south transmission Intertie and the first sales of Northwest power to California and the desert Southwest. But this arrangement was conceived largely as incidental, surplus, and temporary, and not as a fundamental aspect of the Northwest power system, especially in terms of the generation capacity the region would need to develop for peak loads.

Assumption no. 3. In British Columbia there were expectations and even promises from government officials about an increased level of industrial development in the province, including within the Columbia Basin itself. The industry and associated development to come would be powered in part by Canada's share of the low-cost electricity generated on the Columbia, as optimized by the Canadian storage projects. This was one of the benefits British Columbia would realize from the treaty, bringing economic benefits to the province and the basin. This change would also lessen the huge differences in the 1960s between the regional economies, amount of power use, and industrial development of the two nations.

Assumption no. 4. Flood control and power production would remain the dominant concerns from a system perspective. The treaty negotiators apparently shared a little-questioned assumption that the two purposes that drive the treaty— flood control and power generation—would remain the purposes for system water and river management desired by the governments and the people in the two nations, at least in terms of the international river. No treaty provisions recognize any other purpose, or call on or allow the Entities to manage treaty projects, storage operations, river flows, and the downstream projects for any other stated purpose. These people had little or no insight into a dramatic change just over the horizon in western society, which would elevate environmental concerns and values so greatly in the very near future, an evolution in values that would especially and specifically link salmon restoration issues to the heart of Columbia system operations: flows, storage, and generation.

The Pacific Northwest has lived with the treaty for forty-five years now. British Columbia built the dams. Through the Entities and the operating plans, the treaty has operated largely as expected for the purposes of flood control and power generation. In some ways, the treaty is so determined and constrained it could hardly fail to operate as expected. Even so, the essential smoothness of ongoing treaty operations seems remarkable, especially the way the Entities have worked so well and so closely together for efficient implementation with minimal or readily resolved complications and conflicts. This is perhaps not surprising, as it is in the interest of the Entities to closely control treaty implementation, serve the stated purposes, and prevent thorny issues from bothering the governments and the public.

Flood Control: Over-bank flood flows have been almost eliminated in the main stem Columbia, in large part due to treaty dam operations. By one calculation, for example, Treaty storage helped reduce 1997 peak flood flows at The Dalles by 170 kcfs, preventing at least $200 million in direct flood damages in the lower river. The Corps of Engineers has estimated similar benefits from flood control in other flood years, such as 1972 and 1974. The Corps estimates that cumulative flood control damage prevented by Columbia system flood control (total U.S. and Canadian, of which Canadian Treaty storage is a big part) totals nearly $14 billion, $5 billion alone prevented in 1996–97. Prior to the construction of all the Columbia basin storage dams, peak flows at The Dalles exceeded 600 kcfs in more than 40 percent of the years; after completion of the treaty projects, no flows have exceeded 600 kcfs at The Dalles. The benefits in terms of lives, property, and economic activity protected are intuitively obvious. So, too, are the possible implications for the physical and biological conditions of the river.

Power: The downstream power benefits from the operation of the treaty projects have been substantial and increasing over time. For example, the Assured Operating Plan for the 2004–05 operating year (again, adopted six years earlier) determined the downstream power benefits from treaty project operations at approximately 1,070 average megawatts (aMW) of energy and more than 2,300 megawatts (MW) in additions to dependable peaking capability. The Assured Operating Plan for the 2009–10 operating year increased those values to 1,130 aMW of energy and 2,700 MW of capacity. A recent AOP, for the 2012–13 operating year, has a different mix—a little over 1,000 aMW of energy and over 2,600 MW of capacity.

The Canadian Entitlement is half those amounts, or roughly 500 aMW of energy and 1,320 MW of capacity in the most recent AOP. Five hundred average megawatts of energy is enough to power roughly half the city of Seattle. Its value at $30 per megawatt-hour (late 2009 wholesale market prices for power) is about $131 million per year; its value at $75 per megawatt-hour (wholesale power prices in the recent past) is about $330 million per year. The value of the downstream power benefits in 2009 is indisputably significant—well more than anyone expected in 1960—and a substantial economic benefit to British Columbia, although *how* significant that value is (or will be) in a comparative sense and in the context of the entire power system is a complicated and increasingly fractious question, discussed below.

The late 1990s saw the end of Canada's thirty-year sale of its share of the downstream power benefits and the return of those benefits to British Columbia. The two nations had to enter into negotiations as to how the United States would deliver the benefits, as the default provision in the treaty suited neither nation. The treaty provides that if no other arrangement is agreed to, the United States must deliver the increment of power to Canada at the border near Oliver, British Columbia, a small town on the international border in the eastern part of the province, a place lacking suitable transmission and well away from load centers. The negotiation process and the result yielded insights into the present regional situation with regard to electricity. In essence, Bonneville worked out a suitable arrangement with BC Hydro, even as the two countries negotiated at a diplomatic level. The arrangement recognized that while British Columbia still did not need the downstream power benefits to meet its own loads (most of the time, at least) the increment of additional downstream generation represented an additional source of potential revenue to the province from surplus sales to the south, a surplus the province sold too cheaply the first time around as it turned out. BC Hydro was intent not to repeat the mistake of losing control over this resource through a long-term block sale at too low a price.

Thus, the arrangement worked out between the two nations obligates the United States (that is, Bonneville) to deliver the power to BC Hydro at the U.S.-Canada border, most of it at Blaine in western British Columbia and a small portion at Selkirk in the Columbia Basin, where transmission connections already exist. But delivery at Blaine and Selkirk may be at times a formal fiction. Instead, BC Hydro may find a buyer for the power or service and notify Bonneville where to deliver. Even if delivered at Blaine, BC Hydro still largely markets the power somewhere on the grid rather than use it for its local firm-power customers, as I understand matters. BC Hydro often has its own generation to spare as it is, and so its power-marketing arm (PowerEx) has become an aggressive seller in the developing wholesale power market in western North America. One report stated that BC Hydro made approximately $3 billion (in Canadian$) in out-of-province sales in 2003, with 80 percent of those sales to the United States, primarily to customers in California.

So, the first point to emphasize is that the treaty has operated about as expected. The second point is that the future did not. That the future does not turn out as expected is always to be expected, but the people negotiating the treaty seem to have had a different cast of mind in that regard. For an impressionistic look at the present that is not the future people envisioned, I offer the following:

Hydroelectric generation remains the backbone of base load generation in the region, of greater proportion and far more value than expected at the time of treaty negotiations. This came about for a number of reasons. For one, demand for electricity in the Northwest did not continue to rise as dramatically as it was rising in the 1950s and 60s so regional loads are nowhere as high now as they were expected to be from the viewpoint of 1960. The region remained in surplus from hydropower dams and a few other plants far longer than expected. Disastrous forays into large-scale, high-cost thermal plants (for example, the WPPSS nuclear plants) left the region—largely through Bonneville—with a huge financial debt that continues today. What was planned was not needed then, anyway. It left the region with little power to show for it and less from those sources all the time (as these thermal plants, such as the Trojan nuclear plant, are retired). Significant investments in energy conservation (especially by Bonneville and its customers under the Northwest Power Act, guided by the Northwest Power and Conservation Council's regional power plan), a host of smaller-scale thermal plants fueled by natural gas and, in recent years, new wind power developments have taken care of most of the region's additional resource needs—along with imports from the Southwest. The end result is that there has been *far* less large-scale thermal plant development in the region than was expected in 1960, and hydropower generation continues to serve a far higher proportion of the region's base loads than expected. There are also greater, more valuable markets for sale of surplus generation in times of good water than expected (discussed later in this section).

One curious point is that the natural gas–fueled plants that popped up all over the region in the last decade present almost the opposite dynamic than was supposed to result from new thermal plants as envisioned in 1960: Capital investments are relatively small, the main cost is the cost of the natural gas fuel (the operating costs), and the plants are relatively easy to start up and shut down quickly. This makes them much more cost-effective to use for peaking purposes and for market generation at times of higher prices than for base load generation, leaving most of the firm power base load commitment in the region still in the hands of the lower-cost hydrosystem—exactly the reverse of expectations in 1960.

There are a number of other reasons that the value of the Columbia hydrosystem has not only not fallen but instead has grown steadily. These range from the rise of alternatives to oil and other fossil fuels, to the growing demand in a deregulated wholesale power market for responsive resources (and the hydrosystem is or at least can be quite responsive), to the hydrosystem's ability to integrate and firm or balance uncertain renewable resources such as wind and solar.

Yet it is the newest generating resource being added to the region—wind power—that presents even different challenges. Thousands of megawatts of wind turbines are being added across the region, in response to climate-change concerns, related renewable resource portfolio standards imposed by western states, and rising costs for fossil fuels. We are still learning what it means to integrate that much intermittent wind power into the regional power system—lots of energy, with significant variation in the amount; little firm energy and firm power capacity without other resources to firm or balance the variation. The greatest needs right now, it appears, are (1) to develop significantly greater transmission capability to connect wind sites, balancing resources, and loads and (2) to balance wind generation with other resources that can be used very flexibly, in which the operators are able to increase or reduce generation quickly as wind generation does the opposite. An otherwise unconstrained hydrosystem would be perfect for that role, but it is much more of a challenge for a hydrosystem that is used for firm power base loads and demand peaks, a hydrosystem whose flexibility is also highly constrained for biological reasons (a topic discussed at the end of this section). Natural gas turbines and even other wind sites may be better balancing tools in the Northwest, not necessarily an intuitive result. Because we are still learning how to integrate wind power into the regional power supply, it is far from clear how these developments will relate to treaty operations and treaty value.

The one point that is clear is that Columbia hydropower generation, including the downstream power benefits from the operation of the treaty projects, remains of much, much greater value than expected in 1960. For example, a report in the mid-1990s from a multi-agency group in British Columbia called the Downstream Benefits Committee, displayed charts showing that the Canadian Entitlement at the time of its return in the 1990s would be almost three times larger than originally forecast. And after decades of seeing little return from its investments in the treaty projects (from the treaty return itself, that is, and not counting the incidental generation at its Columbia projects), British Columbia now may be benefitting well and alone from the downstream benefits. One recent estimate indicates that the Canadian Entitlement power, delivered based on the optimized power operation hypothesized in the assured operating plans has a current value to British Columbia in today's power market of around $250 million per year. Because the projects in the United States do not actually generate as hypothesized in the AOP, and instead project operations are now heavily influenced in a different direction by the needs of salmon and steelhead (*see* the discussion at the end of this section), staff for the U.S. Entity estimated recently that the United States may get *no* value now from the downstream power benefits.

The conclusion that the hydrosystem, and thus the downstream power benefits, remain of much greater value than expected comes with a number of other cautions or caveats, including:

First, in just the last few years the region is finally seeing the long-expected addition of significant amounts of non-hydropower generating resources. Slowly, and not steadily, increasing demand has been finally outrunning what can be served from the base system of hydropower generation and the few original large thermal plants, stretched for decades by conservation, imports over the Intertie and slower-than-expected growth in demand. The proportion of hydropower generation in total regional generation *is* starting to drop from its still extraordinarily dominant position. Does this mean we will finally experience the expected scenario (expected in 1960 that is) of the treaty-added generation greatly decreasing in significance and value?

This is not likely, for a number of reasons. As noted above, instead of what was expected forty years ago in additional resources—a set of large coal or nuclear plants simply taking over a huge proportion of the base load needs of the region—the added generation has been in relatively smaller units of gas-fired and wind turbines. The new resources have more slowly matched increasing demand, rather than adding a huge increase in capacity in one swoop. This is leaving hydropower generation still largely responsible for its existing service of firm power base loads, while also preserving its value for load following and peaking. Also, these new units are far more capable than big thermal plants of being turned on and off to meet loads, while the fact that they largely still cost more to run than the hydropower plants has meant that it often makes sense (especially with the gas-fired plants) to run the new units only when loads are more than the hydrosystem can serve. And especially at average and above-average water conditions, the hydropower system can still meet a lot of needs, with large amounts of surplus energy left over to market westwide. Another fact of relevance is that most of these units, especially the gas plants, are owned not by regional utilities serving their own retail customers but by independent power producers, another quite unexpected feature of the future not envisioned in the 1960s. So far, the owners have been relatively more interested in leaving these plants free of long-term commitments to regional base loads, preferring short-term and spot-market contracts in short markets and at peak times to maximize revenue against variable operating costs. And finally, as noted above, the increasing proportion of the added generation in wind turbines has simply added another dimension to the hydropower system. Perhaps the region someday will hit a tipping point, but so far the significance and value of hydropower generation remains substantial, and thus so does the continuing high value of the generation benefits added by the Treaty storage.

Yet, second, while the *continuing* nature of the Pacific Northwest power system has meant that the planned use of treaty storage has retained far more value than expected, the *changing* nature of the westwide power system may give still greater value to *different* ways to use the Treaty storage and projects. What could become more and more attractive, for example, could be the revenue value to be realized from using a significant portion of the Canadian storage to optimize Canadian generation, or even both Canadian and U.S. generation, for the purpose of surplus sales into the westwide market in summer, partially undoing the original idea of shifting flows to meet the Northwest's winter peak (*see* below). Summer storage releases for this purpose could also help summer salmon and steelhead migration, another subject discussed below.

If the relative value of the elements of the power system *within* the region did not change as expected at the time of the treaty, the way in which the region's power system relates to the rest of the West took the opposite course. Those negotiating the treaty did not and probably could not have foreseen the degree to which we now have a westwide, inter-connected power system, with a significant expansion of north-south and westwide transmission capacity; significant amounts of power moved between regions; a small but important level of inter-dependency on surplus power from other regions in lieu of developing additional firm power capacity within a region; a developing wholesale power market; and a host of generating utilities and independent power producers selling into that market, either on the short-term or spot market or through longer-term contracts, with no particular allegiance to the region in which the power is generated so long as transmission elsewhere is available. Hydropower generators with surplus generation—either outside of peak loads or at times of higher-than-average flows—are able to take advantage of that westwide market, and to draw on it at times of low flows and peak loads. We also have seen the rocky downside to this dramatic change in the industry in the perfect storm of 2001, combining a severe drought, low runoff, and low flows north *and* south (usually one is up if the other is down) with the wacky experiment in turning California into nothing but a wholesale power spot market without sufficient conditions for such a market. Both California and the Pacific Northwest will be paying for years for that one year. Price and service volatility remain higher than ever in the twenty-first century.

To be sure, most loads in the region are still served by generation in the region, and will be for the foreseeable future. The Pacific Northwest regional peak is still a winter one and will remain so, even as the climate change models project increased winter temperatures and thus reduced generation needs and increased summer temperatures and corresponding increases in generation needs. And transmission capacity between regions is still limited in the overall scheme of things. But even if still relatively small, the extent of the inter-regional connection and the amount of power moved and able to move are now so extensive that they bring a completely new dynamic to the power system in the West. Matters of price, adequacy, reliability, markets, resource availability, and planning for future resource needs are all affected by a westwide power system and must be considered in that context with all that implies for both opportunities and problems. The 2001 debacle highlighted just how dependent the Northwest had become on imports from California for region-wide and system-wide adequacy and reliability in times of low water.

For another example, more closely relevant here, it is no longer clear that the optimal generating use of all the stored water in the Canadian treaty projects will be to augment flows for generation at the U.S. projects to meet the Northwest winter peak, at least not in every year. Perhaps instead, the greatest value would be operations to optimize generation at the treaty projects, or at the treaty projects and the U.S. projects, for sales that capture the most revenue, even if that is in summer. The concept of what optimal should mean with regard to the treaty storage increment is in flux, and no longer obvious: Is it serving Northwest winter peak capacity? Maximizing Northwest reliability? Optimizing contribution to westwide reliability and adequacy? Maximizing revenue? What is the optimal use of this water shift from year to year?

Impacts in the Columbia Basin: British Columbia did not see industrial development to the degree expected in 1964, and certainly not in the Columbia Basin itself. Nor did it see much in the way of other significant load demands. Generation that BC Hydro did develop in the Peace and Columbia rivers has been far more than sufficient for loads in western British Columbia, with surplus left over for sale in the expanding westwide wholesale market. The increasingly cosmopolitan west side of British Columbia has seen significant commercial and residential development, but not anything near the kind of large-load industrial electricity demand envisioned for the future in 1960. Generation at the Columbia projects into a westwide market may be more of a money maker and economic benefit for the province than anything else.

What the Canadian portion of the Columbia River Basin did receive instead of any benefits from the treaty, direct or indirect, was the ongoing adverse economic, social, and environmental impacts of the structures—inundated lands, displaced communities, disrupted river ecosystems, project operations, and locations that yield little recreational potential (especially with regard to Mica Dam), and so forth. A typical accounting of some of the impacts: more than 2,300 residents displaced and a dozen communities lost; 60,000 hectares (231 square miles) of valley bottom land flooded; flooded forests, farm land, and wildlife habitat that limit growth and development and affect all communities; a river system that no longer is pristine, limiting opportunities for recreation and tourism; deep reservoir drafts that interfere with recreation and tourism and cause dust storms that plague communities along reservoirs; and archaeological and natural-heritage losses sustained by First Nations.

These impacts and related resentments festered and increased, not decreased, over the years. The impending return of the downstream power benefits in the 1990s brought all this to a head. The shared depth of feeling among affected citizens that they were owed compensation sparked a momentary political potency that the Columbia Basin usually lacks in a province dominated politically from the west. The result was that province's legislative assembly authorized the Columbia Basin Trust in 1995. The Trust has a board of directors partly appointed by the province and partly by local governments. Initially, it received dedicated funds representing a partial stake in the downstream power benefits to invest in economic development ventures in the basin ($250 million to invest in power-generating projects, and $45 million for other economic development projects), and a mandate to use the proceeds from investments (approximately $4 million annually at this point) to help ameliorate the economic, social, and environmental impacts of the treaty dams and to develop and implement these programs in a highly open, public fashion. A by-product has been the creation of an involved, informed public forum in the basin determined to have a say in what happens to the treaty in 2014 and after. The Trust is well ahead of anyone else in thinking about these things, for obvious reasons.

The Explosion of Environmental Concerns into the Public Sphere: In both the United States and Canada, environmental concerns— including the elevation of fish and wildlife protection and restoration as an important and even equal purpose for decisions on the river—are increasingly shaping public policy decisions. Perhaps more important than any other change—at least those change that are a matter of kind and not just of degree— environmental concerns and values are now as important politically, legally, socially, and economically in both nations as any of the other values and interests in the region and along the river. No issue has dominated system planning and operations within the United States in the last two decades like the drive to find a way to improve conditions for important fish and wildlife species, especially (but not only) juvenile and adult salmon and steelhead that spawn, rear and migrate in and through the hydrosystem.

The United States changed its internal legal framework to bring these interests to the Columbia table, however imperfectly. The impact on U.S. operations of the federal Endangered Species Act after the listing of twelve populations of salmon and steelhead and two species of resident fish (bull trout and Kootenai River white sturgeon) is the most visible today. But a more clear and direct example of how people and political leaders made a conscious choice to integrate these values into the public river is the Northwest Power Act of 1980. This Act created, among other things, a regional interstate planning Council coextensive in coverage with the U.S. portion of the basin, and with two equal mandates: One is to plan for a reliable and adequate power system for the region, and the other is to develop a program out of the recommendations of the fish and wildlife agencies and tribes that will "protect, mitigate and enhance" the fish and wildlife of the basin affected by the hydropower system—a program the federal agencies then have legal obligations to consider and, in the case of Bonneville, to "act in a manner consistent with." The

Act gave those federal agencies actually operating the system (the Corps, BPA and Reclamation) their own explicit and equal obligation under the Act to exercise their responsibilities to "protect, mitigate, and enhance" fish and wildlife affected by the hydropower system in such as way as to provide "equitable treatment" for fish and wildlife with the other system purposes. The ESA-based requirements have built on this regional legal and policy framework.

The annual system operation in the United States is now far from a power-optimized operation. There are several reasons why, beginning with the fact that the United States tries to ensure that its storage projects are at the flood-control level on April 10 but no lower, rather than drafting the reservoirs deeper in the winter and early spring for power operations. The purpose of this change is to allow the system to pass more of the spring runoff at its normal time and shape to mimic to a small degree the conditions under which spring salmon and steelhead migrants evolved. Outflows from Libby Dam differ especially in the spring to assist sturgeon spawning. The United States drafts the storage reservoirs in the summer to try to improve flows for juvenile and adult salmon and steelhead migrants; there is no reason to draft in the summer for an ordinary power and flood-control operation. Probably the most significant change from the power-optimized operation is the large amount of voluntary spill from early spring through the summer required at the lower river generating projects to improve passage survival for juvenile fish.

The legal and political dynamic internal to Canada has taken a different path. This is in part because the particular environmental and fish issues are different, more local, and less systematic, and because the legal framework is not quite so dramatic. But the issues, considerations, values, policies, and laws are there. The most significant for treaty project operations are the biological conditions for whitefish and trout below Keenleyside Dam and the aquatic conditions in Kootenay Lake.

The end result is not just a dynamic that complicates water management for power and flood control, but one that also changes public debates and complicates policy decisions in order to address conflicting environmental desires. Water-management decisions become more complicated because of the need for different operations to benefit salmon and steelhead, including the need to use storage in a different way at different times and to spill rather than generate with significant amounts of the river flow. One obvious example occurs when the needs of salmon management call for system operations to improve conditions for listed and non-listed salmon and steelhead in the lower river (for example, calling for significant on-call drafting of upriver storage reservoirs in summer to increase lower-river flows), and these operations conflict with the operations that upriver fish and wildlife managers seek in order to improve conditions for listed and non-listed resident fish in the upriver storage reservoirs and river stretches (for example, stable, higher reservoir levels and steady river outflows in summer).

Yet all of this is utterly absent from the treaty framework connecting the river across the international boundary. Treaty storage has contributed significantly to the change in the hydrograph and thus in the biological characteristics and communities of the river—from the headwaters all the way down to a major change in the timing and character of river flows through the estuary and out into the plume in the near-shore ocean. Questions as to whether there is a better way to store or release water to improve habitat conditions for fish and other environmental amenities; what the resulting effects might be on the other system purposes; whether reducing generation to benefit fish (such as through spill) is worth more collectively to society than the corresponding reduction in power benefits; and whether and how to decide what the right set of operations might be to balance these needs—questions deeply familiar to anyone working these issues on the United States side of the border—these are questions that cannot even be formally asked or addressed under the treaty, let alone answered. The treaty calls for treaty projects to be operated for power and flood control purposes only, for the generating projects in the United States to be operated in a way that optimizes the generating value of the treaty projects, and for treaty benefits to be determined and delivered to Canada based on optimized power generation. Fish and wildlife do not count in the treaty.

It is true that in the last decade and a half the Entities have found ways to adjust Canadian project operations during the summer to solve particular fish issues in the United States, such as supplemental agreements to draft water from the Arrow Lakes behind Keenleyside Dam in summer that would otherwise (under the treaty) not be drafted at that time. This additional water augments flows in the lower river, so that the United States does not to have to draft the reservoir behind Libby Dam so deeply at that time of the year, yielding localized fisheries and recreational benefits. There have always been ad hoc or supplemental arrangements outside of officially recognized treaty operations. They are entirely voluntary and can always be refused. The Canadians have no legal obligation to operate the storage projects in any way other than to optimize flood control and power generation in the United States and receive the maximum possible downstream power benefits in return (Bankes 1996). In turn, of course, the Canadians have no explicit, recognized, equal, legal ability under the treaty to use even a foot of water in treaty reservoirs in their own country to the benefit of environmental, biological, or recreational qualities in British Columbia. The possibility of ad hoc, voluntary adjustments may be no substitute for the systematic integration of values and interests society holds equally dear into the explicit legal framework of the treaty. This paper discusses this issue in more detail below.

The lack of public process associated with treaty planning and operation is almost as glaring as the lack of environmental purposes. This absence has to be seen in the context of a forty-year transformation of agency decision making in the United States intended to incorporate public involvement, public review, public comment, citizen-suit provisions, and the like. The 1980 Northwest Power Act is, again, a particularly strong example. The Council actually has three mandates, not just two—the third mandate is to do the other two (power planning and fish and wildlife planning) completely in the public view, to maximize public involvement on the Columbia. It is not pretty at times, but it is what the Council does. Underlying this part of the Power Act was a central perception on the part of people in the region and in Congress that the intense power supply and salmon crises of the 1970s that prompted the Act were due in part to the fact that decisions on these matters were made wholly out of the public eye.

Involving the public in river management decisions that cross an international border might be tricky, of course. Simply transplanting domestic public processes probably does not work. It would take us some time and adaptive experimentation to find the right model for public involvement in treaty planning and operations. That is not to say it cannot or should not happen.

What flexibility is there, under the treaty as it is now, to integrate fish and wildlife or other non power purposes, or altered power system purposes, into planning and implementation? The use and potential of supplemental operating agreements.

By now one point to this paper should be obvious—the treaty is in a formal sense an anachronism at best and virtually an embarrassment to public policy in the twenty-first century because it does not include as a purpose for managing the system something that people in both nations desire from the river as much as flood protection and power production— beneficial conditions for fish and wildlife and other environmental qualities and amenities. The laws in each nation that apply to the portions of the river within each nation have long ago incorporated these values, with substantive and procedural expressions and protections. Does it make sense that the treaty that links the two nations of the international river does not? Probably not, no matter how complicated it might be to bring these matters into the relationship.

It is appropriate at this point to note that treaty implementers and supporters, especially (but not exclusively) representatives from the Entities, argue that there is, in fact, no real problem here. They assert that the treaty itself leaves sufficient room for the Entities to agree to project operations that deviate from maximum optimization for power generation and flood control in order to benefit fish and wildlife or realize other desired purposes on the river. And, the argument continues, actual operations over the last decade are proof. For example, a 2002 presentation on "Ecosystem Management and the Columbia River Treaty" jointly from three representatives from Bonneville, the Corps of Engineers and BC Hydro explained how these supplemental agreements come about, provided a number of examples, and concluded with these points, among others (Hyde et al. 2002): (1) The treaty has proved a robust and flexible framework able to address joint stewardship of a major transboundary resource, even if the current context is different from what was expected in 1960s. (2) The treaty allows, by mutual agreement, the inclusion of non power considerations and altered power system considerations in actual treaty storage and release operations. (3) Fishery needs on both sides of the border have been addressed through mutually beneficial supplemental operating agreements, which offer opportunities to operate more broadly for fishery and other environmental considerations along with power and flood control. The track record in broadening stewardship objectives for the river, with regionally developed strategies and solutions, is one of success. (4) While the federal governments have ultimate responsibility under the treaty, challenges have been and can be effectively addressed at the Entity level. The treaty and the spirit of Entity cooperation under it, have enabled resolution of significant issues related to not just non-power expectations but also to the disposal of the Canadian Entitlement and a host of issues related to Libby operations. (5) The Entities and their staffs, by approaching problems with open minds, a spirit of "can-do" cooperation, and some time-tested tools of within-year supplemental operating agreements, have addressed multiple interests in a manner that enhances and broadens treaty benefits to both countries.

How Sound Are These Arguments?

The starting point for me is that it is clear from the treaty that British Columbia cannot be *required* to operate the treaty projects in a way that provides equitable treatment—or any treatment—for fish and wildlife conditions, that takes into account or provides any benefit for fish or water quality or any other biological or environmental qualities, or for any purpose other than realizing the flood control and power generation purposes of the treaty, unless of course those other benefits occur incidentally from flood control or power operations. In plain terms, the treaty does not speak of any purpose for the projects other than flood control and power generation or allow for consideration or inclusion of anything other than power generation and flood control needs in the system regulation studies or the considerations and outputs that make up the Assured Operating Plans. It is the Annual Operating Plans in particular that play the key role in reflecting what is and is not within the purview of official treaty operations. The downstream power benefits, and thus the Canadian Entitlement to a share of those benefits, are calculated on the basis of the studies in the AOPs, which must optimize storage, flows, and U.S. project operations for power generation.

The Entities understand the authority for supplemental operating agreements to come from Article XIV(2)(k) of the treaty. It provides that the Entities' powers include, among a long list, "preparation and implementation of detailed operating plans that may produce results more advantageous to both countries than those that would arise from operation under the [AOPs]." The treaty does not define or explain what is meant by "more advantageous" results. It is quite possible to read this provision in the context of the whole treaty and conclude that what the drafters meant was using current runoff and load information to provide nothing more than "more advantageous" flood control and power operations than could be anticipated in the AOPs, which are produced five years in advance.

That is not how the Entities have chosen to interpret this clause. They have concluded instead that this provision allows the Entities to agree to mutually beneficial supplements or variations to the Detailed Operating Plans for any purpose, power or non-power (the DOPs being adopted annually for that year's operations, based on what it takes to achieve the flood control and power objectives of the treaty and AOP). The Entities occasionally agreed to such variations in 1970s and 80s. The use of supplemental agreements took off in the 1990s, largely due to the expanding need to operate the system in the United States (and to some extent in Canada) for fish and other biological benefits, and have been an ordinary feature of treaty planning and operations for more than a decade now.

The Entities understand that supplemental operating agreements may be used to schedule operations for non-power and power objectives in variation from strict adherence to the AOPs/DOPs so long as the benefits are mutual—that is, advantageous "to both countries." Typical agreements provide non-power benefits for both countries at once or both power and non-power benefits. On occasion the agreements have traded power for non-power benefits, but British Columbia has been resistant to short-term deals that are purely money for water. B.C. has similarly been unwilling to incur significant non-power impacts in Canada to provide non-power benefits in the U.S., no matter what the inducement. What has become the norm is for the Columbia River Treaty Operating Committee, with representation from each Entity, to negotiate one or more supplemental operating agreements just prior to the operating year along with that year's DOP (and it is in the DOP that the Entities authorize the Operating Committee to enter into subsequent or supplemental agreements) or sometime during the year. The agreements allow for an adjustment to strict treaty power operations to meet a defined need in the United States—often in the last decade a fisheries need—coupled with a mutual (usually non-power) benefit recognized for British Columbia at the same time. The supplemental operations are usually expressed as allowing for deviations during the operating year from the "Treaty Storage Regulations (TSR)" set in the DOP, with an obligation to return to specified TSR levels by the end of the operating year.

To illustrate how these agreements work, what follows are examples from two different years. First, during operating year 2004–05, the Treaty Operating Committee agreed to two supplemental operating agreements that revised storage and release patterns from strict treaty operations. The first was a Nonpower Uses Agreement, signed in December 2004. This allowed for the storage by April 2005 of 1 maf more behind Mica Dam than strict treaty operations would allow, storage that the United States could use later in the year for summer flow augmentation to benefit juvenile and adult salmon agreement and to help meet minimum flow requirements at the Vernita Bar in the mid-Columbia region to support fall chinook spawning in the Hanford Reach. In exchange, British Columbia received approval for changes in normal outflow patterns at Keenleyside to improve flows for whitefish spawning in January and trout spawning and emergence in the April-to-June period. The second supplemental operating agreement allowed the United States to store an additional 220 ksfd (approximately 440,000 acre-feet.) in the Arrow Lakes behind Keenleyside Dam in February 2005 under a flow-shaping agreement, with the expectation of release in the May-July period for both power and non-power purposes. This also provided non-power benefits for British Columbia by allowing BC Hydro to store another 220 ksfd in February for release in March to support whitefish flows below Keenleyside.

For the operating year 2008–09, the Treaty Operating Committee agreed to four supplemental operating agreements that will vary storage and release patterns from strict treaty operations pursuant to the Detailed Operating Plan storage regulations:

(1) *Agreement on the Operation of Canadian Treaty and Libby Storage Reservoirs for the period of August through December 2008*, signed in August 2008. Also known as the "Libby/Arrow Swap" or (better) the "Canadian Storage/Libby Swap" and a regular feature of treaty operations in the last decade. The purpose of this agreement is to store more water behind Libby in August (or, more precisely, keep water in storage) than would otherwise occur under the flow requirements of the FCRPS Biological Opinion (remember, Libby is not governed by treaty operations). In exchange, BC Hydro drafted Arrow further during August 2008 than would occur under ordinary treaty operations. The extra draft from the Arrow Reservoir is intended to equal in volume the additional amount stored in Libby. This "provisional draft" of the Arrow Lakes to benefit Libby was then "returned" to the treaty operations by the end of December 2008 by drafting the extra amount stored in Libby at the call of the Entities, each Entity allowed to draft half. This Libby/Arrow swap allows Libby to stay higher and be drafted in a more stable way in August to benefit resident fish in the reservoir and below the dam and to improve summer recreation conditions in the lake, while still allowing the same total amount of flow augmentation to be released from the upper river for summer salmon migration as would normally come just from the U.S. reservoir. British Columbia realizes some non-power benefits (summer lake recreation) and especially power benefits from the ability to schedule the return amounts.

(2) *Provisional Storage Agreement for the period September 2008 through April 2009*, signed in September 2008—also known as the "Fall Storage Agreement." The purpose of this second supplemental operating agreement for the 2008–09 operating year is to alter treaty storage patterns in such a way as "to obtain mutual benefits (both power and non-power) for the [September–April time period] through the shaping of discharge from Arrow reservoir." The agreement allows the Entities to store behind treaty projects additional water from September–December than would be allowed under strict DOP Treaty operations, around 450 ksfd (approximately 900,000 acre-feet) total. Storage is shared equally. The Entities may then schedule the release of their share of the additional storage from November through March, with a release rate limited to no more than 10 kcfs/day over that period and with certain days over the holidays excluded. All must be released by March 31. The United States benefits in that it helps to match releases consistent with winter energy demand, provides additional flows if needed to support chum salmon spawning in the lower river and fall chinook spawning and emergence at the Vernita Bar in the mid-Columbia area, and helps the United States ensure its own storage projects are no lower than the required flood control level on April 10. British Columbia benefits by allowing for releases to protect whitefish spawning below Keenleyside Dam in February and March and by allowing BC Hydro the flexibility to shift water for market generation to periods with higher prices.

(3) *Nonpower Uses Agreement*, signed in November 2008. The Nonpower Uses agreement has become a standard of treaty operation, evolving in several ways since the early 1990s. It again allows for the storage by April 2009 of an additional 1 maf behind Mica and then describes a long list of procedures for changing Mica and Keenleyside operations over the year so as to use this 1 maf for non-power purposes. These purposes include flow augmentation to benefit salmon and steelhead migration from May through July 2009; stable outflows from April through June 2009 to benefit trout spawning below Keenleyside Dam; contributing to meeting minimum stream-flow objectives from January 2009 through May 2009 to cover fall chinook redds at the Vernita Bar and benefit chum salmon spawning below Bonneville Dam; and discharge objectives at Keenleyside Dam during January through March 2009 to protect whitefish eggs downstream.

(4) *2009 January Storage Shaping Agreement*, signed in 2009. New to treaty operations in 2009, this agreement arose out of a mutual desire to reduce outflows from Arrow in January and alter storage and release options in January, February, and March 2009 than as set forth in the "Fall Storage Agreement." The U.S. Entity sought the change to allow it to store additional water as part of its obligation to refill what is known as non-Treaty storage (*see* the discussion of non-Treaty storage in the next section). BC Hydro desired to move what would have been the January flow to February/March to benefit whitefish.

There have been a number of other supplemental operating agreements over the last decade, including agreements such as: (1) *2001 Summer Treaty Storage Agreement*. This was an agreement to refine the operation of treaty storage during the serious drought and power emergency of 2001. The Entities negotiated the agreement in July 2001 (with an addendum in October 2001) to cover storage operations from July 2001 through March 2002. With the drought conditions experienced in 2000 and 2001, conducting 2001 operations to strict treaty rules would have refilled treaty storage to 47.5 percent of full by the end of July 2001. The Summer Treaty Storage Agreement provided for additional storage of water in treaty space to address U.S. fall/winter energy reliability concerns; Canada/B.C. fishery benefits (kokanee access, whitefish spawning) and summer recreation; and U.S. spring flow augmentation for salmon steelhead migration in 2002. By the end of August 2001, approximately 4 maf had been stored in what was called the STS account, equivalent to about 32 feet of additional water at Arrow. Most of this storage was held in Canadian reservoirs until the January-through-March 2002 period. (2) *Arrow local agreements*. These allow the two nations to share additional downstream benefits that arise from implementing what is known as the "Arrow Local" method of computing the refill curve at Arrow, producing power and non-power benefits for both Canada and the United States. (3) *Optimal Balancing of Storage*. A one-time agreement that provided for the optimal balancing of storage and energy between Lake Koocanusa (behind Libby Dam), Kootenay Lake, and the Arrow Lakes during a February 2001 Kootenay Canal Plant outage.

As useful as these annual supplemental operating agreements are, few would argue that they are satisfactory to bear the weight of all the non-power (and additional power) requirements, needs, values, and desires of the two nations into the indefinite future. It is one thing to make use of them when you have to; it is another to continue the regime when the opportunity is presented for an alternative. The non-power use pressures and national legal requirements are only increasing. Why continue baseline treaty operations that are ignorant of these? Also, increasingly complex and changing demands on the hydrosystem's power operations—increased volatility and opportunities in a developing and expanding wholesale power market; power system integration complexities as wind and other renewable resources come on to the system complicate the ability to deal with these considerations in annual and ad hoc arrangements.

On the Other Hand, Is It Necessary to Change the Treaty Itself to Internalize These Considerations?

What about supplemental agreements between the Entities that are long-term and more fundamental? Particular attention has been paid to the way in which the two nations addressed the issue of Libby operations around the turn of the century to protect and improve conditions for Kootenai River white sturgeon. Some see this as a possible model for integrating additional considerations within the existing treaty framework. For others, it is not altogether clear that this story favors the arguments for the necessary flexibility of the treaty, and may instead provide a model for what needs to happen to actually change the treaty. A summary of the details: (1) Kootenai River White Sturgeon live (as long as a hundred years) in the 168 miles between Montana's Kootenai Falls (below Libby Dam) and British Columbia's Kootenay Lake. (2) Population in decline since 1950s. No significant recruitment of new sturgeon adults since completion of Libby in 1974. (3) Sturgeon listed as endangered under ESA in Sept 1994. U.S. Army Corps of Engineers must adjust Libby operations to benefit sturgeon. (4) 1995 Sturgeon Biological Opinion from U.S. Fish and Wildlife Service required the U.S. Army Corps of Engineers to increase spring flows from Libby to enhance sturgeon spawning. Meanwhile, the National Marine Fisheries Service's Biological Opinion on system operations to protect listed endangered and threatened salmon and steelhead required additional Libby storage releases in July and August to augment flows during summer juvenile and adult migration through the lower river. (5) One result: Generation and revenue losses at Kootenay River projects of BC Hydro and West Kootenay Power (now FortisBC). That is because the net effect of these changes has been to decrease the flows from Libby during the fall/winter peak, increasing spring and summer flows. (6) BC Hydro took steps to mitigate the effects, modifying Duncan/Kootenay Lake storage operations to partially reduce adverse power impacts from Libby's sturgeon/salmon operations. (7) BC Hydro also lodged objections with the United States. The Canadian Entity estimated the value of its 1994 to 1999 power losses from Libby's operation at about $12 million Canadian. BC Hydro complained about interference with treaty operations under AOPs and DOPs due to altered Libby operation. British Columbia officials also complained about increased flooding and bank erosion downstream of Libby, as well as adverse impacts to resident fish and summer recreation in the Koocanusa reservoir in Canada. The conflict threatened to prevent the Entities from agreeing to further AOPs. (8) After a lengthy diplomatic dispute, the Entities received permission in January 1999 from their respective governments to negotiate a settlement.

These negotiations led the Entities to such an agreement in February 2000. The *Libby Coordination Agreement*, which achieved the following: (1) Established a framework and procedures for coordinating Libby and treaty project operations in such a way as to allow AOP and DOP planning to continue. Consistent with requirement in Article V of the Treaty Protocol to coordinate the operation of Libby with the treaty projects. (2) Recognized the continued operation of Libby for sturgeon and salmon flows and other non-power requirements. (3) Allows for late summer and fall releases from Arrow and exchanges of power between

BC Hydro and Bonneville to compensate the power losses to British Columbia resulting from Libby flows for sturgeon and salmon. (4) Enabled the Libby/ Canadian Treaty storage swaps (*see* summary of supplemental operating agreements above) that mitigate recreation and power impacts of salmon flow requirements at Libby. (5) Limits the Keenleyside Dam outflow to 80 kcfs during January to provide some protection for whitefish spawning. (6) The United States produces a Libby Operating Plan, and must use that plan to report updated Libby operations (such as resulting from the 2006 Libby Biological Opinion and a recent litigation settlement regarding Libby operations). The treaty AOP is, however, to be based on a Libby operation that is relevant to 2000 operations, *without* updated non-power requirements. (7) The Libby Coordination Agreement is careful to admit no wrong by either Entity and to declare that no precedent is set by the agreement. The agreement may be terminated with but *30* days notice.

The two governments then endorsed the Entities' 2000 Libby agreement. The United States is allowed to operate Libby for sturgeon and salmon flows. BC Hydro received increased operational flexibility at Keenleyside, and mitigation from the U.S. for power losses.

The end result of the Kootenai episode? A relatively long-term agreement between the two nations in which both nations (a) recognized fishery and other non-power, non-flood-control values and purposes as an equally legitimate part of Libby operations with the power and flood control uses of Libby and treaty projects (without, of course, integrating non-power considerations directly into considerations of treaty project operations), and (b) established substantive and procedural elements that attempt to balance, express, and protect these different purposes. This brings the international legal regime into accord with the domestic law and policy of the two nations—for example, this is precisely the type of legal and policy regime already supporting Libby Dam operations under the domestic law of the United States in accord with the disparate policy desires of the people.

This could be a model for a long-term non-power/power integration agreement among the Entities to reshape treaty project operations for the twenty-first century. However, it seems to me that to be successful, such an agreement would so change treaty operations as to largely leave the formal treaty approach a formal fiction that could no longer drive the assignment of responsibilities and benefits (which it still does) any more than it does actual operations. If that is what it would mean, maybe the Libby Coordination Agreement also is a model for precisely how the treaty itself should change?

Two Main Points to Take Away from this Discussion

First, annual supplemental operating agreements must bear the weight of all the modern considerations of fish and wildlife and power. The weight is already too great to sustain much longer, at least in my opinion, and I think everyone would agree with that conclusion if additional weight is added to address flood control and changing hydro conditions resulting from climate change. The status quo is *not* a viable option. Even if the treaty is not changed, the Entities, or the two nations will *have to* negotiate a new longer-term operating agreement or treaty protocol of some sort in the next few years. It is possible such a longer-term operating agreement could address all of these modern issues, solve all the identified gaps, problems and needs, be executed by the Entities with regional input and without the need to crank up the formidable treaty modification/ratification machinery of each nation. Fabulous—if it can work. And there is no need or requirement to wait until 2024 or 2014. We could do this tomorrow.

Second, when the opportunity to begin talking about these matters presents itself, is it possible to imagine the two governments will be able to, or want to, resist a call to bring the non-power values (and the twenty-first century power considerations) systematically into the treaty in some way? After all, these are values expressed in domestic law and policy— values that people on both sides of the border see as equally legitimate with the traditional hydropower and flood control values in the treaty. It seems unrealistic to believe that either country would continue to cast these values as distinctly ad hoc or voluntary or secondary or supplemental—in essence, making them second class treaty citizens. Even so, it is certainly worth exploring whether a supplemental but more comprehensive and durable agreement can be an effective regional solution for bringing the treaty into the modern policy world.

So What Happens Next? Who Knows?

An Assessment of Benefits, Costs, Opportunities, Needs, Desires, Risks, Vulnerabilities, Alternatives and Points of Leverage with Regard to Treaty Operations: The United States and Canada would never negotiate this particular treaty today. It does not fit the present social context and should at least be modified, but to *what* does it make sense to change from the present structure and operations, and more to the point, to *what* will it make sense ten and twenty and forty years from now? Are the interests in change sufficiently mutual to be able to negotiate a different set of operations on these structures—or different structures—that the two countries can agree to as better than the present? And if not, will one or the other nation prefer unilateral treaty termination to continuing as is?

I obviously cannot answer those questions here. What I can do is continue to reflect in a highly impressionistic or speculative way upon current aspects of treaty operations, on how these operations are or might be valued for good or bad by the parties and other interested participants, and how alternative ways of doing things, including the termination alternative, might be accommodated, or desirous, or beneficial, or unpleasant, or disastrous.

What follows is necessarily incomplete, and some of these speculations may be incorrect, or based on incomplete information, while other key topics and issues are missing. The point is just to get them out on the table, to encourage ideas and discussions. I am particularly interested that this be a *wide public discussion within the region*, and not just a closed-door negotiation and decision between the Entities— Bonneville, the Corps of Engineers and BC Hydro—as much as I value the people and principles of all three agencies.

Electricity and Salmon, Flexibility and Constraint

What people think of as the best electricity future for the river is not the river that best suits salmon. This may have always been true to some extent, but after a brief period of possible convergence or better balance, the differences may become far more marked again.

The ability of the Columbia River to generate electricity is not just valuable, it is increasing in value all the time. There are aspects of any hydrosystem and of this hydrosystem in particular that make it particularly valuable in the increasingly volatile westwide power systems and markets not just the sheer bulk of the system, and the fact that in many water years it produces large surpluses of energy for disposal, but also the system's inherent flexibility and responsiveness. There is no reason to believe these factors are going to lessen in the near to middle future; instead they are likely to intensify. There are aspects to treaty operations— set and determined by the treaty provisions and the long advance planning—that actually limit the flexibility of the Columbia system to do what it can do best for power production. The Entities have ways to adjust operations through the year to optimize output to runoff conditions and loads, but the treaty parties will have to think hard about whether and how to introduce more systematic and greater flexibility into project operations for power system optimization.

Changing system management to be more responsive to flexible power system needs may be the exact opposite of the direction required to support fish habitat. If power system considerations move toward flexibility, fish considerations move toward increasing constraints. A biological principle accepted by most of the people working with an otherwise very disparate set of fish and wildlife communities across the basin is that allowing the river to move back toward a more natural hydrograph is an ongoing need (with the possible exception of the need for higher summer flows than natural in the lower river by augmenting from storage if the projects and pools in the lower river remain). This means not drafting the storage reservoirs as deeply in winter; allowing far more of the runoff to pass the reservoirs and move at its appointed time all the way into the nearshore plume; allowing more flood flows, to move sediment and the like; significantly reducing daily and hourly flow fluctuations related to power peaking to improve spawning, rearing and migration conditions; steady or slowly declining releases from reservoirs to protect conditions in river stretches immediately below; steady or slowly declining reservoir levels in summer, with lesser drafts, to maintain natural lake-like biological qualities in reservoirs; spilling more water at run-of-river power dams (and thus generating less) for improved juvenile fish bypass survival; and in general allowing more water to flow in spring and summer to support salmon migration, leaving less water to generate power during winter peak-demand periods. These are constraints that reduce the flexibility to operate the system responsive to regional and westwide electricity demands and that, if taken to their logical extent, call into question— why have the system in the first place?

If we open up the treaty for whatever reason in the coming decades, it is impossible to imagine *not* making the change to incorporate environmental values as an equal purpose of the treaty, among whatever other changes we might make. It is almost as hard to imagine the treaty parties being able to ignore the opportunity to revisit the treaty when it arrives and thus avoid the incorporation of these other values. But in doing so the treaty will have to institutionalize and address the conflict between system management flexibility and constraint, a conflict in which neither perspective has much relation to the well-determined, well-oiled, smoothly operating machinery of traditional treaty project implementation.

Electricity Generation and the United States Perspective (Plus Climate Change?)

Even with these constraints, the United States obviously continues to benefit in terms of increased generation from the fact that flows are stored in and systematically released from storage projects, including the treaty projects in Canada. The long-standing aim of storing the runoff and later releasing the stored water to augment winter flows so as to increase generation in the United States to help meet a winter peak retains its vitality as a power system concept.

Controlled releases from storage for *summer* generation also have increasing revenue value in an increasingly interconnected wholesale power market across the West, presenting both an opportunity and a complication. That value in summer should do nothing but increase as the economy in the Southwest continues to grow, although not much more can be done to take advantage of the opportunity unless the region also significantly increases transmission capacity.

The value of using storage for summer generation will also increase if some plausible scenarios for climate change in the Pacific Northwest come to pass. One likely scenario predicts the region will receive approximately the same average runoff, but more precipitation will come in the form of rain and less in snow, matched by higher temperatures year-round. The obvious effect from less snowpack will be reduced flows and higher water temperatures in summer and fall months. In addition, peak flows are expected to continue the trend of occurring earlier in the spring. Yet with temperatures rising there will be a need for increased summer generation in the Northwest itself (and Southwest), increasing the value of using storage for summer generation. Releasing stored water to generate power in the summer also would improve summer flows for salmon migration, temperature control and other biological needs in the lower river, as compared to holding the stored water into the winter. Salmon managers currently support boosting summer flows in a dammed system. Thus the need for that additional water will become more critical if climate change reduces average summer flows.

In any event, it would seem that right now, in the abstract, the *power* system in the United States could accommodate either the current treaty operation as it is *or* some sort of shift by British Columbia to release more of the storage in summer for the purposes of summer generation. How far the United States (and Canada) could tolerate a shift in treaty storage releases to summer before it saw real effects to the reliability of the winter power supply is not clear, especially given the other dynamic developments in both regional demand and disparate power supply resources. The treaty and the need for winter adequacy/reliability are not the only constraint on any wholesale shift to summer flows, of course. For example, limits on transmission capacity to the south may preclude any radical change in operations to maximize summer generation for sale to the south.

Looked at another way, and assuming the treaty is terminated, would BC Hydro really operate the storage projects much differently from how it operates them now? There is an assumption among many in the United States who like the power benefits of the Canadian storage that when the time comes to terminate the treaty, the United States ought not to fear termination, and should even embrace it, as there are good reasons to believe British Columbia would operate the projects in largely the same way, anyway. The result would be that the United States would still get the altered flows that increase generation and thus the downstream power benefits, and yet would not owe a dime of the benefits north. This is a plausible scenario people will need to grapple with in the next decade. The thinking underlying this scenario assumes British Columbia's interest in removing or changing the structures is negligible, and constrained in any event even with treaty termination, for a number of reasons, including the continuation of the called-upon flood control obligation and because of the value to the province of the hydro generation on the province's Columbia projects. There may not be a different Canadian project operation that has dramatically different effects south of the border that makes sense in the context of British Columbia—that is, nothing of obviously greater benefit to

British Columbia than the present operation, except perhaps a shift of some storage releases to maximize revenue in summer, with obvious limits that presumably could be tolerated in the United States and even be put to profitable use. This scenario also assumes that no large diversions of water are likely to happen in British Columbia (either proposed or that would be approved by the International Joint Commission), and so it remains a fair assumption that water still will be coming over the border.

There are, however, several reasons to wonder whether this scenario really would work out so well for power interests in the United States (let alone other interests, such as fish). First, I suspect that what British Columbia would like to have more than anything else—more than any specific change in operations—is simply more control over how these projects are operated (or perhaps a lot more money, in lieu of that increased control). Presumably this increased control would include how and when water is stored, but variations in the logic of storing water in these structures are few. The key instead would be British Columbia's absolute control over how and when to *release* water and for what purpose or purposes. Also, whether British Columbia would choose power revenue maximization, or improvements in biological and recreational qualities in British Columbia, or some amalgam of both, it is not hard to imagine scenarios where the flows no longer come across the border to the United States at the times and, especially, in the *certain* fashion that provides the reliable power benefits the United States now enjoys. Those who say the Canadians likely will operate the projects largely the same as they do now in the event of treaty termination may be putting far too little weight on how much those benefits are based in the very certainty and reliability of the present operations, a certainty and reliability that stems from the fact that real control over those operations is jointly determined and structured by the planning framework of the treaty.

Second, the "just let the treaty terminate and we'll still get the power benefits in the United States we always have" scenario presumes that this is what the United States will want. It ignores the fact that one of the biggest debates around treaty termination/modification will be a debate *within* the United States about the extent to which the country should negotiate for modifications in treaty operations (and, for some, the removal of structures) to begin incorporating environmental values, especially water management for fish and wildlife needs, as an explicit purpose of the treaty and project operations. Terminating the treaty, or allowing it to be terminated, under the assumption and with the result that the power and flood control operation likely will continue will not be an acceptable result to important political constituencies in the United States. The same can be said for the counter-scenario noted just above; equally unacceptable to this coterie will be treaty termination that results instead in full British Columbian control over the treaty concrete and river flows without some legal infusion at some level of environmental protections for the United States regarding downstream flows. Perhaps if the U.S. fish and wildlife managers knew for certain that British Columbia would change operations to emphasize significantly increased releases in summer to maximize generation revenues from surplus sales to the south, the concept of treaty termination and full Canadian control of the projects may not so greatly concern them or other salmon advocates—but only if they got some solid guarantees that this indeed would happen.

Debates that compare the value to society of predominant or maximized power operations under the treaty (or a variant) versus operations that provide equitable treatment for other purposes will be very real. I doubt we will ever again see the original expectation (that the Treaty-added generation would decline greatly in value as a component of the regional power system). That scenario seemed so dependent on a very particular future of which there is no sign (i.e., huge demand increases, matched by huge investments in numerous large-scale baseload coal and nuclear power plants, with hydropower generation left as a lesser player in the regional system). Will the near future continue the current counter trends, including: more slowly growing loads (especially given the current troubled economy); new resources from a disparate set of generation types and owners (primarily natural gas and, especially, wind turbines, along with conservation) that work relatively well with and prop up the base value of the hydrosystem; perhaps a renewed interest in coal, if carbon sequestration techniques develop, but presumably no more than perhaps one plant in the region over the coming decades; increasing inter-connection and influence of the wholesale power market westwide, yet with transmission constraints that will only slowly ease; and continued uncertainty and some volatility in the power supply market and prices. None of these changes forebode a significant decrease in the use or value of the hydrosystem for power generation over the next twenty years. *Non-power* constraints will be the source of any further de-rating of the system; the power system changes will only complicate how to realize the value of the hydrosystem, not lessen it.

Electricity Generation and the British Columbia Perspective

BC Hydro currently generates power from treaty and nontreaty projects in the Canadian Columbia River Basin. The downstream power benefits are providing real value to British Columbia at the moment; generation from the Columbia projects in British Columbia provides much more value. But because treaty operations are legally dedicated in large part to optimizing generation in the United States, BC Hydro does not have the kind of control over projects in its own province that would allow it to fully optimize *Canadian* system generation, matching that generation best with available loads and sales opportunities. And as noted above, it is a fair guess that if British Columbia could achieve just one thing with treaty termination or renegotiation, it would be to get as much more control as possible over the operation of dams and reservoirs in its own country, for generation and for non-power reasons. And the logic of the present context indicates that what the *province* might like to do with more control—as opposed to, say, what many of the people who live in the Canadian Columbia Basin might want to do—is to optimize Canadian generation to maximize revenue.

Treaty termination would give British Columbia this control. It would cost British Columbia its share of the downstream power benefits from present operations. So, termination and a takeover of control would be worth it to British Columbia only if more value can be realized from control over operations than the present downstream power benefits. If British Columbia terminated the treaty, but then found itself operating the projects largely as it does now, there would have been little point to the province in terminating the treaty and giving up the downstream power benefits for the little gain in flexibility (this is exactly the scenario, of course, in which the United States would gain from treaty termination). And one can imagine that BC Hydro could in fact find itself operating largely the same as at present, even if the intent was to optimize Canadian generation, *if* there is little transmission capacity left to allow for a shift of much of the stored water to summer generation, or *if* operational constraints prevent moving too much water out of Mica and through Arrow in summer, or *if* the growing economy in British Columbia brings even more and more need for dedication of BC Hydro's own resources to a winter peak, or all of these.

But this may not be a correct way of looking at the context. It may be that BC Hydro still would highly value greater control to schedule flows and generation, on a day-to-day, week-to-week and month-to-month basis, even if the overall pattern of operation did not change much. And such a change would require the end of the treaty, or its replacement with something that allowed for more control in British Columbia—and a greater comparative benefit than the present downstream power benefits. Perhaps British Columbia would be happier to swap just *some* of those benefits for just a little more control over operations; whether it can get that kind of incremental change out of treaty re-negotiations is uncertain.

This discussion of generation in the B.C. context ignores for now whether British Columbia may desire more control over project operations to serve purposes other than generation optimization. It seems obvious that the province, in deciding what to do about the treaty in 2014 (if anything), will focus most heavily on comparing the benefits of current project operations against the potential economic benefits, largely from generation, to be gained by securing more control over flows (or as compensation for *not* taking more control). But people in the province and especially in the basin will also desire changes in operations for other purposes, especially to improve recreational and fishery conditions (and any increased economic benefits from both, of course), and these will be important voices to consider, too, if (I believe) less significant to the final outcome than non-power fisheries considerations will be in the United States. There is also the obvious calculation to be made as to what value British Columbia could realize from the United States in return for agreeing to a change in treaty project operations to accommodate U.S. non-power considerations. More on this below, but if taking greater control over the system would allow for a change in operations that accomplished both increased revenue maximization of Canadian generation *and* increased economic activity due to better conditions for recreation and fisheries, then that combination would be very hard for the provincial and national governments to ignore. Whether any change in operations could actually serve both purposes at once is not so easy for me to imagine, perhaps due to my own lack of understanding. But it is just as possible that changes in operations that would optimize Canadian generation for revenue (requiring, for example, increased flow ramping to follow market demands and deeper reservoir drafts in summer), benefits that will accrue (at least at first) to the center of the province, may conflict with operational changes desired by basin residents to enhance recreational and environmental qualities and generate local economic benefits (requiring, for example, higher and more stable reservoir levels and stable outflows).

Downstream Power Benefits

Even if the Entities retain the concept of downstream power benefits upon taking steps either to modify the treaty or change the way it works through a long-term supplemental operating agreement, they will have to reform how the benefits are calculated and the Canadian share determined. There are reasons that it made sense in the 1960s to base the downstream power benefits determination on a system restricted to a 1960s base (to best assure the Treaty storage was valued as the next-added system benefit) and divorce the determination from actual operations (to assure that the United States actually tried to optimize generation with the regulated flows across the border that B.C. sacrificed for). But the method of determination is so divorced from reality now as to seem a weird and formal vestige from a former era. The power system is not just different than what it was in the 1960s; more important, it is vastly different in ways not expected—thermal plant developments have a role in the complicated determinations of downstream power benefits, but wind turbines do not. Does that make sense? And actual dam operations are now so heavily determined by non-power considerations, that to determine power benefits as if an entire project operating purpose (not present in 1960) did not now exist is silly. That does not mean the relative values realized by the countries are wrong or must change, just that the method of determining benefits is unrealistic.

Flood Control and the United States Perspective

Flood control is a touchy, controversial, and emotional subject, as it is a matter of real and perceived risks to life, public safety, and property. The United States receives a real, continuing flood-control benefit from the fact that British Columbia has to make sufficient room in its reservoirs by spring to store that snowpack runoff. Portland is indeed protected by Mica Dam, about 900 miles away. Yet there are some people in the United States— for example, representatives of lower-river tribes calling for improvements in flow conditions for salmon—who argue strongly that the United States may be overprotected vis-à-vis Columbia flood control and could live with a little less protection at little or no additional risk, and gain greater flow benefits in return. The change some have recommended would be not to draft the projects quite so deep in late winter and spring for flood-control purposes, especially in less than average water years. This could be similar to the VARQ flood control operation the Corps has adopted at Libby Dam, which reduces the flood control draft in below average water conditions, leaving less room to store less of the spring runoff, thus improving refill probability and allowing more of the runoff to flow downriver.

Moreover, if climate-change scenarios result in less snowpack and earlier runoff timing, then even if average total runoff doesn't change, even greater changes in flood control operations may need to be made, including flood control operations in late fall and/or early winter months. Knowing when to make the transition will be the trick. The United States conceivably might accommodate even more change in flood control operations through actions either to bolster or build more protective structures, e.g., levees or dikes, in the lower Columbia River *or* non-structural alternatives such as moving more land uses out of the flood plains. Whether much change of this latter type is really available at other than a very high cost is unclear. And at the same time, expected population and economic expansion in the lower river may bring even *more* flood plain development, demanding requiring even *greater* flood control protection.

How much (if any) additional flood risk the people and agencies in the lower river would really be willing to tolerate is debatable. Equally uncertain is whether enough people will agree that some operational changes would bring little additional flood risk, or that the biological benefits of the incremental increases in flow would outweigh the increased risk of flooding of whatever magnitude. And in either event, it would be a long educational campaign to get there. Yet we have less than a decade for people to make any case for reduced flood control.

Also, the relationship of any flood control/flood risk trade-off debate in the United States to actual operations of the treaty projects by British Columbia is tenuous. By this I assume that before decision makers in the United States would voluntarily give up some of the benefits of Canadian project flood control, they would need not only to conclude that the additional flood risk was acceptably minimal, but also that the *United States* would obtain in return very certain high-value benefits from the fact of the reduced flood control operation—that is, that the United States would obtain a very certain, dedicated shift of a percentage of runoff flows to enhance salmon protection in the spring and summer. But if treaty operations go any particular direction from the present regime as far as British Columbia is concerned, it would seem to be in the direction of *more* control of flows by British Columbia for its own purposes, and *less* certain control and flows for the United States. I return to this theme below.

On the other hand, assuming the United States wants to preserve the flood control benefits from the Canadian projects as they are now—and it seems likely that it will—it is important to remember that under the treaty the United States will no longer receive these benefits automatically, as of 2024. I suspect there will be ways of using the replacement called-upon flood control provision to produce the same or similar systematic usual-and-accustomed flood control operations. On the other hand, compensation that better matches the actual costs in Canada of flood control would be due from the United States, and who knows how expensive that will be? Whether British Columbia is able to use the new dynamic to increase significantly the amount of compensation it receives from the United States for continued flood control, and what the United States does in response to significantly increased costs for Canadian project flood control (if they do indeed increase), become new elements to juggle. And where in the United States will the dollars for called-upon flood control come from? The automatic termination of the basic flood control provisions in 2024 will do more by itself than any other dynamic to stimulate negotiations toward a new treaty framework of some sort—at the very least, a long-term supplemental agreement on how to conduct the flood control operations.

Flood Control and British Columbia Perspective

British Columbia does receive minor flood control benefits from the flood control operation, especially in the Trail and Castlegar areas. This seems, comparatively, a minor benefit in the larger scheme, and will be a minor consideration in whether and how to change the treaty or project operations. Even if changes happen to Treaty operations for other reasons, these changes are unlikely to affect flood protection in British Columbia, and even if it does in minor ways, the alternative of paying to move structures and uses away from the flood plain may be acceptable, especially if the costs are borne out of greater benefits received.

But from a British Columbia perspective, why change the flood control operations? Again, changing treaty project operations in a way that provides less flood control would really mean one kind of change—not drafting the projects so deep in winter or early spring, leaving less room for storage, gaining in return higher runoff flows as less is needed for project refill. It is not clear to what end the Canadians would want to keep the projects higher before the runoff period began. Some may see an obvious and overriding biological benefit *in British Columbia* from higher *winter* reservoir levels and/or from greater local spring flows. But I am not aware that this dynamic (especially the latter point) is as central in British Columbia environmental debates as it is in the United States. Controlling winter and spring outflows at Keenleyside Dam for whitefish and trout is the key issue. In many years it is difficult to refill the treaty projects by early summer, which might be a reason British Columbia might desire to reduce how far it will draft the projects to allow for runoff storage. But as I understand it, this is not a significant issue. And as will be discussed below, BC Hydro continues to have a high incentive to draft the projects after refill in both summer and winter for power production and revenue generation. So, British Columbia may likely conclude that it is better off continuing the flood control operation largely as it is, perhaps trying to negotiate either a bit more flexibility in operations, or more compensation for the value these operations bring to the United States, or both.

Any end to flood control that would come from British Columbia *removing* one or more structures would, of course, present the province with a much different benefit-cost consideration. Yet project removal seems a highly unlikely course in the near future, for many obvious reasons, at least three of which are: (1) The provision in the treaty termination clause continuing the called-upon flood control obligation for the "life of" the facilities would seem to prevent removal just because one side (or both) terminated the rest of the treaty under that provision. Ending the flood control obligation, too, and thus allowing for project removal, would thus require the agreement of both parties and a mutual full-stop to the treaty and the projects. It is hard enough to imagine one nation finding that course of action to be within its interest in the next decade, let alone both nations. (2) British Columbia likely may decide that what it desires most to do with this treaty is change it so as to increase the province's *control* over project operations in order to generate power in such a way as to maximize revenue. Removing the projects would hardly be consistent with this aim. (3) British Columbia likely would bear the cost of any project removal, and that is a price the province is unlikely to pay under the present considerations of various cost and benefit scenarios.

All of this weaving speculation arrives at the point that I find it hard to imagine that the flood control considerations, and the flood control operations, are going to change all that much, either in the next decade or after the date of possible termination, whether one or both of the parties terminates the rest of the treaty, or even if the parties modify the treaty by mutual agreement. I can imagine British Columbia seeking somewhat greater control and flexibility in providing the operation (perhaps in changing the way the draft occurs so as to somewhat reduce whatever risk there may be of missing treaty storage refill and somewhat increase the risk of not leaving enough space to contain runoff; that is, shift the risk to the United States somewhat), or a greater monetary return for the operation, or both. But it seems highly unlikely to me that we will see a change in the flood control operations of more than moderate degree. This might be a failure of imagination, but if so I am not sure it is mine alone. On the other hand, how the money flows (and how much flows) to pay for those benefits will change, and that itself will be of *great* significance in overall cost/value considerations of any suite of changes.

Fish and Wildlife and Other Biological and Environmental Concerns and the United States Perspective

You could write a book about salmon and the Columbia hydrosystem. Come to think of it, many already exist. To reduce the particular issue with regard to the treaty to a few essentials: The building of the big storage projects substantially altered the natural flow of the river, changing the physical and biological characteristics of the river across the year at points all the way from the headwaters to the river's plume into the ocean. People strenuously debate every issue possible concerning the effects of human actions on salmon survival, with the possible exception of outright blockages of large amounts of spawning habitat. This is certainly true for the opinions people hold as to whether and, especially to what *extent*, changes in flow from the development and operation of the storage projects has adverse effects on Columbia salmon and other fish and wildlife. People also differ greatly as to whether particular *changes* in current storage project operations will bring improvements in conditions downstream for salmon and resident fish, and whether those benefits would be worth the decrement to other purposes. But there is no doubt that a large and influential group of people and governmental and nongovernmental entities in the United State are persuaded that the storage projects have seriously deleterious effects on salmon and other fish through their alteration of flows. These people believe that changes are needed that will have real benefits, and that sound public policy ought to seek and obtain those changes and benefits. The Northwest Power Act and Endangered Species Act legal regimes and programs would support this conclusion. Any discussions or renegotiations of the treaty must take into account this fact, most certainly the greatest difference from 1960.

Of course, the treaty projects are not alone in changing the hydrograph and otherwise altering the habitat conditions for salmon and other fish and wildlife through the main stem Columbia system, whatever the extent of the biological consequences. All of the storage and generating projects in the United States (and all of the agencies responsible for managing or regulating those projects) have a hard legal requirement to consider the needs of salmon and other fish and wildlife as functionally equal to the other purposes, needs, and values of law and society. We may differ and debate precisely what the effects are and what to do about them, and economic purposes may still carry more weight than the biological in reality, but still the people in the United States have to address these questions and make decisions that try to involve all the interests and purposes and solve the problems—decisions that are subject to public and judicial review. Something similar can be said for decisions wholly internal to British Columbia, from what I understand of the law there.

A similar requirement should apply to the treaty projects. Even the salmon advocates in the United States may not know now precisely in what way the structures or the operations should change to help protect and restore salmon and other biological characteristics in the lower river. But it is an easier proposition to say, now, that any new or revised international agreement about water management on the Columbia *must* include fish and wildlife and other environmental considerations along with other purposes when decisions are made on long-term plans, annual plans, and in-season operations. It is difficult to imagine a United States negotiating position that would not seek to expand the purposes of the treaty in this way, along with some type of process changes to make sure the revised set of non-power purposes is taken seriously into account. It is similarly unlikely that the U.S. government will be able to accept a continuation of the status quo when the window opens to press for changes. Similarly, as stated above, it is also doubtful the United States could simply accept unilateral treaty termination without strenuously working for something different, if termination would mean, in effect, status-quo purposes and operations that would not integrate environmental and biological considerations to some extent.

While the United States may seek to include some consideration of effects on U.S. fish and wildlife and other environmental qualities into treaty principles and operations, there is no particular reason British Columbia has to agree it will operate its projects for that purpose. Thus to obtain a change of this nature in treaty operations, the new arrangement will also have to deliver benefits to British Columbia, in part in the form of non-power benefits and most likely also in the form of a new infusion of power benefits or direct monetary compensation or both, helped along by perhaps a smattering of internal political pressure and moral suasion. That is, if the United States wants this kind of change in treaty project purposes and operations, it will need to devise an operation that has significant environmental or economic benefits or both for British Columbia (the right spring and summer flow operation perhaps that benefits spawning below the dams and maximizes summer surplus sales revenue), or else be prepared to pay handsomely for an operation that provides those benefits in the United States. Whether the result is a better deal for British Columbia than, say, treaty termination and a British

Columbia operation mostly to optimize power generation revenue, may depend on the price?

Fish and Wildlife and Other Biological and Environmental Concerns and the British Columbia Perspective

For all the importance fish have in the debates south of the border over what to do with the river, it is my opinion that the opposite seems the case north of the border. More precisely, fish habitat and fish—native and non-native, riverine and lacustrine—are not such a major driver in British Columbia debates over the effects of the structures and operations in the Canadian arm of the Columbia. There are locally important issues and important legal and policy protections, to be sure, in places quite significant, as much recreational as biological, as much about local water quality as about broadly important fish species. These issues range from stabilizing reservoir levels and nutrients in the Arrow Lakes behind Keenleyside Dam, in the Kinbasket Reservoir behind Mica Dam, or in Kootenay Lake below Duncan Dam, to stabilizing flows and improving water quality below these projects to support local trout and whitefish and other fisheries. There are also obvious wetland and wildlife habitat issues at play. But none of these issues appears sufficient to be a force for significantly affecting system operations, at least not to the extent to be a major driver in Canadian decisions about what to do about the treaty (Columbia Basin Fish and Wildlife Compensation Program).

To a limited extent, people in British Columbia will likely seek and possibly obtain relatively minor changes in treaty project operations to stabilize and thus improve conditions for fish and wildlife in parts of the system, *if* British Columbia were to assume greater discretionary control of the projects through treaty termination or re-negotiation. I can also see a drive simply to achieve recognition that fish and wildlife and other environmental purposes are a legitimate purpose to consider in planning and decisions on project operations, within a changed treaty or added durably in a supplemental fashion to the current treaty. And finally, there is always the need for funding for mitigation or compensation without disturbing overall project operations, which is how I understand the main thrust of the Columbia Basin Fish and Wildlife Compensation Program.

On the other hand, it is hard to see a result in which British Columbia seeks a *major* change in the operation of the treaty projects—either through treaty termination or through a re-negotiated management agreement—to meet an *internal* fish and wildlife purpose. There is not a systematic critique and political force for this particular dynamic. At the moment, I do not see improving fish and wildlife conditions in the British Columbia portion of the Columbia as a major driver in a British Columbia decision to terminate the treaty or negotiate a new one, and I do not see it as a likely driver in actual project operations by the province in the event of termination. (I'm prepared to be proven wrong on this point.) Even when Columbia fish issues are discussed *in* British Columbia in a systematic or broad way, the discussion is almost always about *U.S.* fish issues—e.g., what to do about U.S. desires to alter Treaty project operations (in a systematic way or in particular years) to benefit salmon in the lower river or resident fish in U.S. storage reservoirs, or what to do about the deleterious effects of Libby operations for sturgeon or salmon on Kootenay farm lands just downstream of Libby, or what to do about ocean harvest of reduced Columbia salmon.

Instead, people in the Canadian portion of the Columbia basin have focused more on the social and economic and related environmental effects of the projects, and the desire to see greater economic return in compensation for those adverse effects. Clearly the people would not want to see these benefits come at the expense of further environmental degradation; localized programs and funding for environmental improvement will also be part of the equation. As well, there are people in British Columbia with a stronger position on these issues than I give credit for here. Especially present are people and entities that envision a future without the treaty projects, for social and economic as much as environmental reasons. But I don't see a substantial drive by people in British Columbia to have the treaty include an "equitable treatment" objective for Canadian fish and wildlife with the other purposes of the projects. That seems less likely to be a major issue in provincial decisions on the treaty as compared to a demand that the treaty provide "equitable treatment" for the people in the basin with the people in the western part of the province.

One way this would change, of course, is if Columbia salmon became Canadian, that is, if the idea of allowing salmon back into the Canadian Columbia Basin were to move from the conceptual periphery toward the center. Grand Coulee and Chief Joseph dams in the United States block salmon and steelhead from making their way up the Columbia main stem into Canada to spawn as they once did. The salmon fishery never had quite the importance in Canada as in the lower river— and certainly was not recognized outside of the basin elsewhere in Canada as of much value—but it did have significance, especially locally and especially among the First Nations, and would again if the fish were to return, probably more so. There are people in the United States and in British Columbia who call for salmon passage above Chief Joseph and Grand Coulee. As one response, the Council included a measure in its Fish and Wildlife Program asking for an evaluation of the idea's feasibility. But feasibility is more than suspect at the moment—especially the difficulty of safely moving juveniles downstream through the reservoir and dam—and funds and political will even more so. This is an idea definitely at the periphery of Columbia salmon and river issues at the moment. Grand Coulee and Chief Joseph are also not part of the treaty. Even if British Columbia decided that the return of salmon was something it was so interested in that it became a major issue between the two countries, Canada may have to make it part of the bargain for something the United States wanted from Canada with regard to treaty project operations. It is hard to see what the trade-off of value is here, unless perhaps a significantly revised regime of additional and cooler summer flows.

Control and Autonomy and the British Columbia Perspective

As noted above, I assume, from simple reflection and from comments I have heard from a number of people in British Columbia, that the biggest defect British Columbia sees (or at least that many British Columbians see) in the present treaty is the fact that British Columbia has little control over the operation of large storage reservoirs in its own province. If true, the most desired change would *not* be to some other determined operation, but instead would be to gain a significant measure of control over its own projects—their operation, configuration and even their existence. Whether that control, with all of its continuing limitations, will be worth the price of giving up the downstream power benefits will be one of the singular calculations going into 2014. And with that will be a second calculation to the effect of whether British Columbia places a value on surrendering full control over the projects yet again.

Mica Dam and Non-Treaty Storage

What is called non-Treaty storage behind Mica Dam is another complicating factor, useful again for illustrating a number of themes already brought out. A simplified version of the story begins, as noted above, with the fact that British Columbia built Mica to store far more water than required of Canada in the treaty—5 maf more of active storage capacity. The United States did not pay for this additional storage space; British Columbia did. So BC Hydro has control over this additional storage, except that storage and release of this water is always subordinate to and conditioned on noninterference with treaty operations. For example, the way to fill the non-Treaty storage, if less than full, generally has to be by agreement to a deviation from strict treaty operations to allow for that additional storage.

In the early 1980s, when BC Hydro completed the Revelstoke project downstream of Mica, BC Hydro needed a variance from treaty operations to allow Revelstoke to fill. Part of the deal struck gave Bonneville rights to half of the non-Treaty storage available behind Mica, in what was called the 1984 Non-Treaty Storage Agreement (NTSA). The parties renewed the NTSA in 1991 to expire at the end of June 2003, later extended for a year. Although Bonneville scheduled much of this water for additional generation, use of the NTSA water was not officially limited to power generation in the way treaty storage is. So, the U.S. share has also been available for non-power purposes such as salmon flows in the lower river, without the need for compensation. What this meant in certain years was an agreement with BC Hydro to allow storage in the non-Treaty space in May and June for release in July and August to enhance summer flows in the lower river to improve conditions for salmon migration.

The NTSA expired in June 2004. This time Bonneville and BC Hydro could not negotiate a new NTSA agreement or even agree to extend the old, disagreeing on a host of issues about value, power, control, and more. So, in theory BC Hydro now has full control over this additional storage space, subject (again) to non-interference with treaty operations and to some limitations hanging over from the NTSA: The NTSA space is not full, but the otherwise expired NTSA required the parties to refill the non-treaty storage by 2011. The parties have been working toward filling that storage in bits and pieces every year as they find the opportunity. Refill requires deviations from storage regulations and often energy deliveries by one or both parties to compensate for reduced generation at projects while the non-Treaty storage refills. It will be interesting to see if Bonneville and BC Hydro are ever able to resume negotiations and come to an agreement to share this capacity.

Without the NTSA, the United States has obviously lost access to its half of this additional storage, whether for commercial or biological purposes. This has become an issue in litigation in the federal district court of Oregon over how the federal government operates the Federal Columbia River Power System and the effect of those operations on salmon. Plaintiff environmental and fishing groups and certain of the tribes have asked a federal judge to pressure or perhaps even order the federal government to make greater efforts to re-secure access to the non-Treaty storage and dedicate that water to flows to support salmon migration in summer. (They also question whether the federal agencies could do more to gain access to additional Treaty storage for fish flows through supplemental agreements.) Yet the judge cannot order the Canadians to agree to this or the United States to ensure its delivery—control is in the hands of British Columbia. If the two nations are not able to come to some sort of new agreement over use of the non-Treaty storage in the next few years, this storage will simply become another of many factors for Canada and the United States to talk about at the time of potential treaty termination, when trying to decide what items within the overall scope of Columbia system operations should be subject to an international agreement.

What Is the Vehicle for Integrating Twenty-first Century Non-power and Power Considerations?

As noted above, there are reasons to consider integrating the twenty-first century power and non-power context into the treaty not through a modified treaty but through the Entities developing a long-term, comprehensive, supplemental operating agreement that addresses all relevant issues of operations, purposes, costs, and benefits. If this can work, fine with me. It might avoid formally involving the foreign ministries, the treaty ratification process, and other complicating and ponderous elements of the national governments and national politics. It might help keep control over these developments in the region, where we know more about the issues, the needs, and the desires. It might allow us to better preserve those aspects of current treaty operations that work well.

I and others can also imagine a number of reasons it might not work, or might work but in the end saddle us with entanglements or difficulties, or might work but not be an improvement over actually changing the treaty and embedding these modern considerations directly into the international framework. I am also concerned that while such an approach might serve to integrate substantive purposes into annual planning and day-to-day operations, it is less likely to be useful for reforming treaty governance and opening up treaty considerations and procedures to regional and public involvement, if those interests are also to be served.

Public Process

There are many people in the region in both countries who would like to open up the Treaty box—I know, because I hear from them. I would, too. We would like to see some level of regional public review and input into two different aspects of the treaty: First, into the process for negotiating and deciding on the future of the treaty, and second, into the processes for treaty implementation itself, into long-term and annual planning at the least and into in-season management if possible. Public participation is something we have all grown familiar with over the last thirty years with regard to system operations on either side of the border, especially here in the United States. The fact that this public engagement completely goes away just because another level of the same system operations crosses an international border seems no longer acceptable. The Entities and their respective agencies have promised, going into the "2014/2024 Review," to allow public discussion and input into at least the consideration of what to do with the treaty. Public processes are messy and complicated and frequently ill-informed and fed by emotion. But public access and input into agency action is just as much a part of the modern democratic state as voting, and possibly more important for citizens to feel they have a stake in government policy that affects their lives. And the collective public wisdom can be insightful, astonishing, and at least chastening.

I do not have any particular method in mind right now for how to open the treaty up to the public eye, just ideas. For one, we have a number of modern international agreements to study as models (good and bad) in the integration of public involvement into international agreements, from the Pacific Salmon Commission to the WTO to the European Union to a host of multi-nation agreements regarding rivers, such as the Danube, Mekong, Nile, and Paraná/La Plata. Key avenues to pursue will be to create both a larger public process for review and input and some sort of standing "Interests" section or input group made up of representatives from key regional governmental and possibly nongovernmental entities to advise the agencies. I suspect it will not be necessary to change the basic operating committee structure and implementation methods and procedures. This will remain a nation-to-nation agreement and implementation, and the national-level agencies will do the active work. The question on each side of the border will be how to provide opportunities for input to the agencies as they revise and implement this treaty,

I do know one way this issue will manifest itself—the people of the Columbia River Basin in British Columbia are determined to be heard as to the future of the treaty. The staff and board members of the Columbia Basin Trust have begun floating the idea of establishing an ongoing, cross-border forum as a home for a public discussion of the treaty. Their focus right now is on education and mutual understanding. But clearly the hope is to evolve the forum into a broad public platform for airing policy issues involving the re-negotiation of the treaty. More power to them.

Uncertainty and Flexibility

Finally, I suggest that a little less certainty might be best for us all. One lesson we ought to learn from all efforts at grand, comprehensive public policy—and the treaty provides a particularly good example of this lesson—is that no matter how deeply we think we are looking into our basic natures, the grand schemes are still contingent snapshots in time. We should realize that we have no idea precisely what we will want in the future or what the context will allow or need or demand, whether ten, twenty, or forty years away. With that understanding, we are usually best off with a set of important but general principles and a flexible set of tools. Or, more precisely, we will be better off *if* being able to accommodate the future without needing a new framework is one of the goals of the people making the decision (it isn't always). We must find a way to bring the public's evolving and uncertain interests onto the river, or we will once again find ourselves in the present dilemma of a narrowly determined legal framework and a world that has moved on.

Notes

1. Two significant caveats: First, this is truly a continuing work in progress. Much of the paper is intentionally in outline form to reinforce that point. I assume there are things in here that are not entirely accurate, and no one is responsible for that other than me.

The second caveat is the usual, but still needs to be said and taken seriously: I am the General Counsel for the Northwest Power and Conservation Council, an interstate agency based in Portland that prepares energy and fish and wildlife programs and plans for the

Columbia basin. But, this is an exercise I have undertaken outside of my work. Nothing in here reflects the views or positions of the Council or of individual Council members or the staff. Nor should any of it be attributed to me in my official capacity as legal adviser to the Council.

Two people I particularly need to thank for the time and insights they have given, past and present, are Kindy Gosal of the Columbia Basin Trust and, especially, John Harrison of the Council staff. John has now read and commented deeply on this paper three times, far beyond the call of duty or friendship. I must also acknowledge the knowledge I have gained from and the debt I owe to John Hyde, Tony White, Tim Newton, Kelvin Ketchum and Nigel Bankes.

2. Look at the name closely and you can figure out its origin.

Works Cited

Bankes, Nigel. "The Columbia River Treaty: Responding to Changing Norms." 29 *Curso de Derecho International*, 271–350, Comité Juridico interamericano, 2003

———. "The Columbia River: Multiple Actors in Canada-US Relations." *Presentation to the Canada/U.S. Law Institute.* Cleveland, OH, April 2004.

———. *The Columbia Basin and The Columbia River Treaty: Canadian Perspectives in the 1990s.* Research Publication, Northwest Water Law and Policy Project, 1996.

Barton, Jim. "US Flood Control and Operational Perspective." *Presentation to the Canadian Columbia River Forum.* Vancouver, BC, November 2005.

BC Hydro. *Making the Connection: The BC Hydro Electric System and How it Is Operated.* 1993.

Blumm, Michael. "The Northwest's Hydroelectric Heritage: Prologue to the Pacific Northwest Electric Power Planning and Conservation Act." *Washington Law Review* 55, no. 175 (1982).

Bonneville Power Administration. "Columbia River Treaty Documents." 1979.

Bonneville Power Administration and BC Hydro. "Columbia River Non-Storage Agreement." 1990. Available at *www.bpa.gov/corporate/ntsa/documents/NTSA2011Presentation.pdf*

Bonneville Power Adminstration and U.S. Army Corps of Engineers. "Columbia River Treaty: History and 2014/2024 Review." April 2008. Available at http://www.bpa.gov/corporate/pubs/Columbia_River_Treaty_Review_-_Feb_2009.pdf.

Bonneville Power Administration and United States Entity, Columbia River Treaty. "Pacific Northwest Coordination Agreement." Record of Decision, 1997.

Bonneville Power Administration and U.S. Army Corps of Engineers. "BiOp & Fish Accord Related Briefing on Canadian Storage Operations and Planning." Portland, OR and Spokane, WA, November and December 2008.

Bonneville Power Administration and U.S. Army Corps of Engineers. Columbia River Treaty. Web site. http://www.nwd-wc.usace.army.mil/PB/PEB_08/index.htm.

Bonneville Power Administration and U.S. Army Corps of Engineers. "Columbia River Treaty U.S. Entity Briefing for 2008 BiOp & Fish Accords on Treaty/Non-Treaty Operations and Treaty Operating Plans." *Presentation by Bonneville Power and US Army Corps of Engineers.*

Canada and United States Entities. "Annual Report of the Columbia River Treaty, Canada an United States Entities for the 2002 Water Year, 1 October 2001 through 30 September 2002." 2002 and other years.

Canadian Departments of External Affairs and Northern Affairs and National Resources. "The Columbia River Treaty, Protocol and Related Documents." 1964.

Columbia Basin Fish and Wildlife Compensation Program. Web site at http://www. cbfishwildlife.org/.

Columbia Basin Trust. "Columbia Basin Management Plan." 1997. Web site at www.cbt. org

Columbia River Center of Information. Web site at http://gis.test.bpa.gov/portal/ptk?com mmand=openchannel&channel=29.

Columbia River Treaty Entities. "Annual Report of the Columbia River Treaty, Canada and United States Entities for the 2002 Water Year, 1 October 2006 through 30 September 2007." Annual Report, 2007.

Columbia River Treaty Entities. "Annual Report of the Columbia River Treaty, Canada and United States Entities for the 2002 Water Year, 1 October 2007 through 30 September 2008." Annual Report, 2008.

Columbia River Treaty Entities. "Columbia River Treaty Entity Agreement on Aspects of the Delivery of the Canadian Entitlement for April 1, 1998 through September 15, 2004." 1999.

Columbia River Treaty Entities. "Columbia River Treaty Determination of Downstream Power Benefits for the Assured Operating Plan for Operating Year 2009–10." November 2004 and other years.

Columbia River Treaty Entities. "Columbia River Treaty Entity Agreement Coordinating the Operation of the Libby Project with the Operation of the Hydroelectric Plants on the Kootenay River and Elsewhere in Canada." February 16, 2000.

Columbia River Treaty Entities. "Columbia River Treaty Entity Agreement on the Principles and Procedures for Preparing and Implementing Hydroelectric Operating Plans for Operation of Canadian Treaty Storage." December 2003.

Columbia River Treaty Entities. "Columbia River Treaty Hydroelectric Operating Plan: Assured for Operating Plan for Operating Year 2009–10." November 2004 and other years.

Columbia River Treaty Entities. "Detailed Operating Plan for Columbia River Treaty Storage 1 August 2004 through 31 July 2005." June 2004 and other years.

Columbia River Treaty Entities. "Detailed Operating Plan for Columbia River Treaty Storage 1 August 2008 through 31 July 2009." June 2008.

Columbia River Treaty Entities. "Detailed Operating Plan for Columbia River Treaty Storage 1 August 2009 through 31 July 2010." July 2009.

Columbia River Treaty Entities. "Detailed Operating Plan for Columbia River Treaty Storage 1 August 2009 through 31 July 2010." July 2009.

Columbia River Treaty Operating Committee. "Agreement on Operation of Canadian Treaty and Libby Storage Reservoirs for the period 2 August 2008 through 31 December 2008." August 2008.

Columbia River Treaty Operating Committee. "Agreement on Operation of Treaty Storage for Nonpower Uses for 1 January through 31 July 2005." December 2004.

Columbia River Treaty Operating Committee. "Agreement on Operation of Treaty Storage for Nonpower Uses for 15 December 2008 through 31 July 2009." November 2008.

Columbia River Treaty Operating Committee. "Agreement on Provisional Storage for the Period 1 September 2008 through 3 April 2009." September 2008.

Columbia River Treaty Operating Committee. "Columbia River Treaty Principles and Procedures for Preparation and Use of Hydroelectric Operating Plans for Canadian Treaty Storage." 1991.

Columbia River Treaty Permanent Engineering Board. "Annual Report to the Governments of the United States and Canada." Various Years during the 1970s, 1986–99.

Downstream Benefits Committee. *Columbia Report #1: The Columbia River Treaty.* Victoria, B.C.: Ministry of Energy, Mines and Petroleum Resources, 1993.

Downstream Benefits Committee. *Columbia Report #2: The Canadian Entitlement.* Victoria, BC: Ministry of Energy, Mines and Petroleum Resources, 1993.

Downstream Benefits Committee. *Columbia Report #3: Reservoir Operations.* Victoria, BC: Ministry of Energy, Mines and Petroleum Resources, 1993.

Downstream Benefits Committee. *Columbia Report #4: Alternatives for the Canadian Entitlement.* Victoria, BC: The Ministry of Energy, Mines and Petroleum Resources, 1993.

Engineering Board to International Joint Commission. "Water Resources of the Columbia River Basin." Report of the International Columbia River Engineering Board to the International Joint Comission, 1959.

Goldschmidt, Robert M. "Treaty Implications of Dissolved Gas Management in the Columbia River Basin." Report for the British Columbia Ministry of Water, Land and Air Protection and The Columbia River Transboundary Gas Group, 2001.

Hamlet, Alan. "The Role of Transboundary Agreements in the Columbia River Basin: An Integrated Assessment in the Context of Historic Development, Climate, and Evolving Water Policy." In *Transboundary Challenges in the Americas,* edited by H Diaz and B Morehouse. Spring, 2003.

Hyde, John, Kelvin Ketchum, and Bolyvong Tanovan. "Ecosystem Management and the Columbia River Treaty." *Toward Ecosystem-Based Management: Breaking Down the Barriers in the Columbia River Basin and Beyond.* Spokane, WA, April 2002.

International Joint Commission. "In the Matter of the Application of the Government of the United States for Approval of the Construction and Operation of the Grand Coulee Dam and Reservoir." Order of Approval, December 15, 1941.

International Joint Commission. "Report on Principles for Determining and Apportioning Benefits for Cooperative Use and Storage of Waters and Electrical Interconnection within the Columbia River System." 1959.

Jay, David, and Pradeep Naik. "Historical Changes in Columbia River Hydrology and Sediment Transport." *Toward Ecosystem-Based Management: Breaking Down the Barriers in the Columbia River Basin and Beyond.* Spokane, WA, April 2002.

Krutilla, John. *The Columbia River Treaty: The Economics of an International River Basin Development.* Resources for the Future, 1967.

Martin, Charles F. "International Water Problems in the West: The Columbia Basin Treaty Between Canada and the United States." In *Canada–United States Treaty Relations,* edited by David R Deener. Cambridge University Press, 1963.

Northwest Power and Conservation Council. *Columbia River Treaty.* Web site at http://www.nwcouncil.org/history/ColumbiaRiverTreaty.asp.

Norwood, Gus. *Columbia River Power for the People: A History of Policies of Bonneville Power Administration.* Portland, OR: Bonneville Power Administration, 1981.

Oliver, Stephen. "Canadian Columbia River Basin Forum." *Presentation to the Canadian Columbia River Forum.* Vancouver, BC, November 2005.

Oliver, Stephen (Bonneville Power Administration) and Witt Anderson (US Army Corps of Engineers), Treaty Coordinators, US Entity. "2014/2024 Review: Columbia River Treaty." *Presentation to the Northwest Power and Conservation Council.* Portland, OR, December 2008.

Pendergrass, Rick. "BPA Power and Operations Planning for the FCRPS." Bonneville Power Administration, March 11, 2004.

Shurts, John. "The Columbia River, Regional Government Entities, and Public Participation: The Northwest Power and Conservation Council and the Columbia Basin Trust." *Presentation to the Vietnam–Oregon Study Exchange, Oregon–Portland State University Allliance, Ho Chi Minh Canal and Watershed Restoration Project.* Ho Chi Minh City, Vietnam, June 2004.

Swainson, Neil. *Conflict Over the Columbia: The Canadian Background to an Historic Treaty.* Institute of Public Administration of Canada, 1980.

Swainson, Neil. "The Columbia River Treaty: Where Do We Go From Here?" *Natural Resources Journal* 26, no. 243 (1986).

"The Columbia River Treaty." *Summary Presentation to Canadian Columbia River Intertribal Fisheries Commission Compensation Workshop.* Vernon, BC, October 8, 1993.

U.S. Army Corps of Engineers, Northwestern Division. Water Management Division. *Columbia River Treaty Permanent Engineering Board Website.* http://www.nwd-wc.usace.army.mil/PB/PEB_08/index.htm.

U.S. Army Corps of Engineers and Bonneville Power Administration. "Columbia River Treaty 2014/2024 Review: Phase 1 Technical Studies." April 2009.

U.S. Corps of Engineers, Bureau of Reclamation, and Bonneville Power Administration. "Columbia River System Operation Review, Final Environmental Impact Statement." Final Environmental Impact Statement, 1995.

Vogel, Eve. "Regionalization and Democratization through International Law: Intertwined Jurisdictions, Scales and Politics in the Columbia River Treaty." *Oregon Review of International Law* 9, no. 337 (2007).

Volkman, John. "A River in Common: The Columbia River, The Salmon Ecosystem, and Water Policy." Report to the Western Water Policy Review Advisory Commission, 1997.

William, Sewell. "The Columbia River Treaty and Protocol Agreement." *Natural Resources Journal* 4, no. 309 (1964).

World Commission on Dams. "Dams and Development: A New Framework for Decision-Making." Report, 2000.

The Columbia River Treaty after 2024

Chris W. Sanderson

Introduction

The primary focus of this volume is on the challenges and opportunities surrounding the possible continuation, termination, and/or amendment to the Columbia River Treaty after 2024. Any discussion of what should happen post treaty must be based on a clear understanding of what has happened under the treaty. This chapter will argue that the current treaty served the objectives of its parties admirably and that that success should be recognized as a starting point when considering how best to utilize the resources of the Columbia River going forward. This chapter makes two important assumptions: First, that the primary measure of the treaty's success should be determined by the extent to which it met the ambitions of Canada and the United States, the two sponsoring sovereign states. Second, that bilateralism between Canada and the United States remains sufficiently robust that each country will want to cooperatively manage the Columbia River's water flow in the future to maximize certain objectives, each defined on a sovereign basis, whether or not those objectives continue to be confined to power production and flood control.

Finally, it is important to note that these observations are those of the author alone and do not necessarily reflect the views of nor are based on input from any official participant or stakeholder with the treaty. The value of the observations, if any, stem from the author's past involvement with the treaty and not from any engagement or responsibility he currently has in connection with the treaty.

★★★

In February of 1960, Canada and the United States commenced direct negotiations for the cooperative development and management of the Columbia River. Less than a year later, the Columbia River Treaty was signed. Despite this apparent expedition, the treaty, which was ratified by Canada in 1964, was the culmination of fifteen years of cooperative study and discussion between Canada and the United States.

In its simplest terms, commitments made by Canada pursuant to the treaty enabled the United States to undertake projects on the Columbia River to increase its overall capability for the production of hydroelectric power and to provide the Pacific Northwest with much needed flood protection. In exchange for its commitments, Canada received a lump sum payment for flood control and an entitlement to one half of the extra power benefits produced downstream as a result of the cooperative use of Columbia River water resources.

The treaty serves well as a model for international cooperation. Its underlying principles provide for the equitable utilization of the river's resources, while also providing for the flexibility needed to facilitate desirable project development.

This chapter will outline the historical framework of the treaty, identify some of the issues that have been addressed by Canada and the United States within the treaty context, and highlight the significant aspects of the treaty that make it a model for international cooperation. My thesis is that the existing treaty offers some valuable lessons for structuring the post-2024 arrangement. One is that the treaty permits each nation to cooperate with the other without compromising their sovereignty. Preservation of this feature will be key to successfully addressing the many challenges that face negotiators as they approach 2024 because, as is explained below, it will impose a burden on each of the treaty nations (the parties) to resolve underlying issues of domestic resource allocation in domestic forums. International discussions between the parties can then be left to focus on international issues.

Historical Framework of the Columbia River Treaty
Physical Characteristics of the Columbia River

To appreciate the enormity of the cooperative development facilitated by the treaty, it is useful to have a mental picture of the river itself. The Columbia River and its tributaries drain an area of approximately 259,000 square miles, mostly between the Rocky Mountains and the Cascade Range. The river basin (the main stem of the Columbia River and its major international tributaries) extends 270 miles north into Canada and 550 miles south into the United States. The maximum width is approximately 730 miles. The Canadian portion of the river basin, 39,500 square miles, is situated in the south easterly part of British Columbia. The United States portion of the basin, 219,500 square miles, includes most of Idaho, Oregon, and Washington, all of Montana west of the Continental Divide, and small areas of Nevada, Utah, and Wyoming.

The Columbia River rises in Columbia Lake in the Rocky Mountain Trench and flows a distance of 480 miles in British Columbia before it crosses the Canada–United States border into the northeast corner of the state of Washington. In the United States, the river flows southerly through the central part of Washington to its junction with the Snake River. It then turns and flows westerly and northwesterly to the Pacific Ocean, a total distance of 1,225 miles from its source. The total fall of the Columbia River from its source to the ocean is 2,655 feet (International Columbia River Engineering Board February 1964).

Factors Necessitating Regulation and Management

Regulation and management of the Columbia River have been discussed since 1944 when the Canadian and United States governments first began to consider whether development of the water resources on the Canadian side of the border would be worthwhile (Downstream Benefits Steering Committee 1993). A number of factors served as incentives to work towards cooperative development of the Columbia River. They included: (1) rapidly expanding population and industrial growth within the Pacific Northwest of the United States which created a need for large blocks of electrical power (International Columbia River Engineering Board February 1964, 23); (2) periodic and sometimes devastating flooding in the same area (Downstream Benefits Steering Committee 1993, 1); and (3) a growing need for hydroelectric power in British Columbia.

To appreciate the need for flood control in the Pacific Northwest of the United States, it is important to realize that, in the late 1950s, the Columbia River's average volume of runoff was exceeded only by the Mississippi, Mackenzie, and St. Lawrence rivers. In June of 1894, rapid melting of an above-normal snow pack resulted in severe flooding. That year, the maximum discharge of the Columbia River at the Canada/United States border was estimated at 680,000 cubic feet per second, a truly astounding volume of water. Given the growth in economic, industrial, and residential development alongside the river by the late 1950s, the potential for devastating floods, similar to the one in 1894, was of increasing concern (International Columbia River Engineering Board February 1964, 21, 23, 25).

Process Leading up to the Treaty

In 1909, Great Britain and the United States entered into the Boundary Waters Treaty. Among other things, the purpose of that treaty was to prevent disputes between the two nations regarding the use of boundary waters, including all lakes, rivers, and connecting waterways along the Canada/United States border (Boundary Waters Treaty 1909). The treaty stipulated that except as otherwise permitted by it, or pursuant to special agreement between its signatories, neither party could make use of, obstruct or divert boundary waters in a manner which would affect the natural level or flow of those waters on the other side of the border, unless it first had the approval of the International Joint Commission (Boundary Waters Treaty 1909, Article III). The International Joint Commission (IJC) was an entity formed under the Boundary Waters Treaty, and was granted jurisdiction to decide all cases involving the use, obstruction, or diversion of boundary waters which were specified to require its approval, as well as any other matter referred to it by Canada and/or the United States which raised issues relevant to rights, obligations, or interests in respect of their "common frontier" (Boundary Waters Treaty 1909, Article IX).

In March of 1944, Canada and the United States asked the IJC to look into a number of issues concerning the Columbia River, including: (1) whether further development of the water resources of the Columbia River basin would be practicable and in the public interest from the points of view of the two governments; (2) how the interests on either side of the border would be benefitted or adversely affected by that development; (3) what the cost of the necessary projects would be; and (4) how that cost should be apportioned between the two countries (Departments of External Affairs and Northern Affairs and Natural Resources 1964).

Pursuant to that request, the IJC established an International Columbia River Engineering Board (Engineering Board) to carry out water management studies on the Columbia River. In March of 1959, the Engineering Board presented a report to the IJC which set out alternative plans for water resources development along the Columbia River (International Columbia River Engineering Board February 1964, 19).

The work of the Engineering Board can usefully be considered in two parts. First the Board had to identify what benefits could in theory result from international cooperation that would not be available to either party acting unilaterally. Second, the Board had to apportion the benefits and costs of any cooperative development between the parties.

The Board's conclusions with respect to the first issue are of particular interest. It determined that: "(a) Further development of the water resources of the Columbia River basin is practicable and in the public interest from the point of view of the two Governments. .... (e) The largest and most valuable benefit to be obtained from water resources developments in the Columbia River basin is the production of hydro-electric power. Further, power benefits in both countries can be materially increased by cooperative development and operation of storage and power projects to conform to a plan of basin development" (International Columbia River Engineering Board February 1964, 36–37).

It is clear that the Engineering Board was identifying those characteristics of the Columbia River system that benefitted most from cooperation between Canada and the United States, not which characteristics were of most value. The board clearly understood that the judgments required for valuation were outside its responsibility and it could only help by providing objective information necessary to make those judgments.

Among other things, the Engineering Board concluded that further development of the Columbia River was both practicable and in the public interest. Specifically, as it relates to the fields of water power, flood control, and irrigation, the Engineering Board concluded that greater use of the Columbia River would be made possible by cooperative development of water resources in each of the United States and Canada. It also concluded that the largest and most valuable benefit arising out of cooperative development of Columbia River water resources would be the production of hydroelectric power (International Columbia River Engineering Board February 1964, 36, 37).

It was recognized by the Engineering Board that a key requirement for the development of Columbia River water resources would be the provision of upstream storage in Canada (International Columbia River Engineering Board February 1964, 26). In terms of the production of hydroelectric power, the development and regulation of water storage in Canada would enable the parties to regulate water flow and this would allow both a greater amount of useable energy and a higher level of dependable capacity to be generated at American power plants (Downstream Benefits Steering Committee 1993, 2).

In December of 1959, and again at the request of the parties, the IJC developed a set of principles intended to govern any sharing of benefits between Canada and the United States which might arise as a result of joint development of the Columbia River. To prepare for its task, the IJC reviewed all information available to it on the water resources development needs and possibilities in the Columbia River basin (International Joint Commission 1964, 39). This included the report of the Engineering Board referred to above.

The IJC's work built on the work of the Engineering Board with regard to the first issue and then considered its implications with regard to the second issue—that is, how should costs and benefits be allocated between the parties? Of particular significance, is the IJC's observation:

> *Although other benefits would also be realized from such cooperative use, the outlook at this time is that their value would be so small in comparison to the power and flood control values that formulation of principles for their determination and apportionment would not be warranted. This is not intended to preclude consideration by the two Governments of any benefits, tangible or intangible, which may prove to be significant in the selection of projects or formulation of agreements thereon* (International Joint Commission 1964, 40).

This passage suggests that the IJC was of the view that its job was to advise the parties of where opportunities for intentional cooperation might exist. Each sovereign nation was then expected to determine which opportunities to realize. Only then could the IJC provide advice or begin to develop principles for sharing those benefits.

In making its recommendations with respect to allocation between the parties, the IJC was guided by the basic concept that the principles it formulated should result in an equitable sharing of the benefits attributable to any cooperative undertakings and an advantage to each country, as compared with any alternatives that might be available to it. In total, the commission formulated sixteen principles, thirteen of which applied specifically to questions of power production and flood control. Power Principle No. 6 provided an illustration of the approach. That principle reiterated and elaborated the basic concept outlined above. It stipulated that power benefits should be shared on a substantially equal basis, provided that an equal split of benefits would result in an advantage to each country as compared with alternatives available to it. When an equal split of benefits would not result in an advantage to the two countries, the countries would then have to negotiate and agree upon such other division of benefits as would be equitable to both countries so as to make cooperative development feasible. Moreover, Power Principle No. 6 assumed that each country would bear all capital and operating costs for facilities it would provide in its own territory to carry out the cooperative development (International Joint Commission 1964, 40, 49–50).

In developing its power principles, the IJC recognized the need to incorporate flexibility as a means of accommodating changing conditions. In a general way, consideration was also given to the various practical problems that might arise in the implementation of the IJC principles as a means of ensuring their workability. Although no attempt was made to specifically spell out how the detailed procedures would be worked out, the IJC recognized that the formalities of treaty-making—administrative and legislative actions in each country—may be necessary to work out the procedural details (International Joint Commission 1964, 40, 44).

On February 11, 1960, direct negotiations commenced between Canada and the United States in respect of the selection, construction, and cooperative use of specific projects for the production of hydroelectric power along the Columbia River (Downstream Benefits Steering Committee 1993, 2). The principles developed by the IJC necessarily provided a framework for those negotiations and were the key to the two nations reaching an agreement.

This review of the background to the treaty makes clear that a number of issues that are more prominent today were considered in 1960, but for the reasons elaborated by the IJC were not considered determinative at that time. For instance, those involved in the treaty negotiating process did turn their minds to environmental impacts. Development of the Columbia River for the purposes of hydroelectric power production was of course the focus of most of the studies conducted because the Engineering Board had concluded that this was the largest and most valuable benefit to be obtained through cooperative development of Columbia River water resources. It had further concluded that cooperative development leading to the reclamation of wetlands or joint action to improve domestic water supply and sanitation, navigation, or conservation of fish and wildlife would not add material value to what each country could accomplish on its own (International Columbia River Engineering Board February 1964, 37). However, in the process of looking at the power potential of the river basin in Canada, consideration was given to the beneficial or detrimental impact which hydroelectric development might have on other uses of the river and its valleys, including irrigation, agriculture, forestry, mining, manufacturing, fish and wildlife, recreation, and transportation. Specific to the topic of fish and wildlife, it was recognized that fishing would be affected to some extent by any development of the Columbia River for the purposes of power production. It was also recognized that water diversion plans which gave rise to flooding would likely have an effect on wildlife in the area (Department of External Affairs; Department of Northern Affairs and National Resources 1964, 36, 44). These factors played a role in selecting the most desirable development projects along the Columbia River, and their consideration indicates that the decision making necessitated by the desire for cooperative development between Canada and the United States was neither made in a vacuum nor driven by pure economics.

These observations are not intended to suggest that consideration of noneconomic issues received the same weight then as they might today. Clearly, each country's approach to managing water flows in the Columbia River to meet future electricity loads did not have regard to environmental and other nonpower constraints to the same extent as some (perhaps most) might now wish. However, the focus on power production and flood control was not imposed by the treaty, but rather was a considered decision by each country, prior to entering into the treaty and subsequently in administering the system under the treaty. The best way to view the treaty is as a vehicle that has permitted optimization of the river system as a resource through international cooperation after each party had determined the trade-offs it was prepared to make. Moreover, the treaty did not preclude the parties from varying their preferences as they moved forward. It simply did not offer a bilateral forum for them deciding jointly to do that.

British Columbia's Role

British Columbia was well represented and played an important role during the treaty negotiations. There were a number of issues on which the perspective of British Columbia differed from that of Canada. First, Canada had a long-standing policy of being opposed to electricity exports. The government of then Premier Bennett did not support that policy. Second, Premier Bennett wished to meet British Columbia's long-term domestic needs through development of the Peace River system, both because he saw it as the premier power site in the Province, and because he saw the construction activity which would be connected with dams built there as a key component of his desire to open up the northern part of the Province and foster its economic development.

British Columbia was also concerned with the environmental implications of at least some of the potential configurations of storage development on the Columbia River. In particular, the DorrBull River storage, which was favoured by some

Canadian officials for its economic value, would have meant the loss of significant agricultural land in the East Kootenays. Finally, British Columbia was generally worried about the cost of assuming financial responsibility for the Columbia River, contemporaneous with development of the Peace (Swainson 1979, 115–20).

These differences led British Columbia to withhold its support for the treaty at the conclusion of negotiations. British Columbia's concerns were finally dealt with to its satisfaction in 1964, when a protocol for implementation of the treaty was established and the parties negotiated a sale of the first thirty years of Canada's share of downstream power benefits. Before describing these latter two events, it is necessary to summarize the treaty itself.

The Treaty

The Columbia River Treaty (the text of which can be found in the appendix) was signed on January 17, 1961 in Washington, D.C. It consists of twenty-one articles, two annexes, and one statistical table, which provide detailed information on the cooperative operation of the Columbia River system (Downstream Benefits Steering Committee 1993, 4). I will not review the treaty articles in detail. Rather, I will focus on the primary responsibilities and commitments of each of Canada and the United States.

Power Production

Canada's primary responsibility under the treaty was to provide 15.5 million acre feet of upstream storage by building three dams: Duncan in 1968, Arrow (later renamed Keenleyside after B.C. Hydro and Power Authority's first co-chairman) in 1969, and Mica in 1973. These three dams (together with a fourth built by the United States at the Kootenai River in Montana) more than doubled the water-storage capacity of the Columbia River Basin (Downstream Benefits Steering Committee 1993, 4).

In exchange for providing upstream water storage, Canada received payments totalling US$64.4 million, an amount that was paid out when each of the three Canadian dams became operational, plus a bonus payment for the early completion of the Duncan and Arrow facilities (Downstream Benefits Steering Committee 1993, 4). Canada was also granted entitlement to one half of the extra power benefits produced at hydroelectric plants in the United States as a direct result of the operation of the three Canadian storage dams (Downstream Benefits Steering Committee 1993, 5; Columbia River Treaty 1964, Annex B, para. 5). In a press release issued by Prime Minister Diefenbaker upon the signing of the treaty, the ability of Canada to secure this latter benefit was hailed as a "great achievement" of "basic and far-reaching importance" (Diefenbaker 1964, 82).

The United States accepted an obligation under the treaty to maintain and operate all existing hydroelectric plants in the Columbia River Basin and any new projects on the main stem of the river, in a manner that would make the most effective use of the improvement in stream flow which resulted from operation of the three Canadian storage sites (Columbia River Treaty 1964, Art. III(1). However, Article III(2) makes clear that this is a compensation obligation only. The Canadian entitlement (CE) was to be calculated on the assumption that the most effective use of the surplus for power purposes has been employed. How the United States actually operated the system is up to it. This provision ensured that Canada's interest in the power benefits as contemplated by the treaty would not be jeopardized by changed priorities in the United States. Canada was to be compensated as if maximum power benefits were produced.

The United States also undertook to deliver to Canada at a point on the Canada–United States border near Oliver, British Columbia, or at such other place as the designated representatives of the countries might agree, the power benefits to which Canada was entitled (Columbia River Treaty 1964, Art. V(2)).

Flood Control

On the issue of flood control, the treaty obligated Canada to operate its upstream storage sites in a manner that best ensured protection against downstream flooding. Flexibility was a key element.

In addition to the amount of storage specified in the treaty, the United States was empowered to request during the life of the treaty that Canada provide additional storage within the limits of its existing facilities when extra flood control was needed (Columbia River Treaty 1964, Art. IV(2)(b)). Moreover, the obligation of Canada to provide flood control would continue beyond the life of the treaty. Canada has a continuing obligation to operate upstream storage within the limits of its existing facilities for the purposes of flood control when requested to do so by the United States, for so long as the flows in the Columbia River in Canada contribute to potential flood hazard in the United States (Columbia River Treaty 1964, Art. IV(2)(b)).

In exchange for receiving protection against flooding, the United States undertook to compensate Canada financially. Specific amounts were delineated in the treaty. For providing additional storage upon the request of the United States during the life of the treaty, Canada was to be paid US $1,875,000 for each of the first four flood periods in respect of which a request was made (Columbia River Treaty 1964, Art. VI(3)). Further, and in respect of every request for additional storage made during the life of the treaty, the United States agreed to deliver to Canada an amount of hydroelectric power equal to that lost by Canada as a result of operating its storage facilities to meet the demands for the flood control. For the continuing obligation to provide flood control after the life of the treaty, the United States agreed to pay to Canada the operating costs incurred by Canada in providing the flood control, and compensation for any economic loss to Canada which arises directly as a result of Canada foregoing alternative uses of the storage relied upon to provide the flood control (Columbia River Treaty 1964, Art. VI(4)).

As already indicated, the treaty was signed at the White House in January of 1961. Over the course of the following summer, the United States Congress held hearings in connection with the treaty that ultimately led to its ratification. In Canada, the approval process did not go quite as smoothly. For reasons discussed earlier, British Columbia was reluctant to support the treaty. Canada and British Columbia disagreed on two key matters, and debate on these issues as well as federal preoccupation with other issues delayed ratification. Consequently, at least from the Canadian perspective, the treaty effectively remained in limbo until April 1963. That month a new federal government was formed in Canada under the leadership of Prime Minister Pearson. In May, Pearson met with President Kennedy of the United States at Hyannis Port, and presented him with an action plan for the Columbia River which included suggestions for implementing the treaty in a way which met some of Canada's concerns. The plan also acknowledged the possibility of a sale of Canada's share of downstream power benefits within the United States. Upon his return to Canada, Prime Minister Pearson sent a copy of the action plan he had left with President Kennedy to British Columbia, and suggested that the United States would be prepared to discuss the points raised if Canada and British Columbia could first reach agreement on them. That step facilitated further discussion between the federal and provincial governments aimed at resolving their previous differences (Swainson 1979, 252, 253).

The Post-Treaty Process
The Canada-B.C. Agreement
In 1963, an agreement was entered into between Canada and British Columbia which acknowledged, among other things, that all rights to the downstream power benefits accruing to Canada as a result of the operation of Canadian storage; the proceeds of any sale of those benefits in the United States (the treaty expressly allowed for disposal of Canadian power benefits in the United States; (Columbia River Treaty 1964, Art. VII(1)); and any monies payable and electric power accruing to Canada in return for flood control, belonged absolutely to British Columbia for its own use. In return, British Columbia agreed to construct the three Canadian storage sites at its own expense. It also agreed to operate those facilities (Departments of External Affairs and Northern Affairs and National Resources 1964, 100).

The Protocol

As a result of the May 1963 meeting between President Kennedy and Prime Minister Pearson, tripartite discussions between the United States, Canada, and British Columbia began in August of that same year. These discussions ultimately resulted in a protocol to govern implementation of the treaty (dated January 1964). The protocol made a number of improvements to the treaty from Canada's perspective, including: new procedures for the operation of flood control; reaffirmation and clarification of Canada's right to make diversions for consumptive and other uses; confirmation of Canadian control over the operations of its storage facilities; and an increase in Canada's entitlement to downstream power benefits through the acceptance of more advantageous principles of calculation (Department of External Affairs; Department of Northern Affairs and National Resources 1964, 157; Downstream Benefits Steering Committee 1993, 2). In addition to establishing a protocol for implementation, the parties negotiated the terms of a sale of the first thirty years of Canada's share of downstream power benefits.

The Sales Agreements

As discussed above, Canada's share of the downstream power benefits was not needed in British Columbia in the 1960s as B.C. Hydro and Power Authority (BC Hydro) was in the process of completing construction of the W.A.C. Bennett Dam on the Peace River. Pursuant to an exchange of notes between the two countries dated January of 1964, Canada sold its share of downstream power benefits for US $254 million to a group of American electric utilities for a period of thirty years (Downstream Benefits Steering Committee 1993, 2). At the time of the sale, the best immediate market for the Canadian benefits was California. That market was also of interest to Pacific Northwest producers in the United States, with the result that the sale of Canada's power benefits helped justify the construction of new high-voltage transmission facilities, now known as the Southern Intertie, to interconnect the Pacific Northwest of the United States with California (Pub. L. 88-552, August 31, 1964). The United States also enacted legislation which granted Canada's share of power benefits the same right of access to any and all federal transmission facilities in the United States that was accorded to federal energy— energy generated by the United States at government-owned facilities or otherwise acquired by the United States (Northwest Preference Act, 78 Stat. 756, 16 U.S.C. 837837h 1964)).

Under a second agreement entered into between Canada and British Columbia in January of 1964, Canada agreed to provide to British Columbia the purchase monies generated by the sale of Canada's share in the downstream power benefits (Departments of External Affairs and Northern Affairs and National Resources

1964, 107). Those monies in turn assisted British Columbia with the cost of constructing the storage facilities.

Construction and Operation of the Projects

Once the protocol and the sale of Canada's share in the downstream power benefits were completed, construction and operation of the Canadian storage facilities began. Although this chapter will not explore operational issues, it is important to point out that the treaty requires the preparation of a continuing series of detailed operational plans for power generation six years in advance. Those in British Columbia and the United States who are responsible for the planning and operation of the dams must continually work out a number of points, among them, the river's importance to fish and wildlife habitat; to municipal and industrial water supplies; and to recreation, navigation, and irrigation. (Downstream Benefits Steering Committee 1993, 3).

Issues not related to power have proved important in both countries. Building the Duncan, Arrow, and Mica dams was controversial, and management of the Columbia River by BC Hydro remains a sensitive subject for many British Columbians. Initially, the storage reservoirs gave rise to unanticipated flooding in the area. Moreover, fluctuating stream flows and reservoir levels can affect the aesthetics of the countryside. There are also recreational concerns. Low reservoir levels can make boat access difficult and create significant dust problems for local communities (Downstream Benefits Steering Committee 1993, 1).

There is an important point to keep in mind with respect to these latter issues. The hydroelectric needs of British Columbia would have made construction of similar facilities along the Columbia River highly likely. Accordingly, as much as the significance of the initial problems experienced with flooding and the continuing concerns about the impact of low reservoir levels in the immediate area cannot be discounted, these should not preclude recognition of the long-term economic and developmental benefits derived from Canadian participation in the treaty. This observation is particularly important in light of my earlier comments regarding the extent to which environmental and other nonpower issues were considered by those who negotiated the treaty. The environmental effects resulting from Canadian participation in the treaty are very likely the same as those which would have occurred had development on the Canadian side of the border been undertaken for purely domestic purposes. The economic benefits, however, were much greater.

The same may be true in the United States. That is, there remains considerable controversy in the Unites States over the management of treaty storage and the use of the water in the United States, just as there are on most American rivers. However, those controversies would exist with or without treaty storage and flow regimes. When assessing the treaty the question to ask is whether the intensity of these controversies has been exacerbated or dampened by the treaty.

The 1999 Entity Agreement
The return of power benefits to Canada was staggered because the thirty-year period agreed to as a term of the initial sale in 1964 was calculated in accordance with the intended completion dates for construction of each of the three Canadian storage dams. The result is that the sale of Canada's share in the downstream power benefits to the American utilities expired in tranches on March 31 in each of 1998 (9 percent), 1999 (46 percent), and 2003 (45 percent). On each expiration date, the treaty required the United States to return Canada's power entitlement to a point on the Canada–United States border near Oliver, British Columbia, unless an alternate point of return was agreed upon or a new sale negotiated (Downstream Benefits Steering Committee 1993, 6).

Both Canada and the United States delegated certain decision-making and operational responsibilities to representative entities under the treaty. The United States designated the position of Administrator of the Bonneville Power Administration (BPA), Department of Energy, plus the position of Division Engineer, North Pacific Division, Corps of Engineers, Department of the Army, to form the U.S. entity as its representative. Canada designated BC Hydro as the Canadian representative. Under the treaty, the entities had the responsibility of negotiating the disposal of Canada's entitlement to downstream power benefits (DSBs) (Columbia River Treaty, Article XIV(2)).

Commencing in 1989, the entities began exploring the various options available to Canada and the United States upon the expiration of the sale of Canada's share of downstream power benefits. To allow sufficient time for a deliberation process, they entered into an interim agreement in July of 1992 (Downstream Benefits Steering Committee 1993, 6). While the interim agreement was in place, the entities began formal negotiations to reach a more permanent agreement.

Review of these negotiations is instructive from a number of perspectives. Most fundamentally, the negotiations unearthed some controversies that had long been buried due to the initial thirty-year sale of the entitlement. They also illustrated the increased complexity of modern institutional deal making involving cross border interests. The work started in 1989 did not yield an agreement between the entities until 1999—after ten years of active, sometimes rancorous negotiations. Stakeholder involvement on both sides of the border was more necessary and sometimes more difficult to manage than it had been in 1960, and maintaining a consistency of purpose in the entities and their respective governments over such a long period of time proved difficult.

The most significant controversy that came to light by reintroduction of the obligation to return the entitlement to Oliver was the most fundamental issue of all: What was the Canadian entitlement and how should it be delivered?

From the outset, the treaty had required the entities to prepare operating plans for hydroelectric power generation and flood control (Columbia River Treaty 1964, Art. XIV(2)(h)). These plans were employed to calculate the downstream benefits (DSB) six years in advance, as contemplated by paragraph 5 of Annex B. Prior to 1998, this calculation was of only academic interest in the Canada/United States context because the Canadian entitlement (CE) component in the benefits had been presold and paid for. Beginning in 1998, however, reaching a common understanding of how downstream benefits should be calculated became fundamental to the issue of delivery.

The renewed importance of the DSB calculation predictably put strains on the generally harmonious relationship between the entities. On each side of the border, suspicion developed that those on the other side were seeking to manipulate the formula presented in Annex B to increase or decrease the DSB to suit their respective interests. The result was a tense relationship between the entities in the 1990s that, on at least two occasions, threatened to boil over into an international dispute—a sharp contrast to the spirit of cooperation that had prevailed for the previous twenty years.

Ultimately, the method of calculating the DSB and CE was resolved and an agreement struck in 1999 (the Entity Agreement) that avoided the need to construct new transmission lines in the United States to return the entitlement to the border near Oliver. Instead, the entities agreed that the power could be delivered to the border at the existing points of interconnection at Selkirk/Nelway in the east and Blaine in the west. As contemplated under Article VIII of the Treaty, Canada and the United States also agreed that B.C. could dispose of the entitlement in certain circumstances directly in the United States. In practice, B.C. has not chosen to do that in the ten years since the Entity Agreement was signed.

The awakening of the dormant controversy that occurred in the context of the CE return negotiations also sparked interest in broader stakeholder communities on both sides of the border. In contrast to the 1950s and early 1960s, organized interests in Canada and the United States had found their voices and wished to be heard. Legal requirements imposed by the U.S. National Environmental Policy Act (NEPA, 42 U.S.C. § 4321, 1969), caused BPA to insist on vetting elements of the discussion leading to the Entity Agreement through public consultation processes. In Canada, emergence of the Columbia Basin Trust (CBT) meant a more prominent role for regional stakeholders and their issues. The forty years between 1960 and 1999 had made international deal making a more challenging endeavour.

The public controversy, particularly in Canada, surrounding the original treaty and protocol negotiations was very intense. From the perspective of the day in 1960, the opportunities for public input were unusually extensive. However, there is no doubt that the expectations for stakeholder engagement had grown dramatically by the 1990s and will be significantly greater again in the context of the discussions relating to post 2024 operational issues.

Amendments to the Treaty
Options
The treaty does not contain a provision for amendment. It remains in force until the later of 2024 or ten years after notice of termination is given. Unless the treaty provides otherwise, the basic commitments of its signatories subsist for the entirety of that period. As mentioned above, there are express provisions in the treaty giving the entities the power to modify or supplement some aspects of the relationship between them and empowering them to take various actions to implement the treaty (Columbia River Treaty, Article XIV(2)). To the extent that the parties, going forward, want to carry out modifications expressly contemplated in the treaty, they do not need to concern themselves with international and domestic formalities— the entities are empowered to act. The parties must, however, ask themselves whether the particular subject-matter they want the entities to agree to on their behalf falls within these very specific provisions. We have seen this flexibility exercised in the context of the Entity Agreement, and also in the development and implementation of detailed operating plans for the management of water flow over the treaty dams. However, the basic rights and obligations of the parties can only be altered with formal treaty amendment.

Ten years notice must be given by either party to terminate the treaty and the earliest it can be effective is 2024. Accordingly, the alternatives available to the parties prior to 2014 are to:

(a) give notice of termination to be effective in 2024 and allow the river to be regulated separately in each country subject to international water law obligations, the Boundary Waters Treaty Act, and the few provisions of the treaty that will survive termination;

(b) give notice of termination but seek to amend the existing treaty or draft a new one prior to 2024;

(c) forestall giving notice of termination after 2014 while efforts are made to amend the treaty or draft a new one, thereby extending the treaty beyond 2024 for a comparable period;

(d) forestall giving notice of termination after 2014 while efforts are made to adjust the characteristics of joint operation by agreements that the entities are empowered to make with appropriate exchange of diplomatic notes or as contemplated under the treaty with a view to leaving the treaty in effect.

Each country's representatives can be expected to think carefully about these four options over the next few years. Full discussion of the issues associated with each is beyond the scope of this chapter; however, the following are some observations with respect to these options, and some key principles to be kept in mind in pursuing whatever course is ultimately taken.

Let the Treaty Expire

It is important to note that the treaty resulted from the United States' interest in creating additional storage facilities in the Columbia River in order to better manage flow, and from its willingness to share the benefits that resulted from that storage. The physical characteristics of the river that made storage development attractive in 1960 have not changed, even if perceptions as to how to employ that storage may have. It seems likely that there are benefits to be obtained from a higher level of cooperation made possible by amending or replacing the treaty with or without notice of termination than would occur if the treaty expired.

Amend the Treaty

Formal treaty amendments are nontrivial in both Canada and the United States as each must fulfill the minimal formalities required to conclude an international treaty and have it come into force. In the United States, Congress ratified the treaty and at the time it did so, it made clear its intention that no amendments be made in the absence of congressional approval (U.S. Senate 1961, 40–3). In Canada, Parliament was also called upon to ratify the treaty, and a strong argument exists that any amendment would have to be approved by it. The formalities of treaty-making leave the outcome of any effort to amend the treaty highly unpredictable and dependent on the political will of the parties in 2024. In these circumstances, a treaty amendment is only likely to happen if Canada or the United States feels either sufficiently aggrieved by the terms of the existing treaty, on the one hand, or by the threat of its complete loss, on the other. The parties' perspectives on how best to react in these circumstances may dictate whether notice is given in 2014 or not.

Enter into New Entity Agreements and/or Exchange of Notes

The option of entering into new entity agreements or exchanging notes is attractive to the extent that it avoids the complexity and politicization of full-scale amendment. On the other hand, the range of possible outcomes is considerably narrower if all changes must fit within the words (however broadly they may be interpreted) of the existing treaty.

The parties no doubt will give careful consideration to which options to pursue as they approach 2014. That consideration can usefully be informed by a full understanding of the attributes of the treaty. The next section examines the principles that formed the basis for sharing the benefits created under the treaty with a view to highlighting those attributes.

Principles

There are three basic principles that govern the apportionment of hydroelectric power benefits under the treaty, as well as the responsibility for the costs associated with production of those benefits. They were touched upon earlier:

(1) the power benefits generated as a result of the cooperative development of Canada and the United States are to be shared on a substantially equal basis, provided that an equal division will result in an advantage to each country as compared with the alternatives available to it;

(2) when an equal division of power benefits will not result in an advantage to each country, the countries must then negotiate and agree upon such other division of benefits as will be equitable to both countries and make cooperative development feasible; and

(3) each country is to bear all capital and operating costs for facilities it will provide in its own territory to carry out the cooperative development mandated by the treaty.

Understanding these deceptively simple principles is key to understanding the treaty's success in delivering the value that the parties envisioned in 1960. The principles effectively balance the theoretical potential of international cooperation on the one hand and the need to serve sovereign ambitions on the other.

The case for cooperative development and operation of the Columbia River Basin is too compelling to debate. The highly variable hydrograph, the existing human uses and dependencies on the system, and the migration across borders of nonhuman users of the river, all give the river its international character. On the other hand, the river affects the people, resources, and values of two distinct nations and the internal tensions that are inevitably created within each nation should not unduly inhibit the potential for mutually beneficial international cooperation. The trade-off between local, regional, and national interests, the rights of First Nations or tribes, and the provisions for the protection of the environment in each country may reflect disparate values between countries over time. The prospects for cooperation between the parties will be greatly diminished if success requires both countries to apply a common set of values to those domestic trade-off issues.

This tension was recognized during the negotiation of the treaty and the protocol. The Department of External Affairs, in its official presentation describing the treaty, expressed it this way: "It is, again, an advanced model of bi-national cooperation where the essential independence of both states is maintained within a framework of administrative coordination" (Department of External Affairs; Department of Northern Affairs and National Resources 1964, 14).

The practical effect of the principles was to cause each nation to determine the benefits it believed were attainable through cooperation. A bi-national structure was then developed to provide a mechanism to create those benefits. The principles provided that the benefits would normally be divided 50/50 and each party would bear its own costs. This benefit-sharing formula would be adjusted if the normal approach did not provide a benefit to one of the parties equal to or greater than what it thought it could obtain acting unilaterally.

The great attraction of this approach is that it focuses on gross benefits and eliminates the need for each country to calculate net benefits. It recognizes that determining what the net benefits and costs of a particular project might be in a way that is acceptable to both countries will often be impossible. The wisdom of finessing the need for the parties to agree on valuing intangible attributes such as species at risk or reconciliation with First Nations is amply demonstrated by the difficulty the entities had in agreeing to the quantification of the Canadian entitlement spelled out in the treaty. By allowing each party to assess its own benefits and costs, the treaty avoids the problem of the parties having to agree on these calculations, thereby enabling each country to exploit those opportunities that make it better off than it would otherwise be according to each country's own values. The result is that each country is put in a position to support whatever initiative is being undertaken.

Put simply, the power of the principles which gave rise to the sharing of benefits under the treaty lies in the fact that those principles recognize the benefits in one country, and the costs in the other, without requiring a comparison of the two. In this way, they permit the development of a framework which facilitates a negotiation process that recognizes the legitimacy of the concerns in each country, and introduces a formula that enables both countries to reap benefits from the development.

It is important to recognize that the treaty does not resolve the underlying difficulties associated with reconciling economic and environmental objectives. Nor can it be used to resolve the underlying issues associated with competing water uses that currently exist or will arise by 2024. Resolution of those issues should lie at the heart of what each country undertakes domestically as it contemplates the potential end of the treaty era. The extent to which and the way in which each country resolves those domestic issues is up to that country. Discussion around post 2024 bi-national arrangements should be conducted only once those domestic tasks are completed on each side of the border to the extent each party determines appropriate. At that point, the sharing principles found in the existing treaty can be employed to develop any bilateral agreement that the parties both see as potentially beneficial.

Conclusion: The Challenges Facing Future Negotiation of the Treaty

There can be no doubt that there will be challenges to cooperative management of the Columbia River in the future. The time and effort needed to settle the Entity Agreement attest to the fact that political barriers can stall agreement under the existing treaty principles. The introduction of new objectives beyond those of power and flood control would render those negotiations even more challenging. Based on its current success, there is reason to believe that the treaty is flexible enough to serve the parties as well in the twenty-first century as it did in the latter half the twentieth century. This optimism is tempered by the consideration that the treaty negotiation table is not always the appropriate venue to resolve basin-wide issues associated with competing water uses. The number of current and future social, economic, political, and environmental issues that have the potential to influence water management in the Columbia River basin is formidable. For example, issues relating to: (1) the legacy of the past (and the future need to consult with aboriginal groups); (2) hydroelectricity and the ability to secure energy from other sources; (3) water quality and environmental considerations; (4) water quantity and allocation; (5) industry and livelihoods; (6) salmon and other anadromous fish habitat; (7) public participation and the role of civil society; and (8) climate change, have been identified as driving forces capable of influencing any future water management regime of the Columbia River basin (Davidson and Paisley 2008; McKinney et al. 2010). Behind each of these issues is a legitimate stakeholder interest.

Development of the appropriate mechanisms for resolving competing domestic interests within a nation are the responsibility of each nation. Successful international cooperation depends upon each nation being able to clearly articulate its national values and aspirations. Armed with an understanding of each other's perspective, the two nations will then be able to move forward to benefit the people of both countries through international cooperation. However, that level of cooperation can only be achieved if each country makes its values and goals clear at the international negotiating table.

In summary, the Columbia River Treaty serves as a very useful model for international cooperation; however, it is not the appropriate mechanism to resolve the many legitimate issues and concerns of the various stakeholders in the Columbia River Basin. This chapter has attempted to provide an outline of the treaty's history and content as a means of illuminating the treaty's underlying principles, which are key to its success, and to submit that the success the 1964 treaty is unlikely to be duplicated if the parties move away from these underlying principles.

Notes

The historical review in this chapter is largely based on an earlier presentation of the author to CLE International: The International Energy Exchange Seminar, September 15-17, 1993. It has been updated and adapted for use in this volume with the important assistance of Karin Emond, also of Lawson Lundell LLP. I also wish to acknowledge the authors of the "Columbia Reports," which were prepared by officials at the Ministry of Energy, Mines and Petroleum Resources, the Crown Corporations Secretariat, and BC Hydro and Power Authority, and are cited later in this chapter. These reports were an important source. But this chapter is the sole responsibility of the author and does not purport to set out the view of BC Hydro, British Columbia, or any other party for whom the author has acted or may in the future.

Works Cited

Boundary Waters Treaty. "Treaty between the United States and Great Britain Relating to Boundary Waters, and Questions Arising Between the United States and Canada," 11 January 1909, 36 U.S. Stat. 2448, U.K.T.S. 1910 No. 23.

"Canada–British Columbia Agreement (8 July 1963)." In *The Columbia River Treaty Protocol and Related Documents*, by he Departments of External Affairs and Northern Affairs and National Resources, 100. Ottawa: Queen's Press 1964.

"Canada–British Columbia Agreement (13 January 1964)." In *The Columbia River Treaty Protocol and Related Documents*, by the Departments of External Affairs and Northern Affairs and National Resources, 107. Ottawa: Queen's Press, 1964.

Canada. "From the Canadian and United States Governments to the International Joint Commission (9 March 1944)." In *The Columbia River Treaty Protocol and Related Documents*, by the Departments of External Affairs and Northern Affairs and Natural Resources, 17. Ottawa: Queen's Printer, February 1964.

Columbia River Treaty. "Treaty Between Canada and the United States of America Relating to Cooperative Development of the Water Resources of the Columbia River Basin," 17 January 1961, Can. T.S. 1965 No. 17, 15 U.S.T. 1555.

Davidson, Heather C., and Richard K. Paisley. "Issues and Driving Forces within the Columbia River Basin with the Potential to Affect Future Transboundary Water Management (Draft Discussion Paper)." *Canadian Columbia River Forum*. 2008.

Department of External Affairs; Department of Northern Affairs and National Resources . *The Columbia River Treaty and Protocol: A Presentation*. Ottawa: Queen's Printer, 1964.

Diefenbaker, John G.. "Press Release by the Prime Minister Following the Signing of The Columbia River Treaty," 7 January 1961, in *The Columbia River Treaty Protocol and Related Documents,* 82 by the Departments of External Affairs and Northern Affairs and National Resources Resources, 82. Ottawa: Queen's Press, 1964.

Downstream Benefits Steering Committee. *Columbia Report #1: The Columbia River Treaty.* Victoria, B.C.: Ministry of Energy, Mines and Petroleum Resources, 1993.

Downstream Benefits Steering Committee. *Columbia Report #2: The Canadian Entitlement.* Victoria, B.C.: Ministry of Energy, Mines and Petroleum Resources, 1993.

Downstream Benefits Steering Committee. *Columbia Report #3: Reservoir Operations.* Victoria, B.C.: Ministry of Energy, Mines and Petroleum Resources, 1993.

International Columbia River Engineering Board. "Abstract of Report to the International Joint Commission on Water Resources of the Columbia River Basin" (1959). in *The Columbia River Treaty Protocol and Related Documents,* by the Departments of External

Affairs and Northern Affairs and National Resources, Ottawa: Queen's Printer. February 1964.

International Joint Commission. "Report of the International Joint Commission on Principles for Determining and Apportioning Benefits from Cooperative Use of Storage of Waters and Electrical Interconnection within the Columbia River System (29 December 1959)." In *The Columbia River Treaty Protocol and Related Documents*, 39. Ottawa: Queen's Printer, 1964.

McKinney, M., L. Baker, A. M. Buvel, A. Fischer, D. Foster, and C. Paulu. Managing Transboundary Natural Resources: An Assessment of the Need to Revise and Update the Columbia River Treaty. *West Northwest* 16:307. 2010.

Sanderson, Chris W. Untitled. *CLE International: The International Energy Exchange Seminar.* September 15–17, 1993.

Swainson, Neil A. Conflict Over the Columbia—The Canadian Background to an Historic Treaty. Montreal: McGill–Queen's University Press, 1979.

U.S. Senate, Columbia River Treaty: Hearing before the Committee on Foreign Relations, United States Senate, 87th Cong., 1st Sess., (1961), 403.

Part IV: Governing Transboundary Resources in the Face of Uncertainty

Introduction

Matthew McKinney and Edward P. Weber

Introduction

Before we move to the final chapters in the book, which take a broader and more academic view of governance, it is instructive to clarify the nature of governing natural resource and environmental issues in the twenty-first century. The first two sections of this introduction present our emerging understanding of the types of problems we face and the range of governance arrangements that have emerged to meet the challenges associated with those problems.

Types of Public Problems

Scholars and practitioners increasingly recognize three broad categories of natural resource and environmental issues, ranging from most to least tractable: (1) technical and practical problems; (2) value-laden problems in which people agree on the basic nature of the problem but not on how to resolve it; and (3) value-laden problems in which people disagree on both the nature of the problem and how to resolve it. The last of these—often referred to as 'wicked or intractable' problems—are the problems that grab headlines, generate lots of work for litigators, and pose considerable challenges to policymakers and the administrators responsible for managing them. We believe that efforts to govern transboundary water in river basins fall into this category, and that the uncertainties associated with governing the Columbia River Basin are compounded by five major perturbations—climate change, population increases, changing energy demands, deteriorating infrastructure, a and deteriorating ecological system—noted at the beginning of the book.

Technical and Practical Problems: Technical and practical problems can generally be answered by reasoning and the application of existing knowledge. People are likely to agree on the nature of such problems and on a short list of potential solutions. These problems are susceptible to expert-generated solutions without much consideration of values, and they may not require high levels of involvement by those the problems affect.

Value-laden Problems: A somewhat less-tractable type of problem arises when people generally agree on the nature of a problem but they disagree over the basic direction to take in responding to it. In this type of problem, values and interests begin to pull people in different directions. Even working together in good faith and with reliable information, they are likely to encounter difficulty even in identifying options for consideration. In fact, just acknowledging the need for a solution may raise choices that are too painful even to contemplate, and stakeholders will try to avoid discussing the matter or to dominate any discussion they enter into. Value-laden issues require that the values in tension be given serious consideration, not just by the experts, but also by both those who are interested in and affected by the issues and those who must implement the solutions. Technical experts can help inform possible solutions to these types of problems, but without the participation of those who actually bear the full brunt of the problem (stakeholders), progress will remain elusive. In such cases, people may be wary of each other, and it can take a substantial commitment of time for them to reveal their values, learn about and acknowledge each other's interests, and build trust before they're ready to address specific problems and to seek agreement. Participation by stakeholders in some form of collaborative problem-solving greatly increases the likelihood of success.

What Makes Some Problems 'Wicked': The third category of natural resource and environmental issues consists of problems that are often described as 'wicked or intractable' (Rittel and Webber 1973; Allen and Gould 1986; Heifetz and Sinclair 1988; Susskind and Field 1996; Forester 1999; Mathews 2002; Putnam and Wondolleck 2003). Because they are so difficult to resolve, they warrant a deeper analysis and require more robust tools for responding to them. In contrast to issues that are more readily resolved, issues arising from wicked problems (1) tend to involve many stakeholders with different—often divergent—interests; (2) revolve around complex, sometimes confounding information; and (3) occur in a briar patch of governmental jurisdictions with overlapping and conflicting mandates, laws, policies, and decision-making protocols. In such issues, the power to address the problem is scattered among a host of players. Disputes arise over facts and data, previous or tangential issues, the parties' intentions and "agendas," and over who has the authority to make or implement decisions. Typically, communication fails, trust plummets, and goodwill goes out the door.

More fundamentally, wicked problems are those in which people disagree over not only how to solve the problem and who bears responsibility for doing so, but even over the nature of the problem itself. Such problems involve competing or conflicting values, priorities, ideologies, and worldviews. The deep-rootedness of the sources of the parties' disagreement prevents them from agreeing even on how to characterize, or "name" the problem. The absence of this most basic prerequisite for solving a problem—a shared view of what the problem is—makes it impossible for people to move to the task of identifying options for solving it. (People can't be expected to solve "it" if they can't say what "it" is.) Moreover, precisely because stakeholders see the problem itself so differently, they are apt to see each other as an essential part of the problem. ("The problem is, they don't see the problem accurately; therefore, they must be blind, misguided, or guilty of malign motivation and intent!") Needless to say, when "the people are the problem," working relationships among stakeholders likely will be marked by the inability and unwillingness to cooperate, by the absence of goodwill and trust, and even by a lack of integrity. In such circumstances, it is well nigh impossible for any of the parties to name the problem in a way that others will accept.

In sum, the idea of intractability, or "wickedness," implies that a problem is so freighted with encumbrances that it is extremely difficult, perhaps impossible, for stakeholders to move forward toward a solution. Chief among these difficulties is the inability of the parties to gain sufficient "traction" even to "name" the problem, and hence to "frame" options or potential solutions.[1] This inability is due to several factors. First, there are no clear rules or shared experience for defining the problem. Second, the problems are often so complex that they require a higher degree of abstraction than simpler problems. Third, there is no immediately obvious objective measure of success. Fourth, and perhaps most important, communication channels typically work poorly, if at all—stakeholders are unable to talk with each other without inflaming the situation. What communication exists takes place through lawyers, press releases, and/or symbolic acts (sometimes violent) that are designed to "send a message." By itself, establishing direct, constructive communication between the parties won't solve the problem, but it can help people smooth the way for progress by building a common vocabulary and setting guidelines for civil dialogue.

Addressing intractable problems successfully requires the right political timing— acting when a window of opportunity opens—and a suite of different strategies, tools, and formats that can be used to devise and sustain a disciplined process of naming the problem, framing options, and deliberating about the consequences of different choices (Matthews 1999). As explained more fully below, discussions need to be inclusive, well-informed, and deliberative—artfully organized in such a way that the "forum matches the fuss" (Sander and Goldberg 1994). That is, the forum for deliberation must be tailored to the needs, interests, and purposes of the stakeholders, and also to the characteristics of the problem itself. One size does not fit all. The challenge is to create a public space in which the participants feel unconstrained in applying knowledge, tools, creative intuition, and common sense. They must be unshackled, free to seek unconventional solutions (Hutchinson et al. 2002).

New Models of Governance
In response to the increasing frequency of wicked natural resource and environmental problems, people have invented a variety of ways to match the governance arrangement to the nature of the process. In *Working Across Boundaries: People, Nature, and Regions*, McKinney and Johnson suggest that there is a continuum

of governance approaches that have emerged—from informal networks, to more formal partnerships, to regional institutions (see Figure 1).

Thinking in terms of this continuum helps to recognize that these approaches overlap in some ways, and that the differences among them are often subtle. The distinction between a network and a partnership, or a partnership and a regional institution, is not always clear and clean. These categories are intentionally broad, and within each are various models and approaches that also range from informal to formal. The reality is that these emerging forms of governance are assemblages of cooperating interests and groups, and all have established some type of working arrangement— some more artfully framed than others. The differences appear in aspects such as the range of issues and concerns that bring them together, the size and complexity of the geographical area they are focused on, the strength of the structural relationships they have established in which to function, the type of "official" establishment within recognized public or private organizations, and their method of assuring (or not) a continuing presence.

Key features found in these emerging forms of governance include that they are: (1) collaborative, meaning an attempt is made to include people that are interested in and affected by the issue; those needed to implement any outcome; and those that might oppose the process or outcome; (2) regional, meaning that they revolve more around the "problem-shed" regardless of jurisdictional and institutional fragmentation; and (3) adaptive, meaning that they accept and respond to uncertainty by promoting learning in and through the decision-making process.

The following chapters analyze concrete empirical examples in order to further our understanding of what effective transboundary governance might look like in cases of highly uncertain wicked problems in multi-level, complex governance settings for water resources. More specifically, the authors focus on how we can

Figure 1. Continuum of governance approaches emerging for natural resources and environmental management. An intermediary organization is an agent who acts as a link between parties. © 2012 Center for Natural Resources & Environmenta; Policy, University of Montana

organize and design governance institutions so that they support desirable forms of resilience, adaptability, and collaboration across jurisdictions, groups, and social-ecological systems in the face of the inevitable uncertainty that accompanies such problem-solving exercises.

Eve Vogel's contribution raises the issue of scale in looking at the adaptive capacity of governing institutions in the Columbia River Basin. Vogel challenges our assumptions concerning matching governance to geographic scale using the Columbia River Treaty to illustrate that action at the international scale was required to create a regional dialogue. Vogel furthers the discussion of uncertainty by viewing the time of treaty formulation as one of creative change and adaptation, to be followed by a period of rigid implementation brought on by the high level of international action required for change.

In the first chapter in this section we turn to the increasingly important role of public involvement, as previously noted by Hirt and Sowards in Part I. Then Heikkila and Gerlak look at the adaptive capacity of collaborative institutions by examining the fish and wildlife program of the Northwest Power and Conservation Council. Hill and his collaborators discuss a study in the Columbia River Basin on the use of models or "decision support tools" to inform and involve the public and support decision making where high levels of uncertainty regarding the future exist. They note that (1) use of local knowledge to inform model assumptions that are not scientifically based both increases model transparency and provides a grounded basis for choosing a particular assumption; and (2) use of models to forecast the future may be too constrained to allow decision makers to prepare for surprises. Instead, scenario planning or "backcasting" from a vision of the future may be a more appropriate approach in the face of uncertainty. De Stefano and Schmidt allow us to look at efforts at basin scale public involvement in another setting by examining the efforts of the European Union under the Water Framework Directive. McCaffrey and his collaborators look at broader public involvement through the increasing participation of third party actors in transboundary water negotiations, some bringing the voice of the public to the negotiation and some serving as neutral facilitators of the process. Part IV concludes with a more general overview of public management arrangements concluding that collaborative partnerships are best adapted to wicked problems.

These chapters provide numerous lessons on the need for collaboration, public involvement, and adaptive capacity to govern in the face of wicked problems. First, by examining the evolution of adaptive capacity in the case of the Columbia River Treaty, Vogel shows us that governance highly adapted to the needs of the time can become rigid over time when adaptive capacity is not built in.

Second, when it comes to designing adaptive systems of governance, Heikkila and Gerlak show that governing systems can change and adapt in the face of uncertainty.

Their case suggests that at least some of the necessary conditions for adaptive systems of governance include (1) a constitutional or organizing framework that creates an expectation of learning and adapting through periodic reviews and assessments, (2) the willingness and ability to create new methods of learning, such as scientific review panels, and (3) the ability to integrate this learning into on-the-ground operations. Moreover, the dynamic and rules associated with traditional, dominant decision-making institutions are insufficient for dealing with the new governance challenges, whether in the form of international agreements, fragmented and brittle coupled systems, hierarchical bureaucratic organizations (Thomas), or institutions designed to wrestle with simpler "epistemic" conceptions of problem uncertainty (Hill et al.). Mechanisms to introduce flexibility to decision making are increasingly found in recent transboundary water agreements (McCaffrey et al.). By definition, new decision-making institutions are needed (Kettl 2009). Further, while all of the cases here focus primarily on formal transboundary governance arrangements, in keeping with significant developments in recent years as to the critical importance of informal institutions, norms, and customs to successful governance (North 2005; Putnam 2000; Weber 2003, 2009), each chapter leaves open the question of whether more informal arrangements can also contribute to making the new arrangements more adaptable to the continually changing circumstances arising from wicked problems and multi-level governance.

Third, the failure to build a firm understanding of the types and causes of uncertainty leaves policymakers and practitioners less able to mount an effective governance response. Hill et al. and Heikkila and Gerlak find that uncertainty resides in many places, including: (1) the knowledge of the current state of the environmental conditions and of their relationships to economic and social factors; (2) in the variability of these conditions and relationships over time; and (3) in the projections of the effects of human management interventions on targeted policies and problems.

Hill et al. also recognizes the uncertainty associated with the reliability of the outputs in mathematical models used to assess the given conditions, relationships and projections. In addition, Hill et al. focus attention on the interactive and feedback effects associated with complex coupled social-ecological systems and the increased variability inherent in such settings, especially as it pertains to climate change.

Another common theme addressed by the following chapters is the need to effectively and collaboratively involve both organized stakeholders and the general citizenry. This theme picks up on the growing realization within the larger policymaking literature that many public problems emerge out of "complex patterns of overlapping consequences," including those of social and cultural import, thus effective solutions "will require more than technical solutions" (Lane 1999;

Dietz and Stern 2009). In their review of the European Union's ongoing efforts around watershed planning, De Stefano and Schmidt argue that is it critical to engage citizens and stakeholders early in the process to create trust and legitimacy, presumably by providing a process that is both transparent and accountable. They also suggest that a necessary ingredient for the success of watershed planning is to build the capacity and knowledge of citizens and stakeholders. As mentioned previously, the chapter by Hill et al. suggests one approach to achieve both of these goals. Thomas also makes explicit arguments along these lines. McCaffrey et al. illustrate how third parties are playing a role in this effort in major international negotiations.

As part of engaging a broader slice of civil society, there is also the insight that effective governance for wicked problems and complex, coupled social-ecological systems, involves the successful integration of facts and values (Karl et al. 2007; Adler undated; Leighninger 2006; Brunner et al. 2005; Scholz and Stiftel 2005; Bingham 2003; Fischer 2000; Adler et al. 2000; Ehrmann and Stinson 1999; Lee 1995; Susskind 1994; Ozawa 1991; Jasanoff 1990). Decision systems must have the capacity to work with and apply both traditional science as well as alternative "knowledges" grounded in practitioner experiences in problem solving and culture (Scott 1998; Feldman et al. 2006). Using the Columbia River Basin as a case in point, the chapter by Hill et al. demonstrates the need for a public problem-solving process that integrates scientific and technical expertise with social and political values. The authors suggest that computer models and model-based decision-support systems can mask uncertainty when seeking to address "wicked" public problems. They explore how extended peer review and backcasting from a vision of the future can be used in a deliberative process that incorporaes both facts and values (including ethical and cultural viewpoints).

We encourage the reader to refer back to the articles in Parts I, II, and III in reading the following chapters. Consider the concepts informed by the academic literature in light of the reality of basin-wide water management for competing interests. It is our hope that this collaboration between universities and stakeholders will help inform a productive dialogue among the people of the Columbia River Basin and other populations reliant on transboundary water resources and faced with an opportunity for change.

Notes

1. For more information on intractable problems, see the Web site assembled by Guy and Heidi Burgess at the University of Colorado: http://www.beyondintractability.org/iweb/.

Works Cited

Adler, Peter S. and Juliana E. Birkhoff. Undated. *Building Trust: When Knowledge from Here Meets Knowledge from Away.* The National Policy Consensus Center.

Adler, Peter, Robert Barrett, J. D. Martha, Chris Bean, Juliana Birkhoff, Connie Ozawa, and Emily Rudin. 2000. *Managing Scientific and Technical Information in Environmental Cases: Principles and Practices for Mediators and Facilitators.* Washington, D.C.: RESOLVE.

Allen, Gerald M., and Ernest M. Gould, Jr. 1986. Complexity, Wickedness, and Public Forests, *Journal of Forestry* 84(4): 20–24.

Bingham, Gail. 2003. *When the Sparks Fly: Building Consensus when the Science is Contested.* Washington. D.C.: RESOLVE..

Brunner, Ronald D., Toddi A. Steelman, Lindy Coe-Juell, Christina M. Cromley, Christine M. Edwards, and Donna W. Tucker. 2005. *Adaptive Governance: Integrating Science, Policy, and Decision Making.* New York: Columbia University Press.

Dietz, Thomas, and Paul C. Stern. 2009. *Public Participation in Environmental Assessment and Decision Making.* Washington, D.C.: National Academies Press.

Ehrmann, John R. and Barbara L. Stinson. 1999. Joint Fact-finding and the Use of Technical Experts. In *The Consensus Building Handbook: A Comprehensive Guide to Reaching Agreement,* ed. Lawrence Susskind et al., 375–99. Thousand Oaks, CA: Sage Publications.

Feldman, Martha S., Anne M. Khademian, Helen Ingram, and Anne S. Schneider. 2006. Ways of knowing and inclusive management practices. *Public Administration Review,* 66 (December): 89–99.

Fischer, Frank. 2000. *Citizens, Experts, and the Environment: The Politics of Local Knowledge.* Durham, N.C.: Duke University Press.

Forester, John. 1999. Dealing with Deep Value Differences. In *The Consensus Building Handbook,* ed. Lawrence Susskind et al., 463–94 Sage.

Heifetz, R. A., and R. M. Sinclair. 1988. Political Leadership: Managing the Public's Problem Solving. In *The Power of Public Ideas,* ed. R. B. Reich. Cambridge, Mass: Ballenger Publishing.

Hutchinson, R. W., S. L. English, and M. A. Mughal. 2002. A General Problem Solving Approach for Wicked Problems: Theory and Application to Chemical Weapons Verification and Biological Terrorism. *Group Decision and Negotiation* 11: 257–79.

Jasanoff, Sheila. 1990. *The Fifth Branch: Science Advisors as Policymakers.* Cambridge: Harvard University Press.

Karl, Herman A., Lawrence E. Susskind, and Katherine H. Wallace. 2007. A Dialogue, Not a Diatribe: Effective Integration of Science and Policy through Joint Fact Finding. *Environment* 49: 20–34.

Kettl, Donald F. 2009. *The Next Government of the United States: Why Our Institutions Fail Us and How to Fix Them.* New York: Norton and Company.

Lane, Neal. 1999. The Civic Scientist and Science Policy. *AAAS Science and Technology Policy Yearbook.* Available at www.aaas.org/spp/yearbook/chap22.htm

Lee, Kai. 1995. *Compass and Gyroscope: Integrating Science and Politics for the Environment.* Washington, D.C.: Island Press.

Leighninger, Matt. 2006. *The Next Form of Democracy: How Expert Rule is Giving Way to Shared Governance … and Why Politics Will Never Be the Same.* Nashville: Vanderbilt University Press.

Mathews, David. 2002. *For Communities to Work.* Dayton: Kettering Foundation.

Mathews, David. 1999. *Politics for People: Finding a Responsible Public Voice.* Champaign: University of Illinois Press.

McKinney, Matthew, and Shawn Johnson. 2009. *Working Across Boundaries: People, Nature, and Regions.* Lincoln Institute of Land Policy.

North, Douglass C. 2005. *Understanding the Process of Economic Change.* Princeton: Princeton University Press

Ozawa, Connie. 1991. *Recasting Science: Consensual Procedures in Public Policy Making.* Boulder, Colo: Westview Press.

Putnam, Linda L,. and Julia M. Wondolleck. 2003. Intractability: Definitions, Dimensions, and Distinctions. In *Making Sense of Intractable Environmental Conflicts: Concepts and Cases,* ed. Roy J. Lewicki, Barbara Gray, and Michael Elliott, 35–59. Washington, D.C.: Island Press.

Rittel, H. W. J., and M. M. Webber. 1973. Dilemmas in a General Theory of Planning. *Policy Sciences* 4: 155–69.

Sander, Frank E. A., and Stephen B. Goldberg. 1994. Fitting the Forum to the Fuss: A User-friendly Guide to Selecting an ADR Procedure. *Harvard Negotiation Journal,* January.

Scholz, John T., and Bruce Stiftel, eds. 2005. *Adaptive Governance and Water Conflict: New Institutions for Collaborative Planning.* Washington, D.C.: Resources for the Future.

Scott, James C. 1998. *Seeing Like a State: How Certain Schemes to Improve the Human Condition Have Failed.* New Haven and London: Yale University Press.

Susskind, Lawrence, and Patrick Field. 1996. When Values Collide. In *The Angry Public: The Mutual Gains Approach to Resolving Disputes,* ed. Susskind and Field, 152–97. New York: The Free Press.

Susskind, Lawrence. 1994. The Need for a Better Balance between Science and Politics. In *Environmental Diplomacy: Negotiating More Effective Global Agreements,* ed. Susskind, 66–79. New York: Oxford University Press.

Weber, Edward P. 2003. *Bringing Society Back In: Grassroots Ecosystem Management, Accountability, and Sustainable Communities.* Cambridge: MIT Press.

Weber, Edward P. 2009. Explaining Institutional Change in Tough Cases of Collaboration: "Ideas" in the Blackfoot Watershed. *Public Administration Review* 69 (2) (March/April): 314–27.

Can an International Treaty Strengthen a Region and Further Social and Environmental Inclusion? Lessons from the Columbia River Treaty

Eve Vogel

As we approach 2014 and 2024, when the 1964 Columbia River Treaty (CRT) may be renegotiated and then either terminated or altered, one of the key questions is whether and how a new treaty can incorporate a more diverse range of parties and interests. Since the CRT was ratified in 1964, a host of people, places, and interests have become central participants or considerations in Columbia River management on each side of the international border. Most prominent among these today are Native American and First Nation peoples, native fisheries, and the communities and ecosystems in the Canadian portion of the basin that were negatively affected by the treaty dams. In the years leading up to 1964, however, they were left out from the treaty negotiations—and they have been largely excluded or marginalized from CRT management ever since. Their lack of influence ultimately limits the ability of overall Columbia River management to meet important goals. Ongoing management of the three Canadian treaty dams and the river flows they control must remain within fairly narrow mandates, is inaccessible (except indirectly) to anyone other than two government agencies on the U.S. side of the border and one on the Canadian side, and causes significant social and environmental dislocation (Shurts this volume).

From our current vantage point, it is easy to jump to point to the hubris and lack of accountability of a legal agreement signed by two federal governments. Today, we live in an era when there is a strong movement to devolve natural resource management to smaller-scale jurisdictions, and reorganize it into more natural territories such as watersheds. Many advocate more individualized, even ad hoc, agreements among willing stakeholders rather than formalized, top-down policy (see e.g., Kemmis 1990; DeWitt 1994; Armitage et al. 2007). In the Columbia River Basin, there are thriving local and regional efforts—again, on both sides of the border—to build a collective, participatory approach to understanding and managing the river's diverse processes, interests, and demands (Columbia Basin Trust 2009; Kenney et al. 2000; Northwest Power and Conservation Council 2003). Perhaps this time, some might argue, we need something less formal, less big-government-directed, than an international treaty. Can a new treaty between two huge nations do anything *other* than thwart the kind of thriving grass-roots activity and thinking so apparent on both sides of the border today? Or is there a way a new treaty might actually *help* to democratize participation in Columbia River management, and better distribute the benefits of river management? What can we learn from how the 1964 treaty was negotiated, developed, and institutionalized to try to make sure a new CRT incorporates a full range of people's interests?

It turns out the easy judgment of the CRT as anti-democratic and top-down distorts reality. The 1964 CRT actually had an *empowering* impact on subnational regions on either side of the international border as well as on a kind of transnational region—and this regional empowerment also corresponded to a significant democratization of resource management. The empowerment occurred because the treaty would not have been approved without several associated legal agreements which brought numerous regional jurisdictions and actors on both sides of the border into active participation in Columbia River management decisions, and gave them considerable control over the distribution of the Columbia River's most profitable benefit, its hydropower.

The problem came once the treaty was institutionalized. The international nature of the CRT imposed a permanence that would not have been present in a single-nation law. While the CRT helped to expand participation in Columbia River management and to distribute river benefits more widely, and these results were rooted in regional-scale management of the river basin, only a limited set of regional jurisdictions, actors, and interests profited. Many others were excluded entirely. Viewed from a long historical perspective, then, the CRT initially helped to democratize participation in Columbia River management, but impedes that goal today.

This history seems to have two major contradictions. First, there was a concurrent rise of both international-scale autocracy and regional-scale empowerment and democratization. Second, regional-scale empowerment brought democratization in the past, but presently obstructs democratization. These seeming inconsistencies have major implications for how we should think about the possibilities of further democratization with the treaty's renegotiation. How can we make sense of them?

I suggest that these seem like contradictions because they challenge several standard notions about the relationship among law, jurisdictional level, geographic scale, and democratization, assumptions that turn out to be problematic. If we can break out of these standard notions, we can better make sense of the past, present, and future of the Columbia River Treaty. First, we must realize that international, national, and subnational levels of authority and decision making do not interact in a simple nested hierarchy, in which "higher" levels always trump and control "lower" levels. Second, we cannot make the common assumption that jurisdictional level is congruent to geographical scale. Though it is more complicated even than this, a crucial insight here is that the CRT is international in jurisdiction (that is, it is an agreement between two nations—or actually, between two sovereign "states") but mainly regional in its geography. Third, we need to abandon the wishful notion that resource governance (or governance of other issues) will necessarily be more socially inclusive and balanced if it is devolved or re-scaled to smaller local or regional geographic scales, and will be more attuned to environmental processes and needs if it is re-territorialized along the boundaries of natural systems. These results may follow from devolution of governance—but that depends on other factors besides geographic scale and jurisdictional level. Finally, we have to step back from static, ahistorical and apolitical views of both law and geography. We must recognize that geographical boundaries and institutional forms that structure natural resource governance are political choices, built in particular moments. They can change or be changed—but not entirely from scratch: both challenges and changes are built in part from past institutions, ideas, and constituencies (compare Thelen and Steinmo 1992). The process of constructing new geographical organizations of governance can be a politically open moment in which new actors and interests can break in. However, once new geographical organizations and systems of governance are settled and codified by law, they shape and may impede further political opening.

In this essay, I illuminate the lessons of the 1964 CRT and its associated agreements by analyzing critically the interrelationships among law, jurisdiction, geographic scale and democratization. I draw from recent work in the discipline of geography on the "politics of scale." Geographers have argued that geographic scales and jurisdictional levels are not natural, pre-ordained, nested "containers," each entirely encompassed by the next larger scale, but rather are constructed, dynamic, contested and mutually constituted (Delaney and Leitner 1997; Brown and Purcell 2005; Swyngedouw 2004; Cox 1998a, 1998b; Smith 1992). I suggest this more dynamic, relational, and political conception of scale and jurisdictional level can help to make sense of the seeming contradictions of the CRT history.

The CRT case is a particularly interesting one to examine using the insights from the literature on the politics of scale. A wide range of scholars have now begun to recognize the ways geographic scales and legal jurisdictions can become interrelated and reconstituted. Scholars have tended to see these interrelationships and reconstitutions as recent phenomena, however, brought about by varied processes sometimes identified as "postmodern," or associated with late-twentieth-century and twenty-first-century globalization. These recent phenomena are cast as if they were starkly different from what happened in the early- and mid-twentieth-century "modern" era, when, supposedly, nation-states met other nation-states as unitary actors, with little room for other scales or jurisdictions to challenge the hegemony of the national state (Berman 2002; Glassman 1999; Swyngedouw 1997; Brenner 2004; Keating 1998).[1] Of all the forms of international law, the one that has often been seen as the most traditional, most formal and most state-centric, perhaps, has been the treaty. Yet in the middle of the twentieth-century "modern" era, an international treaty concerning the Columbia River was constituted by complex multi-scalar and multi-jurisdictional politics; and it also reconstituted and empowered smaller scales, with considerable democratization as a result. Far more than recent cases, the history of the CRT proves that the complex and multidirectional relationships we have begun to see among law, jurisdiction, geography and democracy are not new; these complexities are fundamental. But CRT history also provides a cautionary tale. Its longevity allows us to see some of the more problematic long-term effects on geographical and jurisdictional reconfigurations of what were once new international laws and legal regimes.

In the next section of this chapter I draw on current geography and other literature to consider how we can begin to rethink the relationship among international law, subnational and transnational regions, jurisdictions and geographical areas, and democratization of resource management. The third section, "Treaty Construction," examines the multi-layered political contests that shaped the CRT and its associated agreements, and the results in terms of empowerment of regional-scale management and democratization of resource management. The fourth section, "Treaty Codification," outlines the limitations and inflexibilities that resulted from treaty codification. In "Lessons from the History of the Columbia River Treaty," I distill the lessons from the historical treaty to think about the relationship among international law, regional-scale resource management, political openness, and democratization. Finally, "Implications for the Future" outlines the implications of this history on the possible future of the treaty and Columbia River management.

Dis-Ordering Notions of Jurisdiction, Scale, and Democracy

Standard conceptions of international law, jurisdictional level, geographic scale and democratization are built on at least four problematic assumptions. The first two of these assumptions are tightly linked: first, jurisdictional levels are seen as ordered within a fairly clear hierarchy; and second, jurisdictional level is understood to be spatially congruent to geographic scale. Together, then, standard conceptions see a nested series of jurisdictional levels, each tied to a discrete legal territory, each one higher and larger, more powerful and more spatially encompassing than the next. In the United States for example, state law is seen to apply to individual state territories, and to trump local laws that apply only to local areas, while federal law is seen to apply throughout the entire United States territory, and to trump state laws. The relationship between international and national jurisdiction is more complex— some would argue that international authority is dependent on, and therefore lies below, that of sovereign nation-states, despite international law's greater spatial scale. Others would argue that it can at times create standards, opportunities, or strictures to which nation-states must, or at least often do, bend. It seems that whichever stance one takes on the relationship between international and national law, though, a hierarchical view of jurisdiction and geographic scale carries with it the suggestion that international law must reduce the power of regional and local jurisdictions, and de-prioritize the interests that are located within these jurisdictions' territories (Berman 2006; Sneddon and Fox 2007; Center for International Environmental Law 1999; Osofsky 2007).

A further assumption about law, jurisdictional level, geographic scale, and democratization is that governance organized within smaller-scale areas or carried out by lower-level jurisdictions is more democratic and more harmonious. To simplify, the sense is that by organizing governance in smaller-scale areas people can gain collective autonomy, understand interdependencies with each other and their environment, and act to sustain the social and environmental resources and connections on which they collectively depend. This all leads to more sustainable economic growth at the same time it spreads and nurtures wide social and environmental inclusion. Often these ideas are almost philosophical sentiments. But, they have also been translated into a wide range of policies and governmental reorganization efforts around the world, which devolve authority, economic initiative, and resource management to smaller-scale areas (see e.g., Kemmis 1990; DeWitt 1994; Sale 2000; Armitage et al. 1999; Bruch et al. 2005; Kenney et al. 2003; Bradshaw 2003).

Finally there is a tendency to see both jurisdictions and geographical scales as something more solid, permanent and impersonal, than, say, the messy world of "politics" or even environmental change. Based on these assumptions, people often look to "natural" geographies such as those of community and region, watershed and river basin, to transcend or resolve our conflicts. At the same time they fear the inflexibility of fixed jurisdictions like state and nation.

Although these standard notions may make sense in some contexts, a review of recent literature on political and economic geography shows they are often problematic. It also suggests steps toward a reformulation.

The most visible challenges in recent years to static notions of geographic scale and jurisdiction have come from global geopolitical and economic transformations. With the fall of the Berlin Wall in 1989 and the dissolution of the Soviet Union in 1991, the seemingly permanent bipolar geopolitical world suddenly evaporated. Since then, supranational and transboundary regional organizations have increasingly taken over the role of many formerly state functions, and many aspects of the economy are now governed globally. At the same time, states like Yugoslavia have broken apart, and their regions have become states—while others like Somalia have broken into violent stateless fragments (see e.g., Ohmae 1995; Brenner 2004; Keating 1998; Dodds 2007; Demko and Wood 1999). In this context, it is hard for any critical observer of the world to continue to believe any longer that states, regions, or even world hemispheres are permanent.

The same transformations have led to a wealth of writing in the geographical literature that provides critical new insights into the complex interrelationships and fluidities of geographical scale and legal jurisdiction. Literature on Europe has been particularly insightful in illuminating the synergies between international or supranational and regional organization and governance. Authors have linked the development of the European Union (EU) to the re-emergence of subnational and transnational European regions. A European-wide economic and political system has not further disempowered local areas and regions, it turns out; rather, it has freed them from national assimilation policies that often tried to suppress regional differences. In addition, the EU provides funds to regions—not only to the states in which they are located—to further regions' own visions of development. European regions have begun to promote and advertise their own unique cultures, histories, or landscapes, and to see themselves as active participants and competitors in a global marketplace (Keating 1998; Brenner 2004; Murphy 1999a, 1999b). Not only economic regions, but also aquatic regions have emerged or been strengthened: the EU's Water Framework Directive (WFD) has institutionalized river basin management within and across twenty-seven countries. Because river basin management under the WFD is often administered by local and regional institutions, this supranational mandate "herald[s] the regionalisation of water management in the EU" (Moss 2006: 82; see also Page and Kaika 2003).[2] Similarly, in Europe and elsewhere, especially around the Pacific Rim, lowered national trade barriers have been credited for the emergence of thriving economic city-regions, often held up as models and nodes of the global economy (Park 1997; Storper 1997; Yang 2005). Thus, international law and legal regimes in wide-ranging parts of the world are understood to help dismantle the hegemony of the nation-state, and in doing so, to allow for the concurrent rise of subnational and transnational regional power. No longer can we assume that increasing jurisdictional level means a weakening of smaller jurisdictions. Nor can we assume that different jurisdictional levels are nested in a simple hierarchy—for both larger and smaller jurisdictions in some cases seem to be winning out over the traditional state.

A further challenge to the standard assumptions about jurisdictional level comes from a recognition that the process of regional empowerment is not simply about the breakdown of national-level power. In response to proclamations of the end of the nation-state (Ohmae 1995), others have scoffed, arguing instead that it is often the national state itself that is responsible for processes of so-called globalization as well as regionalization and localization—institutionalizing new regulatory systems which allow the increasing transfer of money, goods and information across large distances, for example, or devolve authority to regions or local areas so they can harness new opportunities (Brenner 2004; Mansfield 2005; Swyngedouw 1997). What we begin to see then is that international, national, and regional jurisdiction and authority can all become entwined. We must abandon not only the hierarchical view of jurisdictional level, but also the zero-sum view of power that comes with it. Multiple jurisdictional levels are often both mutually constitutive and mutually supportive. When these multi-jurisdictional systems of regulation and governance are considered in spatial terms, it becomes clear that the same kind of complex intertwining can be seen among different geographic scales as well as among different jurisdictional levels. Different areas become defined and organized as "local," "regional," "national," "international" or "global" in various ways through a complex and often contested interplay of politics at multiple geographic scales and jurisdictional levels (Brenner 2001; Mansfield 2001). Geographical organization, in other words, is "socially constructed," dynamic and contested. This means it is crafted in conception and implemented in practice by human beings. It is subject to change and revision. Furthermore, different people have different notions about how the world should be geographically organized—and they fight over the adoption of particular definitions and practices. This is because some people stand to gain and others to lose from any particular geographical organization. "Democratic decision making" cannot solve this contest, for democratic processes depend on already-delineated spaces, scales, and groups that bound democratic polities. One need think only of the many fights over gerrymandered Congressional districts to realize that geographical delineations of governance have profound implications for decision outcomes (Morrill 1999). However, even once a district is delineated (or gerrymandered), people both within and without can draw on other jurisdictions, other sources of influence, to shape decision making. Think here of external campaign contributions, for example, or a state-level legal challenge to a local town's decision. Drawing from these kinds of insights, we must understand that all levels of governmental organization and all geographic scales overlap and are mutually constituted, and are constituted as well by other axes of social power.

Significantly, much legal scholarship in recent years has developed along similar lines. Legal scholars and historians have highlighted the fact that legal rules and structures, like geographical spaces and structures, are constructed, dynamic, and contested (Horwitz 1997; Nelson 1994).[3]

But if all geographical scales and legal jurisdictions are entwined and interconnected, dynamic and contested—if everything is related to and changeable with everything else—how can we begin to think about the relation among international law, jurisdictional level, geographic scale, and democratic potential in any kind of logical fashion? More specifically for our purposes—that is, as we think about how to build a more inclusive future governance system for the transboundary Columbia River—is the notion of devolution of authority to smaller scales and lower jurisdictions simply irrelevant? Do devolution and region have *any*thing to offer us as we contemplate how to renegotiate the Columbia River Treaty to further wider participation, broader sharing of benefits, and environmental sustainability?

I suggest devolution and region are not irrelevant; rather, the specific character and effect of devolution of authority on wider participation, broader sharing of benefits, or environmental sustainability is indeterminate based on level and scale alone. As geographers Brown and Purcell argue, "There's nothing inherent about scale": just because some resource management program is "local" or some economic initiative "regional," it does not mean that it is necessarily more socially or environmentally beneficial (Brown and Purcell 2005).

However, scale and level are tied to other factors, and together, these influence democratization greatly. Here I define democratization specifically as the broadening of participation in and influence over management decision making and the wider distribution of benefits. More than simply scale and level, it is political conflicts, compromises, and relationships that determine the extent to which democratization accompanies new scales and levels of governance. Devolution of authority can be real, even in the face of complex multi-jurisdictional and multi-scalar interweaving. But exactly who and what a smaller-scale area or lower-level jurisdiction of governance includes and excludes depends on the particular politics and institutions that are embedded within its construction. If national jurisdictions and national-scale interests have traditionally dominated policy making, then empowering other jurisdictions and scales may help destabilize entrenched dominant interests, and therefore empower marginalized groups. Equally, international law can empower particular regions or areas within a nation-state, and particular people or interests within those regions (Berman 2006). The key question is whether international law empowers groups, areas, and interests that were formerly marginalized. If so, it can democratize participation in governance considerably.

But *any* level of law has a quality distinct from other kinds of political and social institutions that makes its influence potentially problematic in the long run. This quality is that law can *codify* geographies and levels of democratic participation, or the interests included within a jurisdiction's or geographic scale's purview. Geographers have been passionate about their insights into the constructed, dynamic, and relational qualities of scale and jurisdiction, because, they say, seeing scales and jurisdictions this way can open up possibilities for political challenge and change (Amin 2004; Massey 2005). The key risk of any legalized geography of governance is that it may diminish subsequent political openness and potential for change. International law may pose an especially high risk, for undoing international agreements can be as difficult as forming them in the first place. In other words, when constituted in part by international law, regional systems of governance may be especially immutable, and particularly resistant to further democratization. In examining how international law may influence democratic empowerment through devolution of authority to smaller-scale areas and lower-level jurisdictions, then, there are two key questions to ask. The first concerns legal and geographical *construction*. What priorities, participants, and political relationships are embedded in a particular geography or jurisdiction of governance? The second is about *codification*. What kinds of policy changes become fixed, limiting further political openness?

Treaty Construction: The Multi-layered Politics of Regionalization and Democratization

How and why did the Columbia River Treaty come to catalyze such strong, clear empowerment and reinforcement of regional actors, interests, and jurisdictions on either side of the border? And how did this help to bring about a kind of democratization—the broadening of participation in Columbia River management and the wider distribution of benefits of river development?

The short answer is that the process of treaty negotiations and construction opened politics in ways that allowed these changes. At a time when the federal U.S. and Canadian governments coveted an international treaty, smaller players and relative outsiders who could prevent the treaty's ratification and implementation had the leverage to win considerable concessions from the two federal governments. The Columbia River Treaty was not enacted until a series of other agreements were concluded. It was these associated agreements that most clearly empowered regional actors, interests, and jurisdictions on each side of the border. Because these smaller-scale players previously had limited influence over Columbia River management, their gains amounted to a widening of participation and distribution of benefits.

Within this over-all dynamic, though, "regionalization" and "democratization" were quite different on the two sides of the border, and there were four distinct conflicts that together resolved into six agreements including the CRT. Quite specific regional actors, jurisdictions, and interests gained greater roles in river management and claimed greater shares of the river's benefits from each conflict; and these differences derived from the distinct jurisdictional and geographical relationships and political and policy contexts, that shaped them.

Table 1 summarizes the six agreements related to the Columbia River that were developed in the process of negotiating the treaty, and their effects on regionalization and democratization. Together, the CRT and its five associated agreements reinforced and empowered two specific existing regions on both sides of the border, and also built a kind of inclusive cross-border region. On the U.S. side,

Table 1: A summary of the agreements associated with the CRT, and their effect on regionalization and democratization of Columbia River management.

Agreement(s)	Effect on regionalization: geography and jurisdiction of river management	Effect on democratization: participation in Columbia River management and distribution of river benefits
1) CRT	Internationalized Columbia River management; required basin-wide coordination of river flows.	Canada became co-manager of Columbia River, and gained benefits even from U.S. portion of river.
2) PNCA	Achieved U.S. Pacific Northwest region-wide coordination of electric power production and distribution—including coordinating Columbia River flows for hydropower.	Diverse U.S. utilities, industries, jurisdictions, and objectives brought into river management decision making and sharing of river benefits.
3) Canada-B.C. Agreement	British Columbia gained full control of treaty implementation, power sale options. Columbia River development became part of province-wide power development and distribution.	British Columbia became a co-manager of the Columbia River; Columbia River power benefits spread widely across province to help residents, farms and industry alike.
4) Terms of Sale Agreement, Sale of Canadian Entitlement to CSPE	Increased integration of B.C. – U.S. Pacific Northwest electric power market system; further strengthened role of collaboration and benefits-sharing among U.S. Pacific Northwest power producers.	B.C. gained control of proceeds from downstream power benefits (Canadian Entitlement). Non-federal U.S. utilities gained a major role as international and inter-regional power sales brokers, and limited treaty participants.
5) Pacific Intertie authorization; Sale of Canadian Entitlement from CSPE to California and Southwest utilities	Interconnected U.S. Pacific Northwest to California and Southwest with high-voltage transmission interties and a large-volume power sale.	California and U.S. Southwest gained influence over the distribution of Columbia River power, and gained large volumes of that power.
6) Pacific Northwest Consumer Power Preference Act	U.S. Pacific Northwest was codified with legally defined territory and preferential access to Columbia River power.	At-cost Columbia River power restricted to U.S. Pacific Northwest.

the region was a Pacific Northwest consisting specifically of Washington, Oregon, Idaho, and western Montana, extending out into small corners of Wyoming, Utah, and Nevada. On the Canadian side, the region was British Columbia.

Conflict 1: Canada vs. the United States (Agreement #1)

In the over-all negotiations between the United States and Canada, Canada won a favorable treaty by bargaining hard at a time when the United States was the more interested party. Downstream U.S. dams and cities needed upstream storage for increased power generation and flood control. But Eisenhower power policy, when superimposed on existing New Deal law, inadvertently precluded the building of storage in the U.S. portion of the Columbia Basin. On the one hand, the Eisenhower administration, in an attempt to support private rather than government development, refused to authorize new federal dam projects. On the other hand, law written during the earlier, more government-friendly New Deal era barred owners of nonfederal storage dams from obtaining compensation for benefits that accrued to downstream federal dams. Since the Columbia River system had several federal dams on the lower river by the 1950s, this meant nonfederal entities that might be interested in building storage dams in the upper river could not easily recover their expenses. Together, two different administrations' policies thus prevented both federal and nonfederal development of storage in U.S. territory. By the late 1950s, U.S. power and flood control interests looked eagerly—almost desperately—to storage in the Canadian portion of the Columbia River Basin (Krutilla 1967; Swainson 1979; Calkins 1970; Blumm 1993).

Canadians, on the other hand, were in a much stronger position. B.C. would soon need new sources of power, but British Columbia's premier of the 1950s and 1960s, W. A. C. Bennett, aggressively pursued power development on the Peace River in northern B.C. (See also Conflict 3 below.) Bennett could thus argue quite convincingly that B.C. did not need Columbia River storage. Canadians could walk away from a treaty if their demands were not met.

Still, a possible U.S.-Canada treaty was a far bigger political issue in Canada than it was in the United States. Once negotiations began, Canadians undertook more extensive analyses, and B.C. policymakers paid closer attention to detail than their American counterparts (Krutilla 1967).

Through hard bargaining and careful analyses, Canada won fifty percent of the power and flood control benefits that would be produced downstream in the U.S. (Krutilla 1967). Although Canada's success did not constitute devolution of governance, it did represent something similar: a jurisdiction with a smaller economy, normally the junior political partner, gained considerable authority and benefits from the jurisdiction that had previously dominated river management. Participation in Columbia River management expanded and internationalized:

Canada joined the United States in coordinated basin-wide management. Canada would now also enjoy a large portion of Columbia River power benefits (Swainson 1979;Volkman 1997; Blumm and Bodi 1996).

Conflict 2: United States: Regional Nonfederal vs. Regional Federal (Agreement #2)

U.S. federal government initiative came not from the national federal government but from the *regional federal* Bonneville Power Administration (BPA), and from the regional offices of the U.S. Army Corps of Engineers, which owned and operated most of the federal Columbia River dams. Here one must understand a critical distinction on the U.S. side: "federal" jurisdictional level in the Columbia River system did not (and does not) necessarily correspond to "national" geographic scale. The BPA was a New Deal agency that had its roots in old visions of regional planning and river valley development. Based in Portland, Oregon, rather than Washington, D.C., the BPA's hydropower program had long been distinctly regional in identity and focus. The BPA transmitted and sold power through its regional federal grid to utility customers in much of Washington, Oregon, Idaho, and western Montana. This three-and-a-half-state "region"—as well as the BPA itself—had been conceived by the Pacific Northwest Regional Planning Commission in 1935, an agency that by 1964 was long defunct. Nonetheless, the commission's institutional heir, the BPA, retained its regional vision. The BPA had from its beginning treated its three-and-a-half-state region as its territorial manifest destiny, and had managed Columbia River power as a regional asset (Vogel 2011; 2008; 2007). The BPA, aided by the regional offices of the Army Corps of Engineers, looked eagerly to "*international*" Columbia River storage to supplement *regional*, not national, electric power.

Despite the regional vision and practice of the BPA, however, power production on the U.S. portion of the Columbia River had retained a federal agency bias. Upper-river federal dams were managed to optimize power production at lower-river federal dams (and flood control) over others' river flow needs. The federal agencies' flow management had for this reason aggravated the Pacific Northwest nonfederal utilities and industrial power customers that relied on Columbia River flows and Columbia River power. This was most clearly an issue for the private and local public owners of dams downriver from Grand Coulee and Hungry Horse dams, but dissatisfaction extended also to their industrial customers and to other utilities and customers to whom they sold power. When discussions began about Canadian storage, the nonfederal dam operators, utilities, and Columbia River power customers on the U.S. side were determined that the management of any new storage dams would not marginalize them (Logie 1993; Northwest Power Pool 1979; Dean and Schultz 1989; Bonneville Power Administration et al. 2001).[4]

These aggrieved nonfederal parties had two sources of leverage over the Columbia River treaty that they did not have over regular ongoing federal river management. First, they could block treaty *ratification* through their influence over the Pacific Northwest Congressional delegation. Second, they could block treaty *implementation*, for they controlled several dams on the mid–Columbia River whose operations had to be coordinated with the Canadian treaty dams to produce the power benefits promised to Canada (Logie 1993; Northwest Power Pool 1979; Dean and Schultz 1989; Bonneville Power Administration et al. 2001). Eager for a treaty, federal Columbia River managers had to appease this group of disaffected utilities and industries.

The result was the Pacific Northwest Coordination Agreement (PNCA) in which nonfederal utilities and industrial power customers became participants in yearly U.S. river planning. These utilities and power customers brought in many local jurisdictions and varying goals, including irrigation, recreation, navigation, and flows for fish ladders. The PNCA also expanded the geography of the Columbia River "system": the Columbia's flows came to be managed not only as a part of an integrated Columbia River hydrologic system, not only as a system for the regional customers of federal Columbia River power, but a multi-party and even multi-river power system that extended throughout the entire U.S. Pacific Northwest (Logie 1993; Northwest Power Pool 1979; Dean and Schultz 1989; Bonneville Power Administration et al. 2001).[5]

The PNCA's Pacific Northwest, it is worth noting, corresponded to Annex B of the Columbia River Treaty, which designated Washington, Oregon, Idaho, and western Montana as the Pacific Northwest states (Columbia River Treaty 1964). Both the PNCA and the CRT followed, that is, the lines of the BPA's longstanding regional vision (Vogel 2011; 2007; 2008).

Conflict 3: British Columbia vs. Canada (Agreements #3 and #4)
The most contentious struggle in the treaty negotiations did not directly involve the United States at all. It faced off the Canadian federal government and the province of British Columbia (B.C.). Within Canada the Columbia River is a single-province British Columbia river, and so, under Canadian federalism, B.C. had the responsibility and right to license and carry out Columbia River development projects on its own. The Canadian federal government, however, had authority over the river at the international border, and had used this authority in 1955 to block independent provincial Columbia River development (International River Improvements Act 1955; Swainson 1979). It was in reaction to this federal veto that B.C.'s premier of the 1950s and 1960s, W. A. C. Bennett, aggressively pursued multiple power development options and was able to bargain during the treaty negotiations from the strong position of not needing Columbia River development

(see Conflict 1.). Bennett met not only U.S. negotiators from this position of strength, but his own federal government. After years of standoff, the Canadian federal government, as well as the U.S. federal government, acquiesced to virtually all of British Columbia's main demands: its preferred dam sites, a large share of the downstream benefits (both in Agreement #1), full control of treaty implementation, the right to export the downstream power benefits (both in Agreement #3), and a single purchaser for the downstream power benefits (Agreement #4) (Canada–B.C. Agreement 1963; Swainson 1979).[6]

Besides clearly devolving management to the province, B.C.'s successes also helped build an international Pacific Northwest. An important step that laid the groundwork for this was the joining in 1961–62 of the three major electric power utilities in British Columbia into a single provincial power agency, BC Hydro (Power Development Act 1961; Swainson 1979; BC Hydro Pioneers 1998). A now province-wide BC Hydro came to work closely with the Bonneville Power Administration, the U.S. Army Corps of Engineers and a host of American utilities, in integrated international management of the Columbia River and a regional power system (Bonneville Power Administration et al. 2001; Bonneville Power Administration 1989; Bankes 1996; NW Water Law and Policy Project 2001; Muckleston 2003; Blumm and Bodi 1996).

To some extent, B.C.'s gains also helped to spread river participation and benefits more widely. The province gained influence it had not had before, and the Bennett government spread power benefits out to the province. While B.C. would not use its share of the downstream power made possible by the dams in the U.S. for the first thirty years, and it would not use the sale of its thirty-year share of the downstream benefits for anything other than building the treaty dams themselves (see next section), BC Hydro was able to build new generation facilities to take advantage of Libby and the Canadian treaty dams, to lower power costs and to improve power distribution to remote parts of the province (Bankes 1996; BC Hydro Pioneers 1998).

Conflict 4: Regional power producers' profit vs. regional power customers' privilege (Agreements #4, #5, and #6)
The final set of agreements came as responses to the British Columbia demand that a single purchaser buy thirty years' worth of Canada's share of the downstream power benefits, known as the Canadian Entitlement. First came two sets of financial agreements, one strengthening Pacific Northwest integration, another one threatening disintegration. In the first financial agreement (Agreement #4), BC Hydro sold thirty-years' worth of the Canadian Entitlement to the Canadian Storage Power Exchange (CSPE), a consortium of the mid-Columbia Public Utility Districts and their customers, backed financially by the BPA. This agreement

actually involved multiple steps: a) Canada and the United States signed a Terms of Sale Agreement to authorize BC Hydro to sell the Canadian Entitlement to a U.S. purchaser; b) The Public Utility District owners of the mid-Columbia dams, together with several of their public and private utility customers, formed the Canadian Storage Power Exchange (CSPE), a nonprofit corporation in Washington State; c) The CSPE sold bonds to finance the purchase of the Canadian Entitlement; d) BPA backed the bonds; e) BC Hydro and CSPE signed the Canadian Entitlement Purchase Agreement on Aug. 13, 1964; f) CSPE delivered payment to Canada on Sept. 16, the day the treaty was ratified; and g) Canada turned the money over to BC Hydro (Swainson 1979; Northwest Power Planning Council and Lesser 1989; Binus 2008; Bonneville Power Administration 1980).[7] The end result was a tight financial interweaving of BC Hydro, the mid-Columbia Public Utility Districts, private utilities, and the BPA—in other words, a large portion of the major power producers and consumers within the bi-national Pacific Northwest—backed by the legal rules and support of B.C, Washington state, and the U.S. and Canadian federal governments. In the process, the CSPE consortium became so central to the treaty that CSPE representatives sat in on the final treaty negotiations, and the CSPE was empowered to use the dispute machinery set up in the treaty (Swainson 1979).

In the second financial agreement, the CSPE found a market for this power among several California utilities (Agreement #5) (Northwest Power Planning Council and Lesser 1989; Binus 2008; Lee et al. 1980; Calkins 1970). While a sale to California could make the treaty possible, it was also a major threat. It would require a high-voltage transmission intertie to California, and Pacific Northwesterners—in particular, large industrial BPA customers—worried that Californians might cut into their access to the nation's cheapest power (Binus 2008; Lee et al. 1980).

The political negotiations focused on how to regulate a transmission intertie to California. This time, democratization reinforced, expanded, and codified regionalization, rather than the other way around. The same array of federal and nonfederal regional power producers, managers, and customers who had resolved their decades-long spat in creating the PNCA and enabling the sale of the Canadian Entitlement now came together in a joint lobbying effort, and won the Pacific Northwest Consumer Power Preference Act. This codified BPA's service region as its old regional territory of Washington, Oregon, Idaho, and western Montana—plus small extensions into the small Columbia Basin portions of Wyoming, Utah, and Nevada, and up to seventy-five miles beyond the hydrological basin to incorporate Montana rural electric coops that straddled the hydrological divide. Customers outside this territory would get BPA power only after regional customers had bought what they wanted (Pacific Northwest Consumer Powers Preference Act 1964; Binus 2008; Bonneville Power Administration 1980; Lee et al. 1980; Swainson 1979; Hanks 1970). The Pacific Northwest had been since the New Deal united in a vision and practice of shared Columbia River development (NRC 1936; Vogel 2011 2008, 2007); now, finally, it was codified as a region with privileged rights to Columbia River power (Agreement #6).

Result: New Democratization and Reinforced Regionalization

These four conflicts and six agreements together brought about significant empowerment of regional actors, interests, and jurisdictions on both sides of the border. They also broadened participation in Columbia River management and distributed the benefits of river development more widely. Finally, they strengthened the coordination and clout of the Pacific Northwest as a region in relation to both the Columbia River and Columbia River hydropower.

They did not achieve these goals, however, because the treaty or regional management of the Columbia River was inherently democratic. Rather, participation was broadened and benefits more widely distributed because the process of treaty negotiations and construction allowed smaller players and relative outsiders to win considerable concessions from the two federal governments. Regional ties and independence were strengthened on the U.S. side because negotiations were built around a U.S. federal Columbia River power system that was already regional. New parties joined and extended this regional system, changing its membership but mainly codifying rather than changing its geography. On the Canadian side, Canada's strong federalism allowed a "maverick" B.C. premier to out-maneuver two federal governments; and the rights and responsibilities that accrued to the province amounted to a kind of regionalization.

Treaty Codification: Political Exclusions and Legal Inflexibility

Ironically, the same dynamic, contested, and multi-jurisdictional politics that opened up the treaty and its associated agreements also built in exclusions and limited flexibility. There were two aspects of the political contests that did this, and two corresponding results. First, the actors, jurisdictions, and interests that had the leverage to fight their way in became participants in river management; management was not broadened further. Management goals were similarly limited; they reflected the goals of these politically successful participants. Second, because the contests were so hard-fought and drawn out, they were resolved by very detailed prescriptions that each party could feel confident would be carried out. The result was that the limited range of participants and goals was locked in for decades.

Exclusions: Limits to Regionalization and Democratization

The limitations of regionalization and democratization under the treaty and its associated agreements can be divided into three categories: limited regional empowerment, limited participation, and limited interests served.

Limited regional empowerment: Regional empowerment and regional-scale manage-ment did not necessarily mean the empowerment of regional jurisdictions—and if it did, that still did not mean the empowerment of those jurisdictions' publics. In Canada, a regional jurisdiction, the province, was empowered, but not all parts or players in the province; rather, only the provincial government's leaders and a new provincial power corporation, BC Hydro. A small group of policy-makers and agency analysts developed B.C.'s position on the treaty essentially by themselves while for thirty years the people in the portion of B.C. most affected by the treaty, the Canadian Columbia Basin, had virtually no voice in treaty development or implementation (Swainson 1979; Halleran 1998; Bankes 1996).

On the U.S. side, regional management remained fundamentally federal; the federal BPA remained at the core of regional collaboration, regional functional interconnection, even regional definition. State governments, the closest thing to regional jurisdictions, gained no authority whatsoever. (The four main Pacific Northwest states would finally gain influence over Columbia River management with the passage of the Northwest Power Act in 1980 [(Pacific Northwest Electric Power Planning and Conservation Act 1980]). Quite a few local jurisdictions in the United States gained authority and benefits through their utilities' participation in the PNCA and CSPE, but these gains related only to these jurisdictions' roles as power producers and dam managers.

Limited participation: Participation in Columbia River management under the treaty and its associated agreements was limited to these governmental units and actors, plus nonfederal utilities and major electricity-consuming industries. On the Canadian side, not even private utilities gained authority and benefits. The most important private utility in B.C. was annihilated in the process of treaty negotiations and bargaining, in order to create BC Hydro (Power Development Act 1961; Swainson 1979; BC Hydro Pioneers 1998). (See Conflict 3.) No other nongovernmental actors or entities participated on either side of the border.

Limited interests served: On both sides of the border, the main interests served were related to power and flood control. Electric power customers in particular—residences, farms and industries alike—of the entire international Pacific Northwest were to benefit from cheap, abundant, and reliable electricity. The goals were to improve people's standard of living and to spread opportunities for economic and industrial growth (Bonneville Power Administration 1980; BC Hydro Pioneers 1998). There were undeniably tremendous benefits for a wide range of people.

However, the ecosystems and wild fish in the upper Columbia Basin, and the dynamic hydrology of the middle and lower basin that was important for downstream river ecosystems, were ignored or willingly sacrificed by the participants who built the treaty and its associated agreements. The treaty dams caused tremendous damage to these, and hurt many people who were dependent on them. Salmon fisheries were not unimportant to the developers of the CRT. Rather, the upper Columbia Basin, blocked to salmon since the construction of the Grand Coulee Dam, was seen by parties on both sides of the border as an area where dams could bring great benefits without harming valuable fish runs—as opposed to alternative hydropower development proposed on the Fraser and Snake Rivers (Krutilla 1967; Swainson 1979; Evendan 2004). The problem was, even though new treaty dams did not create new obstructions for migrating salmon, they dramatically changed downstream flows. Treaty storage dams reduced the spring freshet in order to store water for power generation and flood control. Among other effects, this slowed the migration of juvenile salmon downstream, making them vulnerable to a host of hazards. It also reduced the ability of dynamic river processes to create and sustain important habitat. For example, in Hanford Reach in the mid-Columbia, the most productive salmon area left in the river's main stem (which did not have a dam), vegetation began to encroach on salmon spawning areas because floods no longer scoured them out (ISG 2000). Impacts to downstream salmon hurt fishers from inland tribal fisherman to commercial fisherman who fish for Columbia River salmon along the Oregon, Washington, British Columbia, and Alaska coasts (Shepard and Argue 2005; Committee on Protection and Management of Pacific Northwest Anadromous Salmonids 1996).

Equally ignored and sacrificed were the people and communities that relied on the natural systems of the upper basin. The four treaty dam reservoirs flooded much of the area's already-scarce wetlands and farmlands, and displaced some twenty-three hundred people. Reservoir areas fluctuated seasonally between being huge lakes and huge mudflats; the mudflats became sources of dust storms. Dams, reservoirs, and the loss of wetlands negatively impacted waterfowl, migratory elk and caribou, and resident fish. These ecological losses impacted area residents, sportfishers and hunters and aboriginal people (Halleran; Wilson 1973; Waterfield 1970; Bankes 1996).[8]

Codification: Closure to Further Democratization

All these limitations were locked in place with the codification of the treaty, its minimum sixty-year term, and its imperviousness to national or subnational law or citizen input. Treaty management still follows the stipulations set down in the early 1960s. As Shurts explains in this volume, in recent years management of the treaty dams has been adjusted somewhat in order to provide water for fish and other nonpower needs (Shurts this volume; Hyde et al. 2002). Still, this is secondary to the two fundamental treaty purposes of flood control and optimum power production, and these kinds of adjustments can be made only if both Treaty Entities agree (Bankes 1996; Shurts this volume). Achieving agreement between the entities can be difficult. BC Hydro has no incentive, for example, to allow the BPA and Army

Corps to alter the management of the river in a way that reduces power proceeds to help fish in American waters, unless it receives benefits in return (Bankes 1996).

Two other agreements, though, have allowed some further management democratization around the periphery of the treaty itself. The PNCA, as a United States-only agreement, has been subject to growing legal mandates to help Columbia River salmon and to incorporate the input of states, Native American tribes, and the general public in river management decision making. The result of these legal changes on the PNCA has been that Columbia River flows for fish have become constraints which have to be taken into account before regional system-wide power generation can be planned out. By the 1990s these restrictions had added up to a major restriction on the PNCA's ability to optimize power production—but also reflected a positive ability of this multi-party U.S. agreement to incorporate and meet a growing array of needs and interests (Logie 1993).[9]

More recently, the termination of the thirty-year sale of the Canadian Entitlement also widened participation and spread benefits to previously excluded people and interests. In the early 1990s, as BC Hydro began negotiating a new contract for the sale of the Canadian Entitlement, the people of the Canadian portion of the Columbia Basin, who had gained significant power within provincial politics over the course of thirty years, were able to win for the first time direct control over a significant share of B.C.'s treaty benefits. In 1995, the province committed to providing a portion of the downstream benefits from the treaty dams' second thirty years to a new organization, the Columbia Basin Trust, to be used for economic investments and ongoing economic, social, and environmental programs within the Canadian Columbia Basin. The Columbia Basin Trust has a strong commitment to wide participation in decision making, and has become a leader in discussions about the future of the treaty and the upper Columbia Basin (Bankes 1996; Halleran; Hearn 2008; Columbia Basin Trust).

Lessons from the History of the Columbia River Treaty: International Jurisdiction, Regional Scale and Democratization

To help think through the lessons from the treaty's history, I return to an analysis of the relationship among jurisdictional level, geographic scale, and democratization. I consider the relationship between each pair of factors in turn. I then revisit the assumptions that underlie the common wisdom that formal, "top-down" law is less inclusive and less democratic than more "grass-roots," smaller-scale efforts. Both analyses can now be informed with the specifics of CRT history. From these analyses come more fine-tuned lessons about the specific ways that an international treaty in the Columbia River influences other jurisdictions, other geographic scales, and the extent of democratization of river management.

International Jurisdiction and Democratization

Did the international jurisdiction of the treaty impede democratization? In other words, did it limit participation and inhibit sharing of benefits? Or did it actually help? The answer, in short, is that during the CRT's development, it helped; after it was codified, it limited further democratization.

During its development, the international nature of the treaty contributed to democratization. The treaty's international jurisdiction was not, however, inherently more inclusive. What mattered was how it interacted with the two federal governments and their challengers during the long years of political construction of the treaty. The critical mechanisms whereby negotiations over an *international* law in particular effected democratization, were: first, international opportunity placed the two federal governments in a dependent position, wanting something they could not achieve by themselves; and second, a promised international agreement provided forms of political leverage within domestic political negotiations to nonfederal actors that would not have been available otherwise. The reason this was democratizing—that is, the reason it widened participation and the distribution of benefits—was that U.S. federal agencies had dominated Columbia River policy beforehand, and both federal governments were in many ways more powerful than those who gained against them. It must be noted that were this not the case—that is, if the parties that gained had already had controlling influence over Columbia River management—then their gains against the federal government would not have constituted management democratization.

This merits further comment. Some might legitimately argue that the increased power of private utilities and "direct service industries" in the PNCA and the CSPE (Agreements #2, 4, and 5) was not democratization, but its opposite. No doubt, U.S. private utilities and industries were already politically and economically powerful before 1964, and it was precisely this power that allowed them to gain more influence over Columbia River management through the long negotiation process over the CRT and its associated agreements. How the empowerment of private versus public power producers in the Columbia River (and elsewhere) should be judged in terms of democracy is, of course, a long and storied debate (Dick 1989; Odgen 1949; Robbins and Foster 2000; Brooks 2006). Their gains may be considered democratization in this historical case only because they were not direct participants in river management before the treaty, and because federal government law and policies in the 1930s and 1940s—particularly the public preference clause in the Bonneville Project Act and its implementation by the BPA—had in fact given them (especially private utilities) low priority access to federal Columbia River power. This highlights the fact that democratization is as much a relative notion—that it depends on how groups and interests are arrayed at a particular moment in time and how this changes—as it is an absolute one.

In the treaty's development, international law was democratizing because of the specific ways it interacted with federal law and both regional and national politics at that moment in history. It destabilized existing political hierarchies, allowing new actors, jurisdictions, and interests to break in to become participants in management and recipients of river benefits.

While the international jurisdiction of the CRT helped democratize river management during the period of treaty construction, it had the opposite effect once the treaty and its associated agreements were codified. Unless avenues for challenge are written in—and they were not, in the CRT—international law is not accessible to ongoing political challenge or legislative evolution. Other agreements besides the treaty have proved more flexible and open, such as the PNCA because it is a single-nation agreement which could be influenced by new legal mandates, and the sale of the Canadian Entitlement because it had a more limited term.

The international jurisdiction of the treaty is not the only jurisdiction that has played a role in limiting democratization since the treaty was codified. The entwining of smaller jurisdictions and scales has also reinforced the treaty's immobility. One-sided abrogation is not an option because of federal law and national and regional politics in both countries. On the U.S. side, for example, federal treaties are the supreme law of the land, and one-sided abrogation would require huge compensation to Canada, a partner whom the United States values enough to feel obligated to pay. Further, those who benefit from the large volumes of Columbia River power made possible by the CRT are government agencies and industries with considerable political power within their respective regional and national jurisdictions and alliances. Against this constellation of the politically powerful, those who critique treaty operations cannot leverage the kind of national political agreement that would be required to call for abrogation. Nor is early termination, which would require two-party agreement, possible. This would require wide agreement within regional and national polities on both sides of the border, an impossible achievement for treaty critics.

International Jurisdiction and Regional-scale Management
What has been the relationship between the international jurisdiction of the CRT and regional-scale management in the Columbia River? Has an international treaty undermined regional control? Or, has the CRT, like the EU in Europe, enabled regions to emerge by overcoming the oppressive control of national governments? Alternatively, is there something different or more complex at work?

The CRT strengthened regional control. In some ways, regions were strengthened in the CRT in similar ways to European regions in the EU: international jurisdiction undercut the dominance of the federal governments, and enabled a regional jurisdiction on the Canadian side and regional interests on the U.S. side to claim more decision-making authority and a greater share of the river's benefits. Focusing on this jurisdictional hierarchy, however, obscures the fact that regional, federal, national, and international jurisdictions and scales were fundamentally entwined in CRT negotiations and outcomes, and in the Columbia River "region" itself.

A more accurate statement is that regional jurisdictions and scales were empowered in the course of the CRT negotiations because existing multi-layered political and administrative geographies on both sides of the border were already regionalized under existing national law. Both B.C. and the U.S. Pacific Northwest were preexisting regions whose residents and administrators either had specific regional rights and privileges in relation to the Columbia River under the national constitution (on the B.C. side), or (on the U.S. Pacific Northwest side) were used to acting as if they did, thanks to national legislation creating the BPA and extending its responsibilities to include power from new federal Columbia River dams. Regional empowerment under the CRT and its associated agreements was not so much a rising of the region above other jurisdictions and scales but the entwining of multiple jurisdictions and scales in a way which reinforced, more fully integrated, and codified existing regional management of the Columbia River and Columbia River power.

At the same time the international jurisdiction of the treaty did not create but reinforced regionalization of Columbia River management, regionalization under the CRT and its associated agreements also helped further internationalization. Regionally organized interests were more willing to embrace power distribution, coordination, and financial agreements across international lines, than were the federal governments. This was especially true on the Canadian side. Many in the Canadian federal government feared a loss of Canadian rights and interests against the huge economic power of the United States, which was seen in nationalistic terms as both competitive with Canada and potentially exploitative of Canada (Swainson 1979). British Columbia's Bennett administration, in contrast, felt far more threatened by the ambitions of the federal Canadian government than it did by the United States. Bennett's attitude was not unusual: Canadian provinces have repeatedly chosen to interconnect electric transmission grids with the U.S. rather than with a national or regional Canadian transmission system (Froschauer 1999). Leaders in B.C. also saw regional commonalities that crossed the international border. Development, interconnection, and coordination could bring to B.C. the kind of benefits the U.S. Pacific Northwest had been harnessing from the Columbia River for over two decades (Swainson 1979). "Regional" interests therefore helped to further internationalization.

Regional-scale Management and Democratization

There is a popular notion that regional-scale governance is inherently more inclusive and attuned to wide social and environmental needs than national-scale governance. In the case of regional Columbia River management under the CRT and its associated agreements, however, empowerment and codification of regions improved participation and benefits-sharing only slightly.

On the U.S. side, regionalization did not further democratization; rather, democratization widened and codified existing regionalization. It was only when a host of nonfederal utilities and industries forced their way into participation in river management using their leverage over the CRT that fuller regional coordination in river and power flows developed in the form of the PNCA. After this regional coordination was set up, the regional federal agencies and offices joined with these nonfederal utilities and industries to form a regionally unified political bloc; this was what won regional codification and privilege through the Pacific Northwest Preference Act.

On the Canadian side, regional-scale management helped democratize resource management compared to what might have occurred with federal management, but only to a very limited extent—mainly in helping to distribute the promised benefits of the CRT and in limiting its social costs. Participation in Canadian Columbia River decision making was never wide. The analyses and negotiations over the treaty were run by a small coterie of office-holders and provincial agency analysts. The B.C. government was nonetheless an elected government with representatives from throughout the province, and was more attuned to finer resolutions of needs and interests in the province than was the Canadian federal government. Compared to the federal government's goal of maximum economic development for the province as a whole, the B.C. government consistently prioritized spreading economic opportunity to remote parts of the province, protecting the Fraser River fisheries even if it forced hydropower development into other river basins where it would be more costly, and protecting East Kootenay farmland and communities (Swainson 1979).

The Fallacies of Our Underlying Assumptions: Jurisdictional Level, Geographic Scale, and Democratization in the Columbia River Treaty and Associated Agreements

In Part II I drew on the academic geography literature to reveal the problems with four common assumptions about the interrelationships among geographic scale, jurisdictional level, and democratization. Here I come back to those same assumptions, this time focusing specifically on the case of the CRT. I show that in this case, as well as in more abstract theory, these assumptions profoundly misrepresent the interrelationships among international law, regional-scale management, and management democratization.

First is the assumption that jurisdictional authority is hierarchical, with each higher level trumping the authority of lower level. Thus, international is thought to trump subnational. But in the CRT case, in fact, an international treaty was built largely by regional actors and interests; the treaty in turn further empowered regional authority in the management of the Columbia River. Regionally organized interests and, in Canada, a regional jurisdiction, were also able to trump federal interests and authority on both sides of the border.

Only in one way did international jurisdiction trump others: by codifying the existing management system in a way lower-level jurisdictional law could not. Even here, though, the system it codified grew out of *regional* interests and politics. International law froze time; it did not overpower a smaller space.

The regional federal management of the Columbia River in the United States belies a second problematic assumption: that jurisdictional level is congruent to geographic scale. The CRT case shows this assumption to be drastically flawed. The BPA and the Federal Columbia River Power System are both regional in geographic scale and federal in jurisdiction. The treaty also encompasses a territory quite different from its jurisdiction: management under the CRT is regional in scale, both in terms of its actions and the geographical affiliations of its decision makers and managers, but international in jurisdiction.

The third problematic assumption is that smaller jurisdictions and geographic scales are inherently more democratic. In the development of the CRT and its associated agreements, regional empowerment correlated with more democratic river management, but regionalization did not consistently lead to democratization. On the U.S. side, democratization in the form of increased participation widened and codified already existing regional management, not the other way around. On the Canadian side, regionalization in the form of devolution from the federal to B.C. government furthered democratization in the sense that it helped spread resource benefits, but only to a limited extent; and it did little to expand participation. Regionalization if anything impeded long-term democratization of Columbia River management: regional interests helped to lock in rules to the various agreements that strictly limited subsequent changes in participation, priorities in river management, and benefits distribution.

The final problematic assumption about the relationship among jurisdictional level, geographic scale, and democratization is an assumption about law and geography more generally. Both geography and law are too often seen as timeless and apolitical. The CRT case shows at every step that both law—that is, the authority of different jurisdictions, and the content of legal agreements—and geography—

the geographical organization of management authority and benefits-sharing—were politically contested and dynamic. The treaty and its associated agreements effected considerable change in both geography and law, and were products of these political contests.

How and Why Did an International Agreement Further—and Limit—Democratization of Columbia River Management?

Having problematized the relationships among international jurisdiction, regional-scale management and resource management democratization in the CRT, and having torn apart the common assumptions that underlie standard judgments about how these factors relate, how do we rebuild a conception of their interrelationships that is informed by the history of the treaty? Generally, the CRT case confirms geographers' arguments about the complex, entwined, and fundamentally political relationships among legal jurisdiction, geographic scale, and democratic change. The CRT case shows that even in the "traditional" era of supposedly unitary nation-states, under a customary form of international law like a treaty, different levels of governmental organization and geographic scales were mutually constructed through complex and contested interplays of politics.

But what are the *practical* implications of this history? If these factors are so entwined and complex, how do we conceive of the influence that international jurisdiction in the Columbia River can have on regionalization and democratization? More specifically, under what conditions does international jurisdiction of the CRT further regional-scale management and advance resource management democratization, and in what circumstances does it impede them?

The effects of international law on the authority of other jurisdictions, on interests organized at sub-national geographic scales, and on the breadth of participation in resource management and distribution of resource benefits, must be understood as contingent on the specific way international law interacts with political dynamics among these other factors at a particular historical moment. The case of the Columbia River Treaty reveals one set of circumstances in which international law can lead to both regionalization and (limited) democratization. Once again, by "democratization" I mean specifically the broadening of participation in and influence over management decision making, and the wider distribution of benefits. If, during the development or *construction* phase of an international agreement, national or federal governments need political agreement or cooperation from other parties, and if national or federal jurisdictions or national-scale interests have traditionally dominated policy making, then an international law may help destabilize that entrenched national or federal power and bring new participants into governance. In that case, international law may further democratization. If some of those who gain against national or federal governments and interests are sub-national jurisdictions or regionally organized blocs, then an international law can empower regions as well.

But the CRT case also reveals one set of circumstances in which international law may impede democratization, even when it helps to reinforce regional-scale authority. If an international law *codifies* regional-scale management, and with it, the particular participants or interests included within that region's purview, it can diminish subsequent political openness and potential for change. International law may make regional systems of governance especially immutable, as international agreements are not necessarily subject to systems of challenge inscribed within the law of national and subnational jurisdictions.

Implications for the Future: Columbia River Treaty Reconstruction

What are the implications of this history and analysis for the future? How can it help guide us to think through what we are likely to see, and what we should *want* to see, as we approach 2014 and 2024?

First, it is likely that treaty re-negotiations, if they occur, like the original treaty negotiations, will be a time when the structure of authority is open to transformation and that politics will be most opened. It will be a time of political contest, in which interests that now have limited influence over and gain limited benefit from Columbia River management are able to win a much greater say and share. This is a good thing: it is a product precisely of the rise to prominence of important interests that were excluded from treaty negotiations in 1964, from Native American tribes and First Nations, to wild fisheries, to the communities and ecosystems near treaty reservoirs. A strong role will probably be played this time by those in the Canadian portion of the Columbia Basin, led by the very active Columbia Basin Trust. The next treaty will not so thoroughly marginalize these voices and interests. Good, however, does not mean easy—there will likely be many different conflicts that will have to be negotiated in multiple jurisdictional and institutional settings.

Second, because of this, any future CRT, like the first treaty, will almost certainly be coupled with several associated agreements that address issues that cannot be resolved at the international jurisdictional level. This is the obvious challenge, and the process ahead of us: negotiating a complex and multi-layered array of conflicts into a coherent set of agreements. It is a good thing there will be at least a ten-year window for negotiations, for both conflicts and agreements are likely to be even more complex than those of the 1964 CRT.

There are other challenges that may be less obvious, but which this history suggests are equally important to consider. The third implication is that even as we take to heart the important voices and interests of those who were excluded in 1964, we cannot think that by including them, we will necessarily include all who will be affected by the next CRT. It may be important to seek out other perspectives and needs by, for example, promoting wide media coverage of the options and implications for future CRT management; hosting focus groups in different communities—from southeastern B.C. and northwestern Montana, to estuary towns in Oregon and Washington, to fishing communities north to Alaska; or by studying still too-often ignored interdependencies between people and Columbia River ecosystems, such as the importance of fluvial geomorphological processes that renew habitat important to commercial or sport fisheries, or the continued nutritional importance of subsistence fishing, hunting, and gathering of Columbia Basin species.[10]

The fourth implication is that to truly assess the level of democratization in any set of future agreements, we will need to think in relative terms, asking who and what have gained influence and benefits, and how their gains relate to their existing power relationships. If only the already-powerful gain, we should not pat ourselves on the back too much.

The fifth implication is that we should not assume that *regional* institutions and jurisdictions in the Columbia River are necessarily more democratic and inclusive than institutions at larger spatial scales. They may be, and there are certainly reasons to appreciate particularly the role of new (that is, new since the 1964 CRT) regional players like the Northwest Power and Conservation Council and the Columbia Basin Trust. However, "regional" in the Columbia River is clearly in part about protecting the Pacific Northwest's privileged access to Columbia River resources, especially Columbia River power, and we should not assume this privilege is deserved. Because national governments will necessarily be involved in any treaty renegotiation, there may be some uncomfortable questions raised about the benefits and costs, winners and losers, of regional privileges relative to the other provinces and states. Such questions may be very threatening to central players in the region, but it also could be an opportunity to think critically about what "the region" in the Columbia River is and does, and whether it is in fact as participatory, inclusive, and beneficent as it can and should be. Chances are very good that by the time a new CRT is ratified, the international Columbia River–Pacific Northwest region will be strengthened. However, in the process, it may also be challenged to be more socially and environmentally inclusive.

The final implication is that any inflexibly codified agreements will impede further inclusion later on. No matter how well negotiated or inclusive, no agreements will be able to anticipate all the challenges and changes to come, all the trade-offs. Somehow there will need to be a system set up for ongoing modification of participants and priorities. Of course, constant negotiation among all possible parties and interests would make an international agreement completely unwieldy, and could foster conflict and bureaucracy as much as participation and harmony.

It may be that, for the sake of functionality of management institutions, potential modification should take place only within certain bounds or relatively long time intervals of say, ten or fifteen years. But the inflexibility and inaccessibility of the sixty-year minimum term of the 1964 CRT will not be tolerated in a new 2024 CRT, nor should it be. The creative challenge this time—and the political pressure—will be somehow to institutionalize a system that can be functional, democratic, and both socially and environmentally inclusive on an ongoing basis.

Notes

Acknowledgments: This chapter is adapted from Eve Vogel, "Regionalization and Democratization through International Law: Intertwined Jurisdictions, Scales and Politics in the Columbia River Treaty (in Symposium: Complexities of Scale: The Role of the Subnational in International Law)," *Oregon Review of International Law* 9, no. 2 (2007): 337–88. I thank Hari Osofsky for inviting my participation in the symposium, and the *Oregon Review of International Law* for publishing the original article and permitting this adaptation. Once again, I thank John Shurts and John Harrison of the Northwest Power and Conservation Council for all they have taught me about Columbia River policy, politics, and history, including the Columbia River Treaty. I thank my Ph.D. advisor Prof. Alexander B. Murphy for his help thinking through issues of political geography and scale. John Harrison and Paul Hirt both encouraged me to participate in the conference that preceded this volume, but the difficulties of child care arrangements kept me at home. After the conference, Matthew McKinney encouraged me to contribute to this volume, and Barbara Cosens welcomed me in. Thank you—I am delighted to contribute. The following people met with me to share their knowledge with me: Mike Hansen and Rich Nassief, Northwest Power Pool; Garry Merkel and Josh Smienk, Columbia Basin Trust; Bill Green, Canadian Columbia River Inter-tribal Fisheries Commission; and Anthony White, Bonneville Power Administration. An EPA STAR fellowship enabled me to travel to the Canadian portion of the Columbia Basin in December 2001. All errors are my own.

1. See e.g., Berman 2002; Glassman1999; Swyngedouw1997; Brenner2004; Keating1998; and the special issue on scale and international law, in the *Oregon Review of International Law* 9, no. 1 (2007).

2. The European Commission passed the Water Framework Directive (WFD) in 2000 (European Communities, 2000); European Commission, 2011). Under the directive, all member countries had to assign each river basin within their territory to an individual river basin district, and to designate a "competent authority" to develop and administer river basin management plans. For rivers that cross national borders—and in Europe, there are many such rivers and many such borders—member countries had to coordinate with co-riparians in order to "endeavor to produce a single river basin management plan." (Article 13, No. 3) The resulting map of Europe defies our usual view of Europe, which is often partitioned by national borders (even if those borders keep changing) (UK WRc (2007).

It is worth noting that, parallel to the argument I make in this chapter, the WFD on the surface seems to represent the imposition of a supranational government authority over an entire subcontinent, much as federal law in either the U.S. or Canada might be seen as imposing a mandate across the states or provinces. However, like the CRT, at the local and regional levels, the WFD is in many cases being administered by or else is functioning in coordination with existing or newly developed local and regional institutions— ministries of federal states within Germany, for example, and local farmers' organizations. In negotiating the WFD, Germany—on behalf of its states—insisted that delegation of authority to subnational bodies be allowed. Thus, a supranational decree was both shaped by and empowers subnational and transnational regions (Moss 2006; Page and Kaika, 2003. Again, parallel to the CRT, this was not at the expense of national power—rather, supranational, national, and regional are all mutually constitutive.

3. See e.g., Morton J. Horwitz1977; Nelson1994; the special issue on the work and legacy of Willard Hurst in *Law and History Review* 18 (2000), and the symposium issue on Complexities of Scale: The Role of the Subnational in International Law, in the *Oregon Review of International Law* 9, no. 1 (2007).

4. This section on the PNCA is also informed by conversations with Mike Hansen and Rich Nassief, Northwest Power Pool, in 2000.

5. Logie 1993; Northwest Power Pool 1979; Dean and Schultz 1989); Bonneville Power Administration, U.S. Army Corps of Engineers, and Bureau of Reclamation 2001. The current PNCA is "1997 Pacific Northwest Coordination Agreement: Agreement for Coordination of Operations among Power Systems of the Pacific Northwest," in Fed. Reg. 62 (1997): 43,548, http://www.nwd-wc.usace.army.mil/PB/oper_planning/pnca.html.

6. Canada-B.C. Agreement, July 8, 1963. There was also a second supplementary Canada-B.C. Agreement signed on Jan. 13, 1964. Both agreements are in CRT Documents, at 39 and 44 (respectively).

7. The two central documents from these are: Attachment Relating to Terms of Sale, Jan. 22, 1964, CRT Documents: 117; and Canadian Entitlement Purchase Agreement, Canada-U.S., Aug 13, 1964, http://www.internationalwaterlaw.org/documents/regionaldocs/columbia_river_note1.html

8. In addition to those cited I am indebted to conversations with Garry Merkel and Josh Smienk, Columbia Basin Trust, and Bill Green, Canadian Columbia River Inter-tribal Fisheries Commission, in 2001.

9. Logie1993. A discussion with Mike Hansen of the Northwest Power Pool in 2000 also informed this statement on the PNCA in recent times.

10. This is not to say that there is some foreordained needed level or range of participation during treaty negotiations, that policymakers should not rest until they've talked with every individual and group throughout the basin, or that every single interested party should have veto power over a new treaty. Nor is it to say that a new treaty needs to codify an ever-expanding range of participation directly into ongoing management of the treaty dams. It simply means that, because treaty negotiations represent a temporarily open political moment, there should be an effort to inform and invite the contributions of often marginalized people, places, interests, and groups, even ones we don't necessarily know about. Ultimately the influence of such marginalized people and interests is likely to be felt at least as much in a range of side agreements that are crucial to adjusting how the river is managed, its policies deliberated, and its benefits shared, as within the treaty itself. Of course, all agreements— treaty and side agreements alike—need to have some degree of stability and predictability. The point here is that this stability and predictability (as well as the codified opportunities for change—see continuing discussion) should embed the results of negotiations among an inclusive set of people, places, and interests.

Works Cited

Amin, Ash. "Regions Unbound: Towards a New Politics of Place," *Geografiska Annaler, Series B: Human Geography* 86, no. 1 (2004): 33–44.

Armitage, Derek Russel, Fikret Berkes, and Nancy Doubleday, eds. *Adaptive Co-Management: Collaboration, Learning, and Multi-Level Governance*, Sustainability and the Environment (Vancouver: University of British Columbia Press, 2007).

Armitage, Derek Russel, Fikret Berkes, and Nancy Doubleday, eds. *Adaptive Co-Management: Collaboration, Learning, and Multi-Level Governance*, National Research Council, *New Strategies for America's Watersheds* (Washington, D.C., 1999).

Bankes, Nigel. *The Columbia Basin and the Columbia River Treaty: Canadian Perspectives in the 1990s*. Northwest Water Law & Policy Project (Portland, Ore, 1996).

B.C. Hydro Pioneers, *Gaslights to Gigawatts: A Human History of BC Hydro and Its Predecessors by the BC Hydro Power Pioneers* (Vancouver, B.C., 1998).

B.C. Hydro and Power Authority Act, S.B.C. ch. 8 (British Columbia, Canada, 1962).

Berman, Paul Schiff. "Book Review Essay: Seeing Beyond the Limits of International Law," *Texas Law Review* 84 (2006): 1265–306.

Berman, Paul Schiff. "The Globalization of Jurisdiction," *University of Pennsylvania Law Review* 151 (2002): 311-45.

Binus, Joshua. "Bonneville Power Administration and the Creation of the Pacific Intertie, 1958–1964." (Master's thesis, Portland State University, 2008).

Blumm, Michael C., and F. Lorraine Bodi. "Commentary," in *The Northwest Salmon Crisis: A Documentary History*, ed. Joseph Cone and Sandy Ridlington (CorvallisOregon State University Press, 1996), 125–27.

Blumm, Michael C. "The Northwest's Hydroelectric Heritage: Prologue to the Pacific Northwest Electric Power and Conservation Act," *Washington Law Review* 58 (1983): 175–244.

Bonneville Power Administration, U.S. Army Corps of Engineers, and Bureau of Reclamation, *The Columbia River System: The Inside Story*, Columbia River System Operation Review (2001).

Bonneville Power Administration. "Columbia River Treaty," *Backgrounder*, 3/ 1989.

Bradshaw, Ben. "Questioning the Credibility and Capacity of Community-Based Resource Management," *The Canadian Geographer* 47, no. 2 (2003).

Brenner, Neil. *New State Spaces: Urban Governance and the Rescaling of Statehood* (Oxford, England and New York: Oxford University Press, 2004).

Brenner, Neil. "World City Theory, Globalization, and the Comparative-Historical Method: Reflections on Janet Abu-Lughod's Interpretation of Contemporary Urban Restructuring," *Urban Affairs Review* 37, no. 1 (2001): 124–47.

Brooks, Karl Boyd. *Public Power, Private Dams: The Hells Canyon High Dam Controversy*, Weyerhaeuser Environmental Books (Seattle: University of Washington Press, 2006).

Brown, Christopher J., and Mark Purcell, "There's Nothing Inherent About Scale: Political Ecology, the Local Trap, and the Politics of Development in the Brazilian Amazon," *Geoforum* 36 (2005): 607–24.

Bruch, Carl, Libor Jansky, Mikiyasu Nakayama, and Kazimierz A. Salewicz, eds. *Public Participation in the Governance of International Freshwater Resources* (Tokyo, New York and Paris: Bookwell Publications, 2005).

Calkins, William Clifford. "Some Political and Legal Aspects of the Columbia River Treaty and Related Documents between the United States and Canada." (Master's thesis, University of Oregon, 1970).

Canada-B.C. Agreement, July 8, 1963.

Center for International Environmental Law, *The World Trade Organization and Environment: Technical Statement by United States Environmental Organizations*, vol. 2007 (1999). Available at http://www.ciel.org/Tae/USNGOstatementonWTO.html.

Columbia Basin Trust, *Report to Residents* (Castlegar, British Columbia, 2009). Available at http://cbt.org/uploads/pdf/2009R2R_FINAL.pdf.

Columbia Basin Trust, *Columbia Basin Trust Briefing Book* (Castlegar, British Columbia, 2001).

Columbia River Treaty, Protocol and Related Documents. Departments of External Affairs and Northern Affairs and National Resources (Ottawa, Canada, 1964), Available at http://www.empr.gov.bc.ca/EAED/EPB/Documents/1964_treaty_and_protocol.pdf.

Committee on Protection and Management of Pacific Northwest Anadromous Salmonids, Upstream. *Salmon and Society in the Pacific Northwest* (Washington, D.C., 1996).

Cox, Kevin. "Spaces of Dependence, Spaces of Engagement and the Politics of Scale, Or: Looking for Local Politics," *Political Geography* 17, no. 1 (1998a): 1–23.

———. "Representation and Power in the Politics of Scale," *Political Geography* 17, no. 1 (1998b): 41–44.

Dean, Lawrence A., and Merrill S. Schultz, *Pacific Northwest Coordination Agreement: Background and Issues*. A Report Prepared for the Northwest Power Planning Council (Portland, Ore., 1989).

Dick, Wesley Arden. "When Dams Weren't Damned: The Public Power Crusade and Visions of the Good Life in the Pacific Northwest in the 1930s," *Environmental Review* 13 (1989).

Dodds, Klaus. *Geopolitics: A Very Short Introduction*. New York: Oxford University Press, 2007.

European Commission. Introduction to the New EU Water Framework Directive (2011); available from http://ec.europa.eu/environment/water/water-framework/info/intro_en.htm.

European Communities. Directive 2000/60/EC of the European Parliament and of the Council of 23 October 2000 Establishing a Framework for Community Action in the Field of Water Policy, L372: 1–72 (2000).

Evenden, Matthew D. *Fish Versus Power: An Environmental History of the Fraser River*, Studies in Environment and History ed. Donald Worster and J. R. McNeill (Cambridge, UK: Cambridge University Press, 2004).

Froschauer, Karl. *White Gold: Hydroelectric Power in Canada* (Vancouver: University of British Columbia Press, 1999).

Glassman, Jim. "State Power Beyond the 'Territorial Trap': The Internationalization of the State," *Political Geography* 18 (1999): 669–96.

Halleran, Mike. A video history of the Columbia River development Treaty (WestLand Television, for Columbia Basin Trust: WestLand Television, for Columbia Basin Trust), video (VHS) (Nakusp, B.C., 1998)

Hanks, James Francis. "The Columbia River Treaty." (Master's thesis, University of Oregon, 1970).

Hearns, Glen, and the Canadian Columbia River Forum. *The Columbia River Treaty: A Synopsis of Structure, Content, and Operations* (2008). Available at http://www.ccrf.ca/assets/docs/pdf/hearns_crt_structure_final_20090428.pdf.

Horwitz, Morton J. *The Transformation of American Law, 1780–1860* (Cambridge, Mass. and London: Harvard University Press, 1977).

Hyde, John, Kelvin Ketchum, and Bolyvong Tanovan. "Ecosystem Management and
the Columbia River Treaty." (Paper presented at the conference, Breaking Down the
Barriers: Toward Ecosystem-Based Management in the Columbia River Basin and
Beyond, Spokane, Wash., 2002).
International River Improvements Act, R.S.C. ch. 1–20 (Canada, 1955).
ISG (Independent Scientific Group). *Return to the River: Restoration of Salmonid Fishes in the
Columbia River Ecosystem* (Portland, Ore, 2000).
Keating, Michael. *The New Regionalism in Western Europe: Territorial Restructuring and Political
Change* (Cheltenham, UK and Northampton, Mass., USA: Edward Elgar Pub, 1998).
Kemmis, Daniel. *Community and the Politics of Place* (Norman, Okla. and London: University
of Oklahoma Press, 1990),
Kenney, Douglas S., Sean T. McAllister, William H. Caile, Jason S. Peckham. *The New
Watershed Source Book: A Directory and Review of Watershed Initiatives in the Western United
States* (Boulder, Colo., 2000). Available at http://www.colorado.edu/law/centers/nrlc/
publications/watershed.htm.
Krutilla, John V. *The Columbia River Treaty: The Economics of an International River Basin
Development* (Baltimore: Resources for the Future Press, 1967).
Lee, Kai N., Donna Lee Klemka, and Marion Ernest Marts. *Electric Power and the Future
of the Pacific Northwest*. Public Policy Issues in Resource Management (Seattle, Wash.:
University of Washington Press, 1980).
Logie, Pat. *Power System Coordination: A Guide to the Pacific Northwest Coordination Agreement*,
Columbia River System Operation Review (Portland, Ore., 1993).
Mansfield, Becky. "Thinking through Scale: The Role of State Governance in Globalizing
North Pacific Fisheries," *Environment and Planning A* 33 (2001): 1807–27.
Massey, Doreen. *For Space* (London; Thousand Oaks, CA; and New Delhi: Sage
Publications, 2005).
Morrill, Richard. "Electoral Geography and Gerrymandering: Space and Politics," in
Reordering the World: Geopolitical Perspectives on the Twenty-First Century, ed. George J.
Demko and William B. Wood, second edition (Boulder, Colo.: Westview Press 1999),
117–38.
Moss, Timothy. "Solving problems of "fit" at the expense of problems of 'interplay"?
The spatial reorganisation of water management following the EU water framework
directive," in *Integrated Water Resources Management: Global Theory, Emerging Practice and
Local Needs*, ed. Peter P. Mollinga, Ajaya Dixit and Kusum Athukorala (New Delhi: Sage
Publications India Pvt. Ltd., 2006), 64–108.
Muckleston, Keith. *International Management in the Columbia River* (Paris, France,: UNESCO
2003).
Murphy, Alexander B. "Rethinking the Concept of European Identity," in *Nested Identities:
Nationalism, Territory, and Scale*, ed. Guntram H. Herb and David H. Kaplan (Lanham,
Md.: Rowan and Littlefield, 1999a), 53–73.
Murphy, Alexander B. "The Sovereign State System as Political-Territorial Ideal: Historical
and Contemporary Considerations," in *State Sovereignty as Social Construct*, ed. Thomas J.
Biersteker and Cynthia Weber (Cambridge, UK and New York: Cambridge University
Press, 1999b), 81–120.
Nelson, William Edward. *Americanization of the Common Law: The Impact of Legal Change on
Massachusetts Society, 1760–1830*, 1994 ed. (Athens: University of Georgia Press, 1994).
Northwest Power and Conservation Council, *Columbia River Basin Fish and Wildlife
Program: Twenty Years of Progress* (Portland, Ore., 2003).

Northwest Power Planning Council and Jonathan Lesser, *Review of the Columbia River Treaty and the Columbia Storage Power Exchange* (Portland, Ore., 1989).

Northwest Power Pool, *Position Paper: The Pacific Northwest Coordination Agreement* (Portland, Ore., 1979).

NRC (National Resources Committee), *Regional Planning, Part I—Pacific Northwest* (Washington, D.C., 1936).

NW Water Law and Policy Project: http://www.lclark.edu/dept/water.

Ogden, Daniel Miller Jr., "The Development of Federal Power Policy in the Pacific Northwest." (Ph. D. dissertation, University of Chicago, 1949).

Ohmae, Kenichi. *The End of the Nation State: The Rise of Regional Economies* (London: Free Press, 1995).

Osofsky, Hari. "The Intersection of Scale, Science, and Law in *Massachusetts v. EPA* (in Symposium: Complexities of Scale: The Role of the Subnational in International Law)," *Oregon Review of International Law* 9, no. 2 (2007): 233–60.

Pacific Northwest Consumer Power Preference Act, 837-837h 16 (P.L. 88-552) 1964).

Pacific Northwest Electric Power Planning and Conservation Act, 16 U.S.C. 839-839h (P.L. 96-501) (USA, 1980).

Page, Ben, and Maria Kaika. "The EU Water Framework Directive: Part 2. Policy Innovation and the shifting choreography of governance." *European Environment* 13 (2003): 328–43.

Park, S.O. "Rethinking the Pacific Rim,"" *Tijdschrift voor Economische en Sociale Geografie* 88, no. 5 (1997): 425–38.

Power Development Act, S.B.C. ch. 4 (British Columbia, Canada, 1961).

Robbins, William G. and James C. Foster, eds. *Land in the American West: Private Claims and the Common Good* (Seattle, Wash. and London: University of Washington Press, 2000)

Rodríguez-Pose, Andres, and Nicholas Gill. "The Global Trend Towards Devolution and Its Implications," *Environment and Planning C: Government and Policy* 21, no. 3 (2003):

Sale, Kirkpatrick. *Dwellers in the Land: The Bioregional Vision*, 2nd ed. (Athens: University of Georgia Press, 2000)

Shepard, Michael Perry, and A.W. Argue. *The 1985 Pacific Salmon Treaty: Sharing Conservation Burdens and Benefits* (Vancouver: University of British Columbia Press, 2005).

Smith, Neil. "Geography, Difference, and the Politics of Scale," in *Postmodernism and the Social Sciences*, ed. J. Doherty et al. (London: Macmillan, 1992), 57–79.

Sneddon, Chris, and Coleen Fox. "Transboundary River Basin Agreements in the Mekong and Zambezi Basins: Enhancing Environmental Security or Securitizing the Environment?" *International Environmental Agreements: Politics, Law and Economics* 7, no. 3 (September) (2007): 181–202.

Storper, Michael. *The Regional World: Territorial Development in a Global Economy* (New York: Gullford Press, 1997).

Swainson, Neil A. *Conflict over the Columbia. The Canadian Background to an Historic Treaty* (Montreal: Institute of Public Administration of Canada, 1979).

Swyngedouw, Erik. "Scaled Geographies: Nature, Place, and the Politics of Scale," in *Scale and Geographic Inquiry: Nature, Society, Method*, ed. Eric Sheppard and Robert B. McMaster (Oxford, UK and Malden, Mass.: Wiley-Blackwell, 2004).

———. "Neither Global nor Local: 'Glocalization' and the Politics of Scale," in *Spaces of Globalization: Reasserting the Power of the Local*, ed. Kevin R. Cox (Gullford Press, 1997), 137–66.

Thelen, Kathleen Ann, and Sven Steinmo. "Historical Institutionalism in Comparative Politics," in *Structuring Politics: Historical Institutionalism in Comparative Analysis*, ed. Sven Steinmo, Kathleen Ann Thelen, and Frank Longstreth (Cambridge, UK, and New York: Cambridge University Press, 1992).

Vogel, Eve. "Defining One Pacific Northwest Among Many Possibilities: The Political Construction of a Region and Its River during the New Deal." *Western Historical Quarterly* 42, No. 1 (Spring) (2011): 28–53.

———. "Regional Power and the Power of the Region: Resisting Dam Breaching in the Pacific Northwest," in *Contentious Geographies: Environment, Meaning, Scale*, ed. Michael Goodman, Max Boykoff, and Kyle Evered (Aldershot, UK, and Burlington, Vt.: Ashgate Publishing, 2008).

———. "The Columbia River's Region: Politics, Place and Environment in the Pacific Northwest, 1933–Present." (Ph.D. dissertation, University of Oregon, 2007).

Volkman, John M. *A River in Common: The Columbia River, the Salmon Ecosystem, and Water Policy*. Report to the Western Water Policy Review Advisory Commission (Portland, Ore., 1997).

UK WRc. "National and International River Basin Districts: Submissions in Accordance with Article 3 of the Water Framework Directive,"" (2007). Available at http://ec.europa.eu/environment/water/water-framework/facts_figures/pdf/2007_03_22_rbd_a3.pdf.

Waterfield, Donald. *Continental Waterboy: The Columbia River Controversy* (Toronto and Vancouver: Clarke, Irwin, 1970).

William G. and James C. Foster, eds. *Land in the American West: Private Claims and the Common Good* (Seattle, Wash. and London: University of Washington Press, 2000).

Wilson, James Wood. *People in the Way: The Human Aspects of the Columbia River Project* (Toronto: University of Toronto Press, 1973).

Yang, C. "An Emerging Cross-Boundary Metropolis in China—Hong Kong and Shenzhen under 'Two Systems'," *International Development Planning Review* 27, no. 2 (2005): 195–225.

Institutional Adaptation and Change in Collaborative Watershed Management:
An Examination of the Northwest Power and Conservation Council's Fish and Wildlife Program

Tanya Heikkila and Andrea K. Gerlak

Introduction

Transboundary institutions, such as treaties, compacts, collaborative agreements, councils, and collaborative programs, have long been recognized as valuable mechanisms for addressing and resolving the conflicts and environmental problems that result from the use and allocation of water resources that cross multiple political jurisdictions, both regionally and internationally (Florestano 1994; Lubell et al. 2002; Wolf et al. 2003; Gerlak and Grant 2009). In the United States, such collaborative efforts have emerged in recent years across a number of the largest and most ecologically, economically, and culturally significant watersheds. They can be found in the marsh wetlands of the Florida Everglades, along the coast of Louisiana, in the Midwest's Great Lakes region, and along the Pacific Northwest's mighty Columbia River (Wiley and Canty 2003; Vigmostad et al. 2005; Heikkila and Gerlak 2005; Doyle and Drew 2008; Gerlak 2008).

Such large-scale collaborative efforts bring together federal and state agencies, local agencies, industry, conservation groups, and other resource users to address problems that command-and-control approaches have failed to solve, such as habitat destruction and nonpoint source pollution (Wondolleck and Yaffee 2003; Brick et al. 2001; Karkkainen 2002; Lubell et al. 2002; Koontz et al. 2004; McKinney and Harmon 2004; Sabatier et al. 2005). The scope of these programs can be extensive—often seeking to restore entire ecosystems (e.g., a bay or river basin) while still maintaining economic stability and accommodating growing populations.

A growing body of literature on watershed and collaborative environmental institutions has argued that the robustness of these types of institutions is predicated on their capacity to learn and adapt, thus allowing them to continue to respond to emerging environmental and political challenges (Dietz et al. 2003; Anderies et al. 2004; Scholtz and Stiftel 2005; Brunner and Steelman 2005). Political science scholars who study institutions in the environmental arena have similarly agreed that adaptability over time is necessary to allow an institution to retain its relevance and efficacy in the face of changing external conditions (Steinberg 2009; Scholtz and Stiftel 2005; Brunner and Steelman 2005). Yet, empirical evidence and examples of how such change and adaptation actually takes place is still in its infancy.

In this chapter we examine institutional change within one of the oldest collaborative watershed management institutions in the United States—the Northwest Power and Conservation Council's Fish and Wildlife Program along the Columbia River. The council and its program to mitigate the effects of hydropower on the health of fish and wildlife along the Columbia basin, was established under the congressionally authorized 1980 Northwest Power Act. The council is just one of many transboundary or collaborative institutions engaged in resource management, protection, or planning in the Columbia Basin; the Columbia River Compact, the Columbia Inter-Tribal Fish Commission, and various interagency tasks forces, committees, and councils also include different state, federal, tribal, and local agencies and stakeholders in basin management (United States General Accounting Office 2002). While the council does not represent the entire footprint of collaborative watershed governance in the Columbia River Basin, its long-standing presence and engagements with many of the various stakeholders and players in the basin does provide a well-structured and long-standing institutional setting within which we can examine how institutional change unfolds and how it responds to the complexities of the ecosystem and its larger institutional context.

Analyzing institutional change and adaptation within this context is important not only for the contributions to the academic literature on watershed institutions and their robustness; it also has implications for regional watershed management in the Columbia. In a 1998 discussion paper, eighteen years after the Northwest Power and Conservation Council was established, the council noted: "The issues that face the region now are more complex and important than ever before, and they arise in a much different world than that of 1980 … There was broad agreement on the need for regional solutions to the problems of 1980. Is this still true in 1998 and beyond?" (Northwest Power and Conservation Council 1998). No doubt the challenges that existed in 1998 for regional planning have and will continue to increase, alongside the expected quadrupling of the region's population by 2100 and the increasing competition for water resources, making the recovery of salmon species even more difficult (Lackey et al. 2006). Examining how the Council's institutional setting has evolved may help point to its capacity to adapt in the future.

What Do We Know About Institutional Change in Watershed Management Institutions?

Some experts have argued that many of our long-standing watershed management institutions are incapable of responding and adapting to the challenges of today's physical and socio-economic contexts (Hasday 1997; Ernst 2003; Grant 2003; Sherk 2005), raising the question as to whether such institutions can be strengthened over time and remain responsive to emerging challenges and crises. Others have found that watershed management institutions often do invest in institutional adaptation over time to address emerging challenges and problems (Schlager and Heikkila 2009), often through new operational-level strategies. Many of the "new" types of watershed management institutions have been heralded as examples of "adaptive management" institutions, which formally embrace adaptive change, particularly at the operational level (Lee 1993; Lee 1995; Costanza and Greer 1995; Gunderson et al. 1995). Yet, some of the earlier studies of adaptive management of these programs focus entirely on the technical and scientific approaches for program implementation, without considering how or whether the broader institutional designs have changed, or how or whether the broader institutional design fosters adaptation.

Recent literature has begun to suggest that institutional adaptation, not just technically driven adaptive management, is important for the health of ecological systems, but that the design of institutions matters in fostering such adaptation. For example, scholars who study socio-ecological systems have argued that these systems are more resilient when they include institutional design features that will explicitly identify changing resource conditions and facilitate strategic operational and policy changes in response (Carpenter et al. 2001; Folke et al. 2005; Gunderson et al. 2006; Janssen et al. 2007; Walker et al. 2004). The watershed management literature similarly notes that institutional features that promote both scientific and social learning around ecosystems can support the robustness of collaborative ecosystem and watershed restoration programs (Walters 1996; Gunderson et al. 1995; Gunderson and Holling 2002; Fazey et al. 2005; Van Cleve; et al. 2007). One type of institutional design feature that may foster the capacity for institutional adaptation is the establishment of a "boundary organization" around which actors with diverse knowledge and resources can share information as a source of learning, as well as facilitate the transfer of knowledge between scientific and policy arenas (Guston 1999; Cash 2001; Kallis et al. 2009; Lejano and Ingram 2009).

Given the growing recognition of the importance of the design features that might foster the adaptive capacity of collaborative watershed institutions, we argue that it is important to assess how the design features of watershed-level institutions do change over time. We analyze the types of changes that have emerged within the Northwest Power and Conservation Council's Fish and Wildlife Program, examining changes at the constitutional, collective choice, and operational level of decisions (Ostrom 2005; Crawford and Ostrom 1995). As we have previously identified, the constitutional settings of these institutions includes the basic rules under the signed, formal agreements that establish the collaborative programs; the collective choice settings include the decision-bodies that govern and administer the programs set forth in the agreements; and the operational settings involves the processes or tools for implementing the agreement or other collective choice decisions (Gerlak and Heikkila 2006). In looking at the types of changes that have occurred at different "levels" of institutional decision making, we are able to shed light on some of the design features of institutions that are fostering change and how different levels of changes emerge in a dynamic context.

Examining Institutional Change in a Large-Scale Watershed Collaborative
Constitutional Level Change
The constitutional setting entails the basic rules under the signed, formal agreements that establish the goals and authorize decision-making authority of the collaborative program. The collaborative restoration effort along the Columbia River can be traced to the 1980 Northwest Power Act, which prompted Oregon, Washington, Idaho, and Montana to enter into an interstate agreement for devising basin-wide planning for energy conservation and fish and wildlife protection in the Columbia River basin. The 1970s energy crisis, coupled with growing concerns over the impact of hydroelectric dams on an economically and culturally significant species—salmon—initially brought this issue onto the political agenda (Heikkila and Gerlak 2005). The Act created the Northwest Power and Conservation Council (named the Northwest Power Planning Council until 2003) as a regional decision-making body, which not only brings the four states together, but also many other key actors involved in fisheries management, restoration, and planning in the Columbia basin. These other actors include numerous federal agencies (U.S. Fish and Wildlife, Bureau of Reclamation, U.S. Army Corps of Engineers, Bureau of Land Management), state fish and wildlife departments, Bonneville Power Administration (BPA), irrigation districts, port districts, city governments, the Pacific Fisheries Management Council, and the Columbia Inter-Tribal Fish Commission. The Northwest Power Act, by coordinating regional efforts at fish and wildlife and energy planning was seen as a novel way to address two critical and related dilemmas facing diverse and overlapping jurisdictions at the time (Blumm 1986).

The Council is required by the Northwest Power Act to integrate recommendations from these diverse agencies, while also taking into account the region's needs for an "efficient, economical, and reliable power supply," when developing basin-wide management plans for fish and wildlife. Because BPA (a federal authority) sells power from most of the dams on the Columbia River, the Northwest Power Act tasked BPA with funding the costs for the Council's Fish and Wildlife Program. The Northwest Power Act also requires the Council to make periodic substantive revisions to its Fish and Wildlife Program, following recommendations from stakeholders including federal and state fish and wildlife agencies and Indian tribes in the basin.

Since 1980 the broad constitutional structure of the program—including the fundamental goals of the Council, its authorized members and their authorizing legislation—have remained intact over time. Notably, however, because of the Act's mandate to make periodic revisions to the Council's Fish and Wildlife Program, the institutional design, by nature, is adaptive, within the constraints of the mandates set forth in the Northwest Power Act. The Council's Fish and Wildlife Program, has been modified multiple times since its first program "plan" in 1982, and the goals of those plans and how they are implemented have also changed substantially (Northwest Power and Conservation Council 2003). Yet, the "constitutional level" design of the Council, as established under the Act has changed once, via an amendment to the Northwest Power Act in 1996, which included a requirement that the Council integrate independent scientific oversight to fish and wildlife program funding. One of the outcomes of the amendment was the restructuring of the scientific advisory boards, as part of the Council's collective choice decision structure, discussed below, which further ties into various operational level adaptations in the Council's Fish and Wildlife Program.

Collective Choice Level Change

 The collective choice setting generally includes the decision-making bodies that govern the programs, as well as those that create and approve operational rules for program implementation. The Northwest Power and Conservation Council, as a collective choice body, is made up of two appointed members from each of the four states, who establish the plans for energy and fish and wildlife management that are needed to comply with the Northwest Power Act. In doing so, the Council coordinates the state efforts in their overall hydropower and fish and wildlife planning, as well as the watershed management activities of other entities that participate in implementing the plans. The Council also receives proposals from other entities in the basin (typically by various federal, state, local, and tribal entities) who wish to work with the Council in implementing its plans; proposals are funded according to the Bonneville Power Administration Council's plans, but approved by the Council.

Under the Northwest Power Act, the Council must act to ensure that the Fish and Wildlife Programs, as well as the projects for implementing the program, are scientifically, technically, and economically sound. The Fish and Wildlife Program establishes the overarching goals that the Council agrees are critical for protecting fish and wildlife—such as the goal of improving juvenile fish passage over the Columbia River dams, which was prominent in its first Fish and Wildlife Program—whereas the projects for implementing the program plan will involve the design of a specific fish ladder or water spill schedule over a particular dam, to be developed by a partner agency like the Army Corps of Engineers. The Council's governing body members sit on committees that review and discuss the functional components of the Act's mandate, and devise the program's plans and budgets for program implementation. A staff of about forty-five professionals, including an Executive Director appointed by the Council, help administer the program, coordinate program implementation, and address legal issues and public affairs. The Council meets once per month in an open public meeting to make decisions on program planning and implementation issues. Decisions about program plans are not made by the Council's members and staff alone. Public input and external agency consultation have been formally established within the governance structure of the Council under the Northwest Power Act. The Council is directed to seek recommendations for the Fish and Wildlife Plan from various tribal, state, and federal fish and wildlife agencies, and hearings must also be held in each member state before adoption of the Council's Fish and Wildlife plans.

While the basic structure of the Council, in terms of membership and appointment, has not changed over time, scholars of the Council have noted that the individual members who have been appointed to the Council have had profound effects on the program's goals and strategies. For example, the first chair of the Council in the early 1980s, who had previously served as governor of Washington, Daniel J. Evans, helped bring attention to salmon hatchery issues in the Columbia, which was also a critical issue to tribes (Lee 1993). He also successfully negotiated a water budget with the Army Corps and utilities that would increase spring flows from dams to facilitate juvenile salmon runs. Kai Lee's appointment as a council member was also considered integral to promoting the concept of adaptive management (Cone 1995).

One way that the collective choice structure of the Council (versus the program plans) has changed substantially since its inception is through the establishment of scientific and advisory bodies that inform the Council. The three advisory bodies that provide scientific review and advice for the Northwest Power and Conservation Council came about in 1996, some sixteen years after the start of the program. The creation of three independent advisory boards in the same year can be traced in part to a number of complaints that were lodged against the Council in the early 1990s over how the Fish and Wildlife Program was implemented, as well as broader concerns over the health and status of salmon in the basin (Blumm and Bodi 1996; Dompier 1996; Volkman 1997). For instance, tribes and agencies began to complain that the Council was placing too much emphasis on the power side of its mandate, with the program's plans focusing its mitigation efforts on hydropower operations, such as spills and flow augmentation from dams, while relying less on hatcheries. Still, the best approach to restoring fisheries populations in the basin was open for debate. Conservation groups and fisheries experts had begun to recognize the need for more flow augmentation from dams to protect native fish and raised concern over the impact of hatcheries and increasing harvests on wild fish (e.g., see Williams et al. 1991). The emerging science supported the arguments for petitions for listing Snake River salmon under the Endangered Species Act. These listings challenged some of the fundamental assumptions of the Northwest Power Act, namely its emphasis on hydropower impacts and the reliance on agency and tribal recommendations that focused on hatcheries as restoration strategies.

In 1992 the Council devised the "Strategy for Salmon" to deal with the endangered species issues, but lawsuits were brought against the Council by environmental groups for the plans' limitations in adequately addressing the detriments to salmon in the basin. In 1994, the Council revised the way it had been developing its fish and wildlife plan when a federal appeals court required that the Council give a "high degree of deference" to the fish and wildlife agencies and tribes (Northwest Power and Conservation Council 2003). Native fish concerns, which were fundamental to the ESA listings, were not, however, addressed by the Court. The 1994/1995 Fish and Wildlife Program was developed based on recommendations by the Court and regional input from federal agencies, state water and land managers, and numerous tribes. It was set forth as a long-range, comprehensive plan for fish and wildlife recovery in the basin (Northwest Power and Conservation Council 1994).

Despite these new plans, the ESA listings continued to raise questions as to whether the restoration program was working and the extent to which fisheries science was being adequately incorporated into the process. The Council was to some extent paralyzed in its ability to implement its plan and needed to get the science right (Shurts 2009). The Council then decided to create the Independent Science Group (ISG) in 1994. Both the ISG and the National Academy of Sciences, commissioned by Congress, were asked to provide independent reviews of the underlying science of fisheries restoration. The National Academy report in 1995 and the ISG's "Return to the River" report in 1996, both criticized the notion that technology could compensate for the loss of ecosystem function. The ISG report further provided the "state of the science" on the main-stem management issues and gave recommendations for an ecological "framework" for future fish and wildlife planning (Merrill 2005). That same year, NOAA Fisheries produced a report that recommended the creation of a board to advise both the Council and NOAA Fisheries on Columbia basin fisheries management issues. The Council decided the ISG members could provide the foundation for a new board—the nine-member Independent Scientific Advisory Board (ISAB).

The work of the ISAB generally follows the path set by the ISG in the mid-1990s by continually reviewing the state of the science related to fish and wildlife management in the Columbia basin. The Council's 2000 Fish and Wildlife Program provided more explicit direction for the ISAB to review existing research and summarize the state of the science in key areas (Northwest Power and Conservation Council 2008), further institutionalizing the formal review of existing data, knowledge, and research on the basin's fisheries and ecological status. In 2002, the Columbia basin Indian tribes were added to the list of partner agencies served by the ISAB. The Council, NOAA Fisheries, and the Tribes each have an ex-officio member on the board who interfaces between the agencies and the group. For the Council specifically, this board focuses on the scientific merits of the Council's long-term plans for fish and wildlife recovery.

A second advisory body, the Independent Scientific Review Panel (ISRP), was set up under the 1996 amendments to the Northwest Power Act (Section 4H10D, Public Law No. 96-501) to review the operational projects, such as a habitat restoration, which partner agencies propose for implementing the Fish and Wildlife plan. In addition to the program's struggles to adequately deal with emerging endangered species issues, there were growing concerns in the 1980s and 1990s that the program and its process for reviewing restoration and mitigation projects were too political (Dompier 1996). Prior to the 1996 amendments to the NWPA, the BPA was in charge of reviewing proposed projects, which were not subject to any scientific review (Heikkila and Gerlak 2005). To help de-politicize the process, the 1996 amendments required projects to be reviewed by the eleven-member ISRP, selected from a list of scientists provided by the National Academy of Sciences. The ISRP mandate was to determine if projects are "based on sound science principles; benefit fish and wildlife; and have clearly defined objective and outcome with provisions for monitoring and evaluation of results" (Northwest Power Act, 94 Stat. 2710, as amended by Pub.L. 104-206, § 512(4)(h)(10)(D)(iv), September 30, 1996, 110 Stat. 3005). The 1996 amendments also required the Council to respond in writing to the ISRP before recommendations go to the BPA for project sponsorships. Initially, the ISRP went through massive numbers of projects (over four hundred in six months) (Merrill 2005). The first review process initially found 40 percent were not technically sound. The second review process found numerous projects inadequate, which prompted a review of the project-funding process. As a result, ISAB established more specific guidance for the entities that implement these projects. It also provided a preliminary progress review, which implementers can then respond to and document project changes (Merrill 2005). The 1996 amendment further directed the ISRP, with assistance from scientific peer review groups, to review annually the results of prior-year expenditures based upon the project review criteria and submit its findings to Council. Together, the creation of the independent science groups and the science reviews in the mid-1990s were a paradigm shift for the Council (Shurts 2009). First, science became an institutionalized feature of the decision-making process and second, fish and wildlife restoration would be guided by a framework focused on ecological principles (Hill et al. 2012).

The mandate under the 1996 amendments for the ISRP to review expenditures reflects another issue of concern in the early 1990s—rising program costs, which at the time were estimated to be about $40–$60 million per year of out-of-pocket costs paid by BPA, equivalent to about $50 per salmon (Lee 1993). Since one of the mandates of the Northwest Power Act was to ensure that fish and wildlife recovery efforts were economically sound, the Council decided to create a third advisory body in 1996, the Independent Economic Analysis Board (IEAB) as part of the Act's broader requirements to establish scientific advisory committees (Northwest Power and Conservation Council 2009a). It is comprised of eight economists, nominated and screened by a group of peer economists, who are supposed to help assess the cost-effectiveness of fish and wildlife recovery projects (Northwest Power and Conservation Council 2009a). The Council can also request the IEAB to provide economic advice on a variety of other fish, wildlife, and/or energy issues as needed.

Operational Level Changes

The "constitutional level" agreements outlined earlier, and the decisions made by the collective choice bodies, set the stage for the Fish and Wildlife Program's implementation and operational-level decisions. While there have been a couple of major shifts in the constitutional-level agreements and structure of the governing bodies, especially the introduction of science groups and their ecological focus, the operational-level plans and processes for implementing the program have gone through more frequent changes. These changes are often driven by, or at least linked to, changing scientific and stakeholder advice, as well as external reviews from oversight agencies or court challenges.

As previously noted, at the operational level, the Council receives proposals from state, local, and tribal governments, as well as universities and conservation organizations, who want to obtain funding to undertake fish and wildlife recovery projects in the Columbia River Basin (such as restoring habitat along the riparian areas of streams, improving fish passage in streams, or conducting population studies of species in the basin). Prior to the creation of the ISRP, only the Bonneville Power Administration reviewed these proposals for funding absent any review of their technical characteristics. Now, with the advice of the ISRP, the Council reviews and approves hundreds of these projects each year if they meet the goals of the program. The changes in implementation are directly tied to the Council's periodic adaptations to its Fish and Wildlife Program, which sets the fundamental objectives for the program and are reviewed and influenced by the scientific advice of the ISAB. This approach to program implementation is currently guided by the ecosystem assessment framework set forth by the science groups (Shurts 2009). The Fish and Wildlife Program incorporates the framework into a "vision" through its planning assumptions, biological objectives, scientific principles, and basinwide strategies, which are then further incorporated in the sub-basin plans (Northwest Power and Conservation Council 2009b).

Initially, in the 1980s, the Fish and Wildlife Plan focused heavily on improving passage for salmon at main stem dams but the Council was criticized for this narrow focus and for failing to take into account many stakeholder interests. Early efforts at program implementation were also marked by the recognition that the program needed to be a learning process. Council member Kai Lee, noted for his fostering of adaptive management in the Council, helped shepherd an explicit recognition of monitoring and learning from implementation in the 1980s, which was formally recognized in the 1987 Fish and Wildlife Program (Northwest Power and Conservation Council 1994).

The Council began a series of revisions to its Fish and Wildlife Program in the 1990s that resulted in various other operational level modifications. In 1992, the Council's "Strategy for Salmon" established rebuilding targets for salmonid populations in the basin and set forth a number of operational strategies, such as increasing releases from reservoirs in order to improve habitat, using water more efficiently, assisting the passage of fish passage over dams, and implementing more efficient energy operations to reduce demands on dams run by BPA (Northwest Power and Conservation Council 1992). Legal challenges to the 1992 Strategy, however, resulted in a shift toward more broad-based approaches to fish and wildlife protection in 1994/95 that recognized sources of fish and wildlife loss in the basin beyond the hydropower system, i.e., hatcheries, harvest, and habitat (or the four "H's"). However, as noted earlier, the strategies established under the 1994/95 plan were criticized for not being firmly grounded in the current science of the time.

The shift in the Council's advisory structure toward more independent scientific review in the 1990s led to an even more comprehensive and detailed plan in 2000 that emphasized the systemwide nature of the resource management problem more explicitly and included more specific biological objectives and strategies to meet those goals (Northwest Power and Conservation Council 2000). The previous plans also had very specific steps assigned to different actors rather than comprehensive recovery efforts. The 2000 plan established sub-basin plans for the sixty-two tributary sub-basins and it is supposed to be updated in phases, consistent with ecological principles and scientific information (Northwest Power and Conservation Council 2003). Sub-basin plans have now been established to help direct BPA in their funding of projects, as well as to help other partnering federal agencies to meet endangered species recovery obligations (Northwest Power and Conservation Council 2001). Sub-basin plans can be developed independently by stakeholders in the sub-basins as long as they are consistent with the Council's Fish and Wildlife Program, and the Council reviews these plans for inclusion.

Since 2000, the Council has amended its program twice, once with the main-stem amendments in 2003, and again with the sub-basin plans in 2005. In preparing the main-stem plan, the Council solicited recommendations from the region's state and federal fish and wildlife agencies, tribes, and others. The Council conducted an extensive public comment period on the draft main-stem plan before finalizing these program amendments. The Council also solicited recommendations for program amendments, in the form of sub-basin plans, so that they could adopt more specific biological objectives and measures for tributary sub-basins and specific main-stem reaches.

Some critics of the efforts to restore salmon and other fish and wildlife in the Columbia River Basin have noted that the efforts to adapt have been weaker than expected, resulting in continual declines in the health of the ecosystem, particularly in salmon populations. As Blumm and Bodi (1996) have found, the difficulty of adapting to changing problems in part stems from the limited scope of the Northwest Power Act, or the constitutional design of the program, which "focused on the hydroelectric system while overlooking harvest regulation, federal land management, and state water law." Volkman (1997) has noted that the adaptive management approach of the Council in fact has not been implemented because of an unwillingness or inability to learn from the process. He notes: "Organized learning requires an enormous effort of coordination…Collaboration can be urged but not enforced, and dissatisfied parties can erect obstacles to implementation… Merely calling something a learning process doesn't absolve it from a burden of history, mistrust and politics" (Volkman 1997).

Notably, the history of mistrust that Volkman refers to in the basin, evidenced by lawsuits over endangered species protection, has been mitigated somewhat by the May 2008 Record of Decision by BPA for Fish Accords between BPA, the Columbia basin tribes, the Army Corps of Engineers, Montana, and Idaho. The Accords settle litigation that dated back to 2000 and authorizes projects to mitigate the effects of BPA's power production on endangered species. These "Accord" projects are subject to scientific review by the ISRP, which under the Northwest Power Act is tasked with reviewing projects funded by BPA. The Council's 2009 Amendments to its Fish and Wildlife Plan, as a result, has incorporated some of these strategies to work in tandem with the Accords.

Table 1 provides a brief snapshot of the major institutional design features of the Northwest Power and Conservation Council's Fish and Wildlife Program.

Table 1: Northwest Power and Conservation Council Institutional Design Features

Institutional Level	Institutional Change	Year
Constitutional (Authorizing Policy)	Northwest Power Act of 1980	1980
	Northwest Power Act Amendment	1996
Collective Choice (Governing Body)	Northwest Power and Conservation Council	1980
	Northwest Power Planning Council (name change)	2003
Collective Choice (Advisory Bodies)	Independent Science Group (ISG)	1994 (ended 1996)
	Independent Science Advisory Board (ISAB)	1996
	Independent Economic Analysis Board (IEAB)	1996
	Independent Scientific Review Panel (ISRP)	1996
Operational (Plans & Strategies)	Strategies focus on mainstem fish passage, flows, hatcheries – research, action plans, goal setting	1980s-1990s
	Plan Amendment	1985
	Plan Amendment	1986
	Plan Amendment	1987
Operational (Plans & Strategies)	Broader strategies toward "4Hs" and broader stakeholder input	1990s-2000
	Strategy for Salmon	1992
	Plan Amendments	1994/95
	New Program	2000
Operational (Plans & Strategies)	Basin-wide approach, sub-basin plans; ecological principles	2000-Present
	Mainstem amendments	2003
	Sub-basin amendments	2005
	Plan Amendments	2009

Discussion and Conclusion

Our analysis of the Northwest Power and Conservation's Fish and Wildlife Program illustrates that collaborative institutions at the large watershed scale are capable of change in dynamic ways, demonstrating the adaptive capacity of these institutions to respond to emerging ecological, political, and economic concerns. As we argue, it is critical to understand the ways in which these types of watershed governance institutions change at different levels of institutional choice, beyond the operational level, which has been the focus of much of the early analysis of such institutions. Doing so illustrates not only the depth and diversity of changes that can emerge, but the connections and interplay between the different levels of institutional design. Through understanding the interplay of the changes at these levels, scholars and practitioners interested in adaptive governance can better understand how the design of an institution matters in shaping adaptive capacity.

Starting with the constitutional level, we find that the Northwest Power and Conservation Council by design, through its enabling legislation, is structured to periodically update its operational strategies under its Fish and Wildlife Program plans with review and approval of its collective choice decision body, the Council, and input from partnering agencies that implement the program. While the constitutional design encourages adaptation, the mandates of the Northwest Power Act also pose limits on the types of adaptation that can take place, as previously noted. Where we have seen further adaptive capacity demonstrated is through the program's changes to the collective choice structure when the program created its three current advisory boards (supported by an amendment to the enabling legislation). These three advisory bodies expanded the technical and scientific authority, scope, and networks of the program. They arguably created new ways of understanding the problems of fish and wildlife protection, as well as venues for agreement on the new strategies for addressing these problems (e.g., moving away from single species, main-stem approaches to ecosystem-based planning efforts). Such changes align with the literature on "boundary organizations," which are supposed to bridge divergent experts and interests to create new ways of thinking and problem solving (Guston 1999). At the same time, new program strategies that were supported by the advisory processes also required direction and support from program leaders and partnering agencies.

Changes to the program obviously have not only been driven by internal forces. External criticism, notably lawsuits and challenges to the legality of elements of the program, particularly around the protection of endangered salmon species, has occasionally stepped up the internal assessments as to whether the program's design and strategies are working, based on the operational outcomes. As noted earlier, these criticisms and external factors were instrumental in pushing the creation of the 1996 advisory boards. Concerns over the politics and economics of the program further fed into the creation of the advisory boards and the Amendments to the Northwest Power Act. As such, the performance of the operational level strategies, particularly as they interplay with the broader ecological, political, and economic environment, can feed into changes at the constitutional and collective choice levels as well.

In sum, the design of an institutional arrangement plays an important role in watershed management adaptation. As we have seen in this case, institutional adaptation emerged through the complex and dynamic interplay of different institutional design features (goals and mandates, decision structures, and operational strategies) along with external forces. While this ability to respond to emerging challenges over time is likely to be critical for the success of the institution, it is not without costs. As others have noted, the creation of new operational plans by the Council involves substantial time and effort to negotiate, as well as the costs of shifting or finding resources and staff for implementation, which can make it difficult for partnering agencies to sustain a coordinated basin-wide management approach (Schlager and Blomquist 2008). How large-scale collaborative programs, such as this one along the Columbia River, continue to adapt and respond to political, technical, and management challenges will undoubtedly be of interest and importance to scholars and practitioners in the coming years. Seeing how the adaptive capacity of this institutional arrangement was established provides valuable insights into the ways in which adaptation can be encouraged in the future.

Works Cited

Anderies, John M., Mark A. Janssen, and Elinor Ostrom. "A framework to analyze the robustness of social-ecological systems from an institutional perspective." *Ecology and Society* 9, no. 1 (2004): 18.

Blumm, Michael C., and Loraine F. Bodi. "Overhaul of Columbia River Operations Needed: Commentary." In *The Northwest Salmon Crisis: A Documentary History*, edited by Joseph Cone and Sandy Ridlington, 262–64. Corvallis,: Oregon State University Press, 1996.

Blumm, Michael. "Reexamining the Party Promise: More Challenges than Success to the Implementation of the Columbia Basin Fish and Wildlife Program." *Environmental Law*, 1986: 227–357.

Brick, Phillip, Donald Snow, and Van Wetering. *Across the Great Divide: Exploration in Collaborative Conservation—and the American West*. Washington, D.C.: Island Press, 2001.

Brunner, Ronald D., and Toddi A. Steelman. "Beyond Adaptive Governance." In *Adaptive Governance: Integratng Science, Policy, and Decision Making*, edited by Ronald D. Brunner, Toddi A. Steelman, Lindy Coe-Juell, Christina M. Crommley, Christine M. Edwards and Donna W. Tucker, 1–47. New York: Columbia University Press, 2005.

Carpenter, S., B. Walker, J. Anderies, and N. Abel. "From Metaphor to Measurement." *Ecosystem*, 2001: 765–81.

Cash, David W. "In Order to Aid in Diffusing Useful and Practical Information: Agricultural Extension and Boundary Organizations." *Science, Technology and Human Values* 26, no. 4 (2001): 431–53.

Cone, Joseph. *A Common Fate: Endangered Salmon and the People of the Pacific Northwest*. New York: Henry Holt and Company, 1995. Reprinted 1996, Oregon State University Press.

Costanza, Robert, and Jack Greer. "The Chesapeake Bay and Its Watershed: A Model for Sustainable Ecosystem Management?" In *Barriers and Bridges to the Renewal of Ecosystems and Institutions*, edited by Lance H. Gunderson, C. S. Holling, Stephen S. Light, 169–213. New York: Columbia University Press, 1995.

Crawford, Sue, and Elinor Ostrom. "A Grammar of Institutions." *American Political Science Review* 89, no. 3 (1995): 582–98.

Dietz, Thomas, Elinor Ostrom, and Paul C. Stern. "The Struggle to Govern the Commons." *Science* 302 (2003): 1907–10.

Dompier, Douglas W. "Commentary." In *The Northwest Salmon Crisis: A Documentary History*, edited by Joseph Cone and Sandy Ridlington, 203–5. Corvallis: Oregon State University Press, 1996.

Doyle, Mary, and Cynthia A. Drew. *Large-Scale Ecosystem Restoration: Five Case Studies from the U.S.* Washington, D.C.: Island Press, 2008.

Ernst, H. *Chesapeake Bay Blues: Science, Politics and the Struggle to Save the Bay.* Lanham, Maryland: Rowman and Littlefield, 2003.

Fazey, I., J. A. Fazey, and D. M. A. Fazey. "Learning More Effectively from Experience." *Ecology and Society* 2, no. 4 (2005): art 4.

Florestano, Patricia S. "Past and Present Utilitization of Interstate Compacts in the United States." *Publius* 24, no. 4 (1994): 13–25.

Folke, C., T. Hahn, P. Olsson, and J. Norberg. "Adaptive Governance of Social-Ecological Systems." *Annual Review of Environment and Resources* 30 (2005): 441–73.

Gerlak, Andrea K. "Today's Pragmatic Water Policy: Restoration, Collaboration, and Adaptive Management Along U.S. Rivers." *Society and Natural Resources* 21 (2008): 538–45.

Gerlak, Andrea K., and Keith A. Grant. "The Emergence of Cooperative Institutions Around Transboundary Waters." In *Mapping the World Order: Participation in Regional and Global Organizations*, edited by Thomas J. Volgy, Zlatko Sabac, Petra Roter, and Andrea K. Gerlak, 114–47. London: Blackwell, 2009.

Gerlak, Andrea K., and Tanya Heikkila. "Collaboration and Institutional Endurance in U.S. Water Policy." *PS: Political Science & Politics* 40, no. 1 (January 2007): 55–60.

Gerlak, Andrea K., and Tanya Heikkila. "Comparing Collaborative Mechanisms in Large-Scale Ecosystem Governance." *Natural Resources Journal* 46 (2006): 657–707.

Grant, Douglas L. "Interstate Water Allocation Compacts: When the Virtue of Permanence Becomes the Vice of Inflexibility." *University of Colorado Law Review* 74 (2003): 105–80.

Gunderson, L., and C. S. Holling. *Panarchy: Understanding Transformations in Human and Natural Systems.* Washington, D.C.: Island Press, 2002.

Gunderson, Lance H., C. S. Holling, and Stephen S. Light. *Barriers and Bridges to the Renewal of Regional Ecosystems.* New York: Columbia University Press, 1995.

Gunderson, Lance H., Steve R. Carpenter, Carl Folke, Per Olsson, and Gary Peterson. "Water RATS (Resilience, Adaptability, and Transformability) in Lake and Wetland Social-Ecological Systems." *Ecology and Society* 11, no. 1 (2006): art 16.

Guston, David H. "Stabilizing the Boundary between Politics and Science: The Role of the Office of Technology Transfer as a Boundary Organization." *Social Studies of Science* 29, no. 1 (1999): 87–112.

Hasday, Jill E. "Interstate Compacts in a Democracy Society: The Problem of Permanency." *Florida Law Review* 49 (1997): 1–47.

Heikkila, Tanya, and Andrea K Gerlak. "The Formation of Large-Scale Collaborative Resource Management Institutions: Clarifying the Roles of Stakeholders, Science, and Institutions." *Policy Studies Journal* 33, no. 4 (2005): 583–612.

Janssen, Marc A., John M. Anderies, and Elinor Ostrom. "Robustness of Social-Ecological Systems to Spatial and Temporal Variability." *Society and Natural Resources,* 2007: 307–22.

Kallis, Giorgos, Michael Kiparsky, and Richard Norgaard. "Collaborative and Adaptive Management: Lessons from California's CALFED Water Program." *Environmental Science Policy,* 2009: 631–43.

Karkkainen, Bradley C. "Collaborative Ecosystem Governance: Scale, Complexity, and Dynamism." *Virginia Environmental Law Journal,* 2002: 189–243.

Koontz, Tomas, M. Todi, A. Steelman, JoAnn Carmin, Katrina Smith Kormacher, Cassandra Moseley, and Craig W. Thomas. *Collaborative Environmental Management: What Roles For Government?* Washington, D.C.: Resources for the Future Press, 2004.

Lackey, Robert T., Denise H. Lach, and Sally H. Duncan. "Policy to Reverse the Decline of Wild Pacific Salmon." *Fisheries* 31, no. 7 (2006): 344–51.

Lee, Kai. *Compass and Gyroscope: Integrating Science and Politics for the Environment.* Washington, D.C.: Island Press, 1993.

Lee, Kai. "Deliberately Seeking Sustainability in the Columbia River Basin." In *Barriers and Bridges to the Renewal of Ecosystems and Institutions,* edited by L. H. Gunderson, C. S. Holling, and S. S. Light. New York: Columbia University Press, 1995.

Lejano, Raul P., and Helen Ingram. "Collaborative Networks and New Ways of Knowing." *Environmental Science and Policy* 12, no. 6 (2009): 653–62.

Lubell, Mark. "Collaborative Watershed Management: A View from the Grassroots." *Policy Studies Journal* 32, no. 3 (2004): 341–61.

Lubell, Mark, Mark Schneider, John Scholtz, and Mete Mihrye. "Watershed Partnerships and the Emergence of Collective Action Institutions." *American Journal of Political Science* 46, no. 1 (2002): 48–63.

McKinney, Matthew, and William Harmon. *The Western Confluence: A Guide to Governing Natural Resources.* Washington, D.C.: Island Press, 2004.

Merrill, Erik. July 20, 2005, personal communication with author.

Northwest Power and Conservation Council. "Strategy for Salmon, Vol. II." 1992. Available at http://www.nwcouncil.org/library/1992/2_0_Summary.pdf.

Northwest Power and Conservation Council. "1994/1995 Columbia River Basin Fish and Wildlife Program." 1994. Available at http://www.nwcouncil.org/library/Default.htm.

Northwest Power and Conservation Council. "Discussion Paper: The Role of the Northwest Power Planning Council." 1998. Available at http://www.nwcouncil.org/LIBRARY/1998/98-8.htm.

Northwest Power and Conservation Council. "Northwest Power Planning Council Columbia River Basin 2000 Plan." 2000. Available at http://www.nwcouncil.org/library/2000/2000-19/FullReport.pdf.

Northwest Power and Conservation Council. "Council Briefing Book." 2001. Available at http://www.nwcouncil.org/library/2001/2001-1.pdf.

Northwest Power and Conservation Council. "Northwest Power and Conservation Council." 2003. Available at http://www.nwcouncil.org/library/2003/2003-20/report.pdf.

Northwest Power and Conservation Council. "Briefing Book." 2005. Available at http:// www.nwcouncil.org/library/2005/2005-1.pdf.

Northwest Power and Conservation Council. "Independent Scientific Advisory Board Background." 2008. Available at http://www.nwcouncil.org/fw/isab/background.htm.

Northwest Power and Conservation Council. "Independent Economic Analysis Board Background." 2009a. Available at http://www.nwcouncil.org/fw/ieab/background.htm.

Northwest Power and Conservation Council. "Northwest Power and Conservation Council Columbia River Basin Fish and Wildlife Program 2009 Amendments" 2009b. Available at http://www.nwcouncil.org/library/2009/2009-2.pdf.

Ostrom, Elinor. *Understanding Institutional Diversity.* Princeton, N.J.: Princeton University Press, 2005.

Sabatier, Paul A., Will Focht, Mark Lubell, Zev Trachtenberg, Arnold Vedlitz, and Marty Matlock. *Swimming Upstream: Collaborative Approaches to Watershed Management.* Cambridge, Mass.: MIT Press, 2005.

Schlager, Edella, and Tanya Heikkila. "Resolving Water Conflicts: A Comparative Analysis of Interstate River Compacts." *Policy Studies Journal* 37, no. 3 (2009): 367–92.

Schlager, Edella, and William Blomquist. *Embracing Watershed Politics.* Boulder, Colo.: University Press of Colorado, 2008.

Scholtz, John T., and Bruce Stiftel. *Adaptive Governance and Water Conflict: New Institutions for Collaborative Planning.* Washington, D.C.: Resources for the Future, 2005.

Sherk, George W. "The Management of Interstate Water Conflicts in the 21st Century: Is it Time to Call Uncle?" *New York University Environmental Law Journal* 12, no. 3 (2005): 764–827.

Shurts, John. Personal Interview. April 4, 2009.

Steinberg, Paul F. "Institutional Resilience Amid Political Change: The Case of Biodiversity Conservation." *Global Environmental Politics* 9, no. 3 (2009): 61–81.

United States General Accounting Office. *Columbia River Basin Salmon and Steelhead: Federal Agencies' Recovery Responsibilities, Expenditures and Actions.* GAO-02-612. July 2002.

Van Cleve, F. Brie, Thomas Leschine, Terrie Klinger, and Charles Simenstad. "An Evaluation of the Influence of Natural Science in Regional Scale Restoration Projects." *Environmental Management* 37, no. 3 (2007): 367–79.

Vigmostad, Karen E., Nicole Mays, Allen Hance, and Allegra Cangelosi. *Large-Scale Ecosystem Restoration: Lessons for Existing and Emerging Initiatives.* Washington, D.C.: Northeast Midwest Institute, 2005.

Volkman, John. "A River in Common: The Columbia River, the Salmon Ecosystem, and Water Policy—Report to the Western Water Policy Review Advisory Commission." Western Water Policy Review Advisory Commission, Springfield, Va., 1997.

Walker, B., C. S. Holling, S. R. Carpenter, and A. Kinzig. "Resilience, Adaptability and Transformability in Social-Ecological Systems." *Ecology and Society* 9, no. 2 (2004): 5.

Walters, Carl "Challenges in Adaptive Management of Riparian and Coastal Ecosystems." *Ecology and Society* 1, no. 2 (1996): art 1.

Wiley, Helena, and Dennis Canty. "Regional Environmental Initiatives in the United States." Report to the Puget Sound Shared Strategy, Seattle, Wash., 2003.

Williams, Jack E., Willa Nehlsen, and James A. Lichatowich. "Pacific Salmon at the Crossroads: Stocks at Risk from California, Oregon, Idaho and Washington." *Fisheries* 16, no. 2 (1991): 4–21.

Wolf, Aaron, Shira B. Yoffe, and Mark Giordano. "International Waters: Identifying Basins at Risk." *Water Policy* 5, no. 1 (2003): 29–60.

Wondolleck, Julia, and Steven L. Yaffee. *Making Collaboration Work: Lessons from Innovation in Natural Resource Management*. Washington, D.C.: Island Press, 2003.

Uncertainty, Society, and Resilience: A Case Study in the Columbia River Basin

Gregory Hill, Steven Kolmes, Eric T. Jones, and Rebecca McLain

Uncertainty, in its many manifestations, is a central factor in the management of complex systems, whether environmental (van der Sluijs 2007), climatic (Risbey and Kandlikar 2007), medical, or financial (McDaniel and Driebe 2005). Nowhere is this more apparent than in the management of ecosystems heavily affected by human use. True to the characteristics of complex systems, the distinction between cause and effect is blurred as environmental degradation drives changes in economic factors and social-cultural relationships, which in turn influence the state of the ecosystem (Liu et al. 2007). In such a system, uncertainty exists in many attributes: in the knowledge of the current state of environmental conditions and their relationships to economic and social factors; in the variability of these conditions and relationship over time; in the projections of the effects of human management interventions; and in the reliability of the mathematical models used to assess these conditions, relationships, and projections (Walker et al. 2003). The different actors involved in planning, each having a unique point of view, see uncertainty differently—for example regarding a high level of uncertainty either as a reason for action or as a factor supporting inaction. We will argue in this chapter that the way in which uncertainty is managed in a planning and policy process has important implications in terms of framing the deliberations and influencing the balance of power among the various entities involved in system governance.

It is worth taking a moment to note that at the inception of the development of American environmental thought, consideration of values and ethics was an explicit keystone. Concerning "The Land Ethic," Aldo Leopold wrote (Leopold 1949):

All ethics so far evolved rest upon a single premise: That the individual is a member of a community of independent parts. His instincts prompt him to compete for his place in that community, but his ethics prompt him also to co-operate (perhaps in order that there may be a place to compete for). The land ethic simply enlarges the boundaries of the community to include soils, water, plants, and animals, or collectively: the land . . . No important change in ethics was ever accomplished without an internal change in our intellectual emphasis, loyalties, affections, and convictions. The proof that conservation has not yet touched these foundations of conduct lies in the fact that philosophy and religion have not yet heard of it. In our attempt to make conservation easy, we have made it trivial.

Today, philosophy and religion are engaged with the ethics and values of conservation, but ethical analysis has generally vanished from scientifically based conservation planning. Discussions of environmental planning inherently contain ethical dimensions, because they are discussions about the relationship between humans and the planet. The more complicated the circumstances, the more uncertainty in possible future scenarios, and the greater the stakes to more stakeholders in a society, the greater the need for ethical analysis, if we wish to avoid making conservation planning "trivial."

The authors of this chapter are investigators in an on-going research project to study the use of computer modeling and decision support systems in multi-stakeholder environmental problem solving processes. Our study focuses on the effectiveness of these tools to address issues of equity and breadth of public participation and the quality of planning outcomes. The context of our study is the Columbia River Basin and the public deliberation process for the recovery of species of Columbia River salmon and steelhead as required under the Endangered Species Act. The perspectives expressed in this chapter are based in part on ethnographic interviews conducted in 2008 throughout the Columbia River Basin. Interviews were conducted with a wide range of managers, scientists, and public stakeholders involved in, or concerned about, planning for salmon recovery.

The Nature of Uncertainty

The way in which uncertainty is perceived, understood, and managed plays a crucial role in the public participation process for salmon recovery in the Columbia River Basin. We introduce three distinct varieties of uncertainty, labeled as "epistemic," "stochastic," and "irreducible," and then describe the roles they play in this process.

Most narrowly conceived, uncertainty pertains to the quantitative data and model output involved in salmon recovery planning: spawner recruitment, smolt survival rates, critical habitat and spawner capacity, etc. These uncertainties, sometimes called "epistemic uncertainty" (Walker et al. 2003) can be effectively described in the language of statistics. The result of imperfect knowledge, these uncertainties may be reduced through additional research.

But upon broader analysis, we find ourselves in a more complex situation. Many of the ecological phenomena in the Columbia River Basin, such as annual rainfall, or the pattern of the Pacific decadal oscillation, have a high level of variability. These uncertainties, which we will call "stochastic uncertainties" (Refsgaard et al. 2007), can still be described by statistics, but are not amenable to reduction through further research. The potential for clarification is limited to the likely endpoints or limits of the range of variability. They represent uncertainties to which we must adapt. This is the situation that the practice of adaptive management is designed to address, but as the complexity of the social-ecological system increases, uncertainties rise to yet another level.

At this third level, uncertainty cannot be described statistically and "irreducible uncertainty" appears. The changes can be discontinuous and abrupt, "surprises," in the language of Holling and others (Holling 1986; Kates and Clark 1996; Carpenter and Gunderson 2001) are the rule and not the exception. Although there is no method to decide exactly when a system achieves the level of complexity that exhibits this chaotic behavior, there are accepted characteristics that such a complex dynamical system must possess (Holland 1995). An important aspect of such a system is that the sub-systems are tightly linked through a network of feedback loops and interacting factors. Negative feedback loops (which resist movement away from a normative state) will tend to reduce the dominance of chaotic events in the system; positive feedback loops (which are self-accelerating) will tend to enhance the dominance of chaotic behavior in the system. For the Columbia River Basin, the salmon themselves, with their wide-ranging life histories, tie together the ecological, bio-physical, cultural and economic systems very tightly, across temporal and geographical scales. Uncertainty in this context cannot be captured by statistics alone. It is the theory of resilience (Berkes et al. 2000) that proposes a set of principles and methods for management. We will use the term "irreducible uncertainty" (Gallopin et al. 2001) to describe uncertainty arising as a consequence of this highest level of system complexity.

Although we have described three levels of system complexity and associated types of uncertainty, it is important to note that all three contexts may well be present at different locations and at different times within the same complex system. The abrupt changes that are endemic to complex systems occur at thresholds— boundaries between two possible stable states of the system (Carpenter and Folke 2006). These "regime shifts" (Scheffe et al. 2001) are distinguished from merely large changes within one system—they represent a "rapid reorganization of a system from one relatively unchanging state to another" (Carpenter and Folke 2006). As an example of such a regime shift, consider the construction of the Grand Coulee Dam and Bonneville Dam, which induced abrupt changes in the ecosystems of the upper Columbia River Basin. In the early 1930s, as the decision to construct the first large hydro-electric dams on the Columbia drew near, the social system in particular was in a particularly high level of uncertainty. Dam construction plans were tied to massive unemployment and a national agenda to generate a sense of hope in the midst of the Great Depression. Walker and Meyers' (Walker and Meyers 2004) classification of such thresholds would describe these examples as a "shift in the ecological system, driven by the social system." Far away from thresholds, where responses are proportional, systems uncertainties are likely to be manageable, but near thresholds, small changes can produce large unexpected results, bringing the

Figure 1: Complexity and decision stakes domains

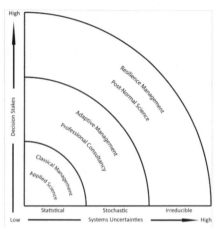

system into the zone of irreducible uncertainty. The construction of The Dalles dam and subsequent main-stem Columbia River and lower Snake River dams from the 1950s onwards were superficially similar events, but occurred in a period of considerable affluence. They were quite predictable on a social level given the dominant paradigm privileging economic development over ecological concerns and native cultures, and the decisions to construct these facilities would result in a very different ethical analysis than the dams begun in the 1930s.

Today, scientists are asked to participate in contexts that differ qualitatively from what Kuhn (Kuhn 1962) called "normal science,"which is curiosity-driven,"puzzle-solving" science whose outcomes are characterized by certainty and accuracy. Ravetz and Funtowicz (1994) proposed an alternative methodology for problem solving at the science-policy interface, designed to deal with contexts in which "facts are uncertain, values in dispute, stakes high and decisions urgent." Crucial to their theory is an analysis of the levels of uncertainty present in complex systems. Several typologies of uncertainty (Walker et al. 2003; van Asselt and Rotmans 2002; Refsgaard et al. 2006) have been proposed and some have been implemented in planning processes (Wardekker et al. 2008; van der Sluijs et al. 2005) but for the purposes of this discussion, we will use the three levels of uncertainty that are closely tied to the degree of complexity of the system under consideration. In a now classic diagram Functowicz and Ravetz (1994) developed a three-level classification of scientific problem-solving approaches: applied science, professional consultancy, and post-normal science (as distinguished from Kuhn's "normal science"). Figure 1 is our adaptation of the original diagram that maps types of management systems onto the three contexts.

The horizontal axis of the diagram denotes the degree of system complexity with three zones indicating the three levels of increasing complexity described above. The vertical axis designates the level of decision stakes. Defining risk as the combination of the level of consequence of an event and the probability of its occurrence, the three rings then represent three levels of increasing risk.

In the inner domain, with both decision stakes and uncertainty low, traditional applied science with its emphasis on reduction of uncertainty and linear modeling of cause and effect is an appropriate methodology and classical management practice is an effective method for translating that science into policy and action. In this context, decisions are simple enough and stakes low enough that they do not require deeper levels of public involvement and deliberation. Note that even if the complexity of the ecological system is low, high complexity in the social system (for example, multiple ethical or cultural perspectives in the stakeholder population) would move the problem to a higher level. Knowledge quality assessment, typically kept within the scientific and management communities at this level, involves formal or informal peer review and evaluation by managers, and is aimed at judging the quality of the scientific process and ultimate product.

In the intermediate domain, risk has risen to a significant level, either through higher levels of uncertainty, or consequence, or both. If the uncertainties have risen to the stochastic level, they may no longer be reducible to achieve the certainty required for the applied science approach. In the environmental context, adaptive management was developed to cope with this situation (Walters and Holling 1990; Lee 1993; Gunderson and Light 2006). If decision stakes are significant, multiple perspectives, perhaps in the form of additional professional opinions, are appropriate. Here, knowledge quality assessment involves a wider range of professional participants, representing differing standpoints, engaging in a deliberative process to assess the quality and reliability of scientific input to planning.

But if negative consequence, uncertainty, or both rise to the highest levels these approaches may not be sufficient. Functowicz and Ravetz propose a problem-solving methodology for this context (Funtowicz and Ravetz 1995) that they call post-normal science (PNS). Two striking innovations in the post-normal framework are the approach to uncertainty and the role of nonscientists in knowledge quality assessment. When system complexity reaches a level at which uncertainties become irreducible, the issue becomes not how to reduce uncertainty, or even how to describe it statistically, but rather how to live with irreducible uncertainty.

Most environmental problem-solving contexts involve a negative consequence of an external impact (often a human-induced impact) to an environmental system. As the time period for reversibility of such an impact extends further and further into the future, the need to utilize the techniques of post-normal science increases. Figure 2 shows that in the case of a relatively quickly reversible process, post-normal science is only needed when decision stakes are high and systems uncertainty is high. However, when reversibility of a process would take a longer and longer time (moving from left to right in Figure 2), the cumulative effect of even a modest

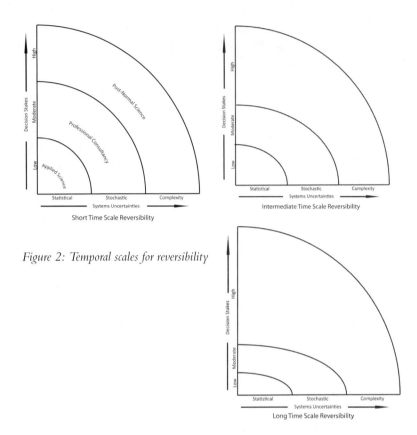

Figure 2: Temporal scales for reversibility

mistake becomes very much greater, and the complex and inclusive processes of post-normal science are appropriate at lower and lower immediate decision stakes. This can be thought of as akin to the need for more societal consultation and planning when preparing to spray a highly persistent organochlorine pesticide like DDT compared to an organophosphate pesticide with similar immediate effectiveness but far less persistence. The compression of the curves for statistical and stochastic uncertainties as the time scale of reversibility increases represents the increased seriousness of persistent errors. The decision stakes get higher more quickly with increased uncertainty as the time frame lengthens.

Models and Uncertainty

Increasingly, computer models and model-based decision support systems (DSS) are playing a central role in environmental assessment and planning (Matthies et al. 2007). Environmental problems in particular give rise to some of the most complex problem-solving processes our society faces. In these processes "where facts are uncertain, values in dispute, stakes high and decisions urgent" (Functowicz and Ravetz

1995), multiple stakeholders seek to further highly contested objectives and critical variables can only be estimated, making projections uncertain at best. A complex network of influences, competing objectives, and uncertainties can overwhelm the cognitive capacity of even the most sophisticated decision maker (McDaniels and Gregory 2004). Increasingly, computer-implemented decision support systems are being introduced to assist stakeholders and decision makers in coping with these complexities. From local watershed councils (Independent Scientific Advisory Board 2001) to the Environmental Protection Agency to the Nature Conservancy, organizations are using DSS in a wide range of environmental problem-solving scenarios. The use of these DSS tools has increased rapidly in recent years. In a study by the National Commission on Science for Sustainable Forestry (Gordon et al. 2005), over a hundred different decision support systems were evaluated in the field of sustainable forestry alone. Within the Columbia River Basin, models play central roles in the planning processes directed by the National Oceanic and Atmospheric Administration (Fullerton et al. 2009) and the Northwest Power and Conservation Council (Williams 2006). But with the increased use of these technological tools come risks, especially when they are used in the post-normal domain. Writing of the need for natural resource managers to come to grips with complex social-ecological systems, Westley et al. (2002) conclude that the key factors in the failure of technological approaches are that "they focus on the wrong types of uncertainty and on narrow types of scientific practice," assuming that the system under consideration is near equilibrium and paying little attention to "uncertainties that arise from errors in tools or models." We now examine these considerations.

Knowledge Quality Assessment and Extended Peer Review

Developing guidance to detect and manage model uncertainty has been an active area of research, particularly in Europe (Refsgaard et al. 2007; Wardekker et al. 2008; Pereira and Quintana 2009). Following a controversial public case in which the reliability of "data" reported by the Netherlands National Institute for Public Health and the Environment (RIVM) was criticized so heavily for its reliance on models rather than observations that it made front page news (van der Sluijs 2006), the national government of the Netherlands developed and implemented sophisticated protocols for managing model uncertainty (Janssen et al. 2005). In the U.S., examples also exist of methodologies for assessing the quality of model output (Environmental Protection Agency 2009; National Research Council 2007), but no protocol has been mandated as in the Netherlands. In a recent National Research Council report (Dietz and Stern 2009), the authors summarize a wide range of research on public participation in environmental decision making, addressing the role of uncertainty issues extensively. With respect to model output, they emphasize the need to communicate uncertainties in a transparent manner and in a way that the lay public can understand. Other researchers (Kallis et al. 2006; Pahl-Wostl 2002) often working in the post-normal science tradition, formulate a more expansive role for stakeholders, especially in the context of irreducible uncertainties and high decision stakes. These authors hold that in such a problem-solving context, the quality of planning processes and outcomes is best assessed by an "extended peer community" including scientists, managers, and policy makers, but also including other stakeholders from a wide range of perspectives. In particular, the role of extended peer review in assessing the use of model-based decision support is well described in this literature (Refsgaard et al. 2007; van der Sluijs 2006).

Members of an extended peer community have an important role to play in revealing sources of uncertainty, often ones that go unrecognized by scientists and managers. Scientists are usually aware, to varying extents, of the implications for uncertainty of the complexity of ecological systems, but less often appreciative of the implication of social, cultural, or bureaucratic complexity. For example, one commonly used form of model-based decision support in use in the Columbia River Basin (EDT) is particularly demanding for managers to populate. With over forty input variables consisting of numerical data or "expert opinion," sometimes hundreds of stream reaches to evaluate, and a demanding time-line, populating the model puts a high demand on managers and field biologists. In one ethnographic interview with a manager directing one of the sub-basin evaluations, we discussed the output of the model. In a table of model output describing the optimal achievable salmon populations (historical abundance) for a watershed, a four-digit number was reported. When asked how many of the digits he thought were significant, this interview subject reported, "None of them. I don't believe that number to an order of magnitude." This perceived high degree of uncertainty about model output was not related to the ecological complexity, or to the inner workings of the model, but rather to an informed opinion of the social complexities of populating the model with the input values required. In fact, the input data was often in the form of the "expert opinion" of biologists familiar, to varying extents, with the stream reaches in question. Subsequently these data were transformed by the mathematical equations of the model into numerical output, giving the impression of an unwarranted level of accuracy. Taken on their own terms, these qualitative data (expert opinions) may well have been reliable, but transformed into numerical outputs with the illusion of accuracy, they can be quite misleading. This kind of evaluation of uncertainty is often best performed by the holders of local knowledge and is one of the valuable contributions that an extended peer review can make to the quality of a planning process.

Ethics and Uncertainty

The transition from applied science to post-normal science can also be thought of in terms of Type I and Type II errors. If we use the analogy of a scientific hypothesis to a defendant on trial, then a Type I error would be the conviction of an innocent defendant. The scientific analogy would be to accept as true a hypothesis (or the existence of a relationship) that is actually false (does not exist). In contrast, returning a not guilty verdict for a guilty defendant is an example of a Type II error. In a scientific context, it is the error of dismissing a hypothesis that is in fact true. Stated differently, it is the error of dismissing a relationship when one in fact exists. It is customary in traditional science to prioritize the avoidance of Type I errors over Type II errors, often by setting explicit acceptable levels of probability for these errors— often 5 percent for Type I and 20 percent for Type II (Lemons et al. 1997). The Precautionary Principle would assert that, as uncertainty, and hence risk, increases it becomes more and more important to readjust these priorities by putting more emphasis on the avoidance of Type II errors (Marshall and Picou 2008; Kriebel et al. 2001). We can therefore think of the post-normal domain as the region in which the Precautionary Principle is most appropriate, the applied science domain as a place where it is less relevant and the intermediate domain as a transitional zone (Ravetz 2004). Thus, as systems uncertainties increase, the way in which uncertainty is managed and communicated has important ethical implications.

An example of these ethical implications, as described in the following paragraphs, can be observed in the salmon recovery planning process in the Columbia River Basin, particularly in the use of model-based decision support. The mathematical models that scientists create to organize information, make projections, and help choose future actions are often assumed to be impartial, scientifically rigorous, and an analytical way to quantify nature into a manageable set of variables. Although traditional scientific peer review process generally ensures that the models are internally consistent and predictable in their operations, the issues of underlying assumptions and data transformations (such as the qualitative to numerical example above) are not well controlled by this process. Models involved in salmon recovery planning in the Columbia River Basin often possess characteristics that are invisible except with very detailed analysis, which renders them either far less scientifically based than they seem, or else far less impartial to the interests of different stakeholders than they appear.

For example, the Viable Salmonid Population (VSP) methodology, first described in "Viable Salmonid Populations and the Recovery of Evolutionarily Significant Units" (McElhany et al. 2000) has a goal of first identifying genetically distinct populations of salmonids listed under the Endangered Species Act, and then to establish guidelines for population levels and growth rates designed to minimize the threat of extinction to these populations. The quality of the model's mathematical operations and its consistency with the contemporary fisheries literature is unimpeachable. However, the VSP definition of a minimized threat of extinction is problematic. The model sets a goal of limiting the risk of extinction for genetically distinctive populations to a 5% risk over a period of 100 years. Neither the extinction risk level of 5%, nor the choice of a century, has any basis in scientific thought or analysis.

The 5% risk owes its origins to the 5% probability level (p-value) used to demonstrate statistical significance between the treatment or observational groups, conventional in the biological sciences for interpreting experimental results. This decision statistic level has no logical or mathematical connection to a "minimized extinction risk" under the VSP, it simply adopts a scientific-sounding number that biologists are familiar with seeing in abstracts, to a novel context where it has no support. In fact, in experiments where the potential consequences of error are considered more grave (e.g., clinical experiments on drug efficacy) an acceptable p-value below 5% (1% or 0.1%, or even 0.01%) is considered prudent. Extinction, it might be argued, is also a grave risk, and so selection of an arbitrary value of 5% is ethically questionable in this context. There is no natural non-arbitrary value to use in this case, but rather than raising a 5% risk (or other risk levels) for discussion by a wide array of stakeholders, it quietly becomes an assumption of the VSP model, making this value seem more impartial, scientifically rigorous, and certain than in reality it is.

Similarly, the choice of a 100-year time period is arbitrarily chosen, appearing to the human time scale as a "long time." From the perspective of the salmonid populations, 100 years is an eye-blink in the geological history of the Pacific Northwest. Salmonids have flourished in this river basin for millennia, and planning for 100 year periods is related neither to some number of generations of fish, nor to a number of human generations, nor to the chaotic behavior, and irreducible uncertainty, introduced by high levels of system complexity. Our conventional use of base 10 may make this 100 year period appear more tidy than (say) 82 or 437 years, but it adds an appearance of certainty to the VSP projections that is out of keeping with what is in reality an unsupported assumption. Given that a 5% risk is another unsupported assumption, setting the VSP goals as a 5% extinction risk over 100 years compounds both the arbitrary nature of these standards and the appearance of rigor and certainty. Given these starting points the model operates well. Traditional peer reviews of models, examining model construction and internal consistency, may not provide the perspectives needed to address the larger systemic and ethical considerations we raise. Extended peer review, including multiple stakeholders able to bring situated local knowledge to bear, can add valuable perspectives on plausibility, sources of uncertainty, and ethical considerations that would ultimately enhance the salience and quality assessment of models and model output.

A second example of the importance of distinguishing between arbitrary values and empirically derived values is raised in McElhany et al. (2003, 2006)'s interim viability reports for Willamette and Lower Columbia River salmonids. These reports introduce and develop an approach to salmonid recovery planning called Stratum Persistence Probability Categories. To explain this in brief, salmonids are divided into a list of historical populations and life histories in specific ecoregions that are referred to as strata. For lower Columbia River Chinook, in the Columbia River Gorge, the Fall and Spring runs of returning fish constitute largely genetically distinct groups. There were four historical populations in the fall and two populations (when water levels made different parts of the river accessible) in the spring. These constitute two separate chinook salmon strata (one with four populations and one with two), with a shared geographic location but distinctive temporal pattern.

For each population in a stratum, a population persistence probability is assigned on a 5-point scale, ranging from 0 (either extinct or at very high risk of extinction, 0-40% probability of persistence over the next 100 years) to 4 (very low risk of extinction over 100 years, greater than 99% probability of persistence). The goal of this exercise is to allow salmonid recovery planners to reconstruct something like the historically stable salmonid populations using the Columbia River Basin in a manner distributed over space and time, to provide more resilience in the face of unknown environmental fluctuations and challenges in the future. However, the way these scores are determined and utilized is a maze of arbitrary assumptions.

The actual persistence scores are determined by "professional judgment," meaning that local fisheries biologists score the health of local populations using the 5-point scale. Ideally the individuals doing this scoring would be working off at least partial numerical data sets for spawner abundance, productivity, and recruitment, and a good estimate of likely habitat modifications on each area of stream over the next century. In reality, people providing scores do their best, but often in the face of missing numerical data and little ability to predict the future environment of any specific stream. Whether a population in a specific location spawning at a specific

Table 1: Stratum persistence categories based on averages of individual population risks. The category thresholds are based on professional judgment.

Stratum Persistence Probability Category	Average of Population Risks
Low persistence	Average < 2
Moderate persistence	2 < = Average < 2.25; at least two populations > = 3
High persistence	Average > = 2.25; at least two populations > = 3

time of year is evaluated correctly or not depends on so many variables that the quality of these assessments is surely uneven and sometimes doubtful.

To compound the way these scores are used, a stratum persistence probability is calculated based on the averages of the individual population risk scores determined by professional judgment. Averaging ordinal data is an unusual exercise at best, but the averaged stratum scores thus determined are then compared to a table for their interpretation that ranks the probable persistence of the stratum. Table 1 is from the 2003 report.

The recovery goal for each stratum, in order to have a high probability of persistence, is to have salmon recovery efforts achieve a stratum persistence probability of 2.25, with at least two populations over a value of 3. The 2003 interim viability report acknowledges that there is a nonlinear relationship between population persistence scores and population persistence categories since some scores have wider ranges (40 percent) than others (10 percent). The 2006 interim viability report expresses concern that people not be misled by an implied precision of a score that goes out to two decimal places (e.g., 2.25) and which originates in averages of ordinal scores from 0 to 4. However, these caveats aside, the interim viability report of 2006 continues to promote this approach. Other scoring methods that would be too lengthy to discuss in detail here are also discussed in these documents. These include quantification on scales of 0 to 4, based on expert opinion, for attributes as diverse as: the risk of fish declining to a four-year annual average population of fifty spawners within one hundred years; whether juvenile outmigrant fish are declining, stable or increasing in numbers; loss of within-population genetic and life history diversity; and present and projected future habitat quality. In all of these cases, there would be a privilege extended to people deemed to possess expert knowledge to populate data sets with these scores, and then scores would presumably be subjected to mathematical operations such as the averaging of population persistence categories. It is difficult to overstate how disenfranchised stakeholders not deemed to possess expert opinions would be, were this sort of process to be implemented. Again, an extended peer review process, in which a wider range of stakeholders were charged with evaluating the uncertainty and ethical loading of these numerical criteria, might make significant contributions to the quality of planning.

Many examples of the value of extended peer review can be found in the scientific literature. Vogel et al. (2007) suggest an alternative approach to the science-policy interface in which "different experts, risk-bearers, and local communities are involved and knowledge and practice is contested, co-produced and reflected upon." Based on a case study of southern Africa related to vulnerability assessments, the authors describe an "extended peer" approach in which an "expert" identification of an issue as a "food" problem was expanded, based on suggestions from a broader range of stakeholders, to include a larger socio-ecological perspective. Informed by this broader perspective, plans and actions were altered.

Hartley et al. (2006) examine a case study of the Northeast Consortium, a group supporting cooperative fisheries research in the Gulf of Maine and on the Georges Bank, involving fishermen, scientists, and managers as partners in knowledge production and management. The authors find that "the scientific results from cooperative research were considered more credible and were likely to produce outcomes of direct value to the fishing industry and fisheries, such as better and more selective fishing gear technologies." Moller et al. (2004) consider two case studies, one from Canada and one from New Zealand, that demonstrate the value of integrating traditional monitoring methods into scientific monitoring of wildlife species. They find that, "although traditional monitoring methods may often be imprecise and qualitative, they are nevertheless valuable" resulting in an improvement in the overall quality of knowledge and hence decision making.

Work by Coburn (2003), illustrates how input from an extended peer community, in this case low-income residents of Brooklyn, New York, can contribute to better scientific understandings of asthma, air toxics, and risks associated with eating locally caught fish. In a plan to assess community pollution exposure, the Environmental Protection Agency (EPA) planned to model the dispersion of hazardous air pollutants, and later present model output to the community. Activists shifted the process to a collaborative investigation, arguing that finer level detail was needed than could be obtained from modeling the air pollutant, ultimately producing a detailed community map of polluters. The EPA revised its dispersion model approach to take these data into account.

From our case study in the Columbia River Basin, an example of the value of extended peer review for knowledge quality can be found in the Walla Walla watershed of southeastern Washington. Diverse groups of local watershed residents, informally organized, collaborate with federal agencies in all levels of planning and assessment. One of our ethnographic informants, a high-level manager with a federal agency, described a scenario in which a federal agency planned to issue fishing take restrictions on a local creek. In the words of our informant, "there's this thing called the 'Walla Walla Way,' when there's a threat of something bad, they get together and start talking." A plan was worked out with the agency to augment the flow of the creek, averting the fishing restrictions. Broadly based local knowledge had an impact on the implementation of the proposed plan as one participant observed "if you put more water in and ran it down the cement ditch in the blazing sun you'd be putting hot water in cold pools. It would have killed the fish." Our informant described this process as "a feedback loop with local knowledge. It's their culture and every year they've managed to keep the water flowing."

The Forecasting Frame

Uncertainty, and the method by which it is managed, can play an important role in the framing of public discourse. Following Lakoff (Lakoff 2005), we understand a frame as "a cognitive structure, one that is necessary for understanding and reasoning," and like Lakoff, we are particularly interested in frames that are largely unacknowledged, but nonetheless play a significant role in guiding and possibly limiting discourse and deliberation (Pinker and Lakoff 2007). We begin by describing the "forecasting frame."

Using methods rooted in the culture of analytical modeling, forecasting is a standard framework for exploring the effect of policy scenarios on future states. This modeling culture looks temporally forward, beginning with causes then proceeding to outcomes. Typically, models are built from a small range of values for the explanatory variables, have fixed assumptions, and a defined set of inputs. Modelers, following best-practices, are hesitant to extrapolate far from the model design when exploring alternative policy scenarios, and will not modify assumptions or add input variables. While this is good practice from a modeling perspective, it discourages or prohibits exploration of substantial departures from status quo, even when such departures may offer the only opportunities of achieving long-term aspirational goals. This framework promotes a disciplined approach of representing the mechanisms moving from cause to effect, an approach that serves science well. However, the traditional modeling approach and associated focus on incremental variations from existing data (status quo) limits policy discussions to "the path" and not to the ultimate destination. Overall, disciplined analytic modeling encourages sound, data-based policy decisions. However, in the context of a participatory public process, this modeling culture, combined with the extra legitimacy inferred from computer-based DSS's, discourages full participation of stakeholders with viewpoints based upon different assumptions or ones with aspirations requiring significant components not included in the design of the model. This can limit the consideration of a full range of policy alternatives.

In the applied science domain of Figure 1, uncertainties can be measured statistically, and reduced through further research. In this context, forecasting is a particularly effective tool. Uncertainties in forecasted effects of policy-based actions can be described well using standard methods of statistics (but often are not) and the actions can be evaluated in terms of their effectiveness in addressing a particular goal. In the stochastic range, the uncertainties in model output used in forecasting are harder to quantify, and may not be reducible through further research. Climate change and the impacts of invasive species are good examples of contexts in which stochastic effects are prominent and uncertainty needs to be handled with great care (Pahl-Wostl et al. 2000; Cooney and Lang 2007). For example, in the work of the Intergovernmental Panel on Climate Change, uncertainties are not reported numerically, but rather qualitatively to avoid giving the impression of overly accurate analytical assessments to policy makers and the public (Risbey and Kandlikar 2007). To the extent that irreducible uncertainties are involved, forecasting is called into question as an effective planning tool. Thus, in a post-normal context, the use of models must be treated with great care. Some researchers (Gunderson et al. 2008) recommend that in this context, models be used solely to develop hypotheses or to investigate systemic causal relationships, avoiding any predictive use. In writing about the need for Knowledge Quality Assessment and the role of modeling in a post-normal context, van der Sluijs (2006) writes: "A widening in focus from 'reducing uncertainties' to 'coping with untamable uncertainties and complexities' is needed. This can avoid misunderstandings and undue expectations of the role and competence of science in complex environmental problems."

While avoiding "misunderstandings and undue expectations" is crucial, a lack of a robust and transparent system for managing and communicating uncertainties, appropriate to the complexity of the problem context, can also play an important role in limiting the use of relevant planning modalities. Without such a system, uncertainty is directed away from risks associated with continuing the "business as usual" scenario and the forecasting frame is reinforced. As a result, planning principles that do not rely on incremental changes away from the current trajectory are excluded. In the next sections we will discuss two such planning principles: scenario planning and the precautionary principle.

Scenario Planning and Aspirational Goals
While the use of decision support, coupled with a lack of a robust practice of uncertainty evaluation and communication, may restrict the decision space and privilege the current trajectory, this outcome is not inevitable. In this section, we describe several instances of "scenario planning" processes used in the Columbia River Basin, as examples of modalities that foster a more transparent treatment of uncertainties and a broader range of future trajectories.

Peterson et al. (2003) describe scenario planning as "a structured account of a possible future" to create "alternative, dynamic stories that capture key ingredients of our uncertainty about the future of a study system." Wilkinson and Eidinow (2008) argue that scenario planning, coupled with the involvement of the wider range of epistemologies represented by a greater diversity of stakeholder involvement, may be particularly well-adapted to the post-normal context. An example of such a scenario planning process was the Northwest Power and Conservation Council's basin-wide multi-species framework (MSF) planning effort of the late 1990s: In 1998, the NPCC initiated a multi-species framework planning process to establish a framework for making science-based decisions about fish and wildlife recovery efforts in the Columbia River basin (NPPC

1998a). As with sub-basin planning, broad-based stakeholder involvement and the use of decision support tools were key elements. The Council began the process by sending out approximately 1500 letters to a wide variety of stakeholders soliciting concept papers. Proposers of concept papers were asked to "formulate a broad vision for the Columbia River basin that reflects the biological/ecological, cultural, social and economic priorities" (NPPC, 1998b:18). Based on these visions, proposers completed the concept papers by developing objectives, strategies and management actions to realize their vision for the basin. The proposers of the 27 concept papers represented a wide range of standpoints including those of organizations such as the Save Our Wild Salmon Coalition, the Columbia River Inter-Tribal Fish Commission and Reynolds Aluminum as well as a number of unaffiliated individuals (BPA 2003).

The concept papers were fleshed out in a workshop organized by the NPCC and reviewed in two public meetings. Council staff distilled the twenty-seven papers into seven well-defined alternatives including Alternative 2 calling for the breaching of 1 dam on the John Day and 4 dams on the Lower Snake river to Alternative 7, envisioning a river managed for maximum economic benefits (NPPC 2000b). Only after stakeholder participation framed the alternatives to be considered was the decision support tool EDT brought in to evaluate the plans for biological benefit. The Council used a separate process to evaluate the plans for social and economic impacts. Although the process was not intended to select a preferred alternative, participants produced rankings for fish, wildlife and social/economic benefits. (McLain et al. publication pending)

The comparison of the MSF process with sub-basin planning is particularly revealing since both planning processes used the same decision support tool (EDT) but quite different use methodologies to arrive at substantially different outcomes. Perhaps the key difference was the timing of the use of the decision support tool. The MSF process began with an open-ended, stakeholder-led process to develop scenarios and only after scenario development was complete did the decision support tool make an appearance. By delaying the involvement of modeling until after the human-centered visioning process was complete, the MSF process afforded a much broader decision space, avoiding much of the framing imposed by the use of decision support tools in the sub-basin planning process.

The model analysis of the scenarios did not identify a preferred alternative, but rather scored the alternatives according to a range of ecological, social, and economic factors. According to McLain et al.:

In terms of benefits for salmon recovery, the analysis found that, 'Alternative 2 performs better for chinook population recovery under the Technology Pessimistic worldview and poorer under the Technology Optimistic view. Alternative 2 is projected to produce a larger increase in chinook abundance from current levels, than either of the other two alternatives regardless of the worldview (Marcot et al. 2002).' The human effects analysis ranked Alternative 2 highest in monetary costs and lowest in nonmonetary costs (NPPC 2002c). The multi-species framework's planning process paid much greater attention to uncertainties than the sub-basin planning process. In spite of using the same decision support tools, the MSF scenario planning process, combining a high level of extended peer involvement and scientific review, yielded a product with significantly different outcomes from the sub-basin planning process. Key contrasts in the MSF planning outputs include the more sophisticated approach to the treatment of the uncertainties; a much more detailed social and economic evaluation; and the delivery of a range of options for decision makers to consider (Marcot et al. 2002; NPCC 2002c).

Backcasting, a specific form of scenario planning, is the name given by Robinson (1982) to describe a method of analyzing future options in which the concern lies "not with what futures are likely to happen, but with how desirable futures can be attained. It is thus explicitly normative, involving working backwards from a particular desirable future end-point to the present in order to determine the physical feasibility of that future and what policy measures would be required to reach that point." Thus, backcasting contrasts with forecasting by the adoption of an explicit focus on desired outcomes, in advance of an emphasis on the process of modeling cause and effect (Figure 3).

Backcasting can be viewed as a particular approach to "scenario planning," one in which normative goals and underlying principles are made explicit and are fundamental to the planning process. In their publication "Scenario Planning: a Tool for Conservation in an Uncertain World" (Peterson et al. 2003), the authors identify scenario planning as particularly appropriate in the face of uncertainty, stating that "unlike forecasts, scenarios stress irreducible uncertainties that are not controllable by the people making the decisions," and that "Scenario planning is

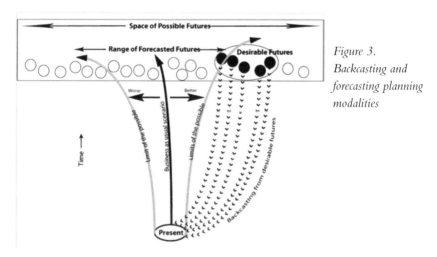

Figure 3. Backcasting and forecasting planning modalities

most useful when there is a high level of uncertainty about the system of interest and system manipulations are difficult or impossible."

Backcasting has been used most often in planning scenarios requiring a substantial revisioning of the current state of affairs to achieve "desirable futures" that accommodate a particular goal. Robinson credits Amory Lovins with developing backcasting as a planning tool in his work on "soft energy paths" and the method continues to be used in sustainability planning (Jablonowski 2007; Wilson et al. 2006). The "Natural Step" method for sustainability planning, developed by Karl-Henrik Robert (Robert 2000; Broman, Holmberg, and Robèrt 2000), uses "backcasting from principles" as its main method for breaking from familiar practices and moving to a different paradigm. In these and other examples, backcasting is a method that accommodates aspirational goals, often ones derived from deeply held beliefs and ethical concerns.

Such aspirational goals are not hard to find in the planning arena for recovery of endangered runs of wild salmon in the Columbia River Basin. In a document composed by the Catholic bishops of the Columbia River watershed, "The Columbia River Watershed: Caring for Creation and the Common Good" (Catholic Bishops of the Pacific Northwest 2001), ethical and spiritual beliefs are brought to bear to describe the "Rivers of Our Vision." The process of writing this document began with fifteen "Listening Sessions" held from 1997 to 1999 across the Columbia River Basin (including much of Oregon, Washington, British Columbia, and a small portion of Montana). At these sessions people from many walks of life were asked to explain what the Columbia River meant to their beliefs, livelihoods, and communities. This rich quilt of testimonies helped the bishops envision a future in which the dreams and aspirations of these people, as well as basic requirements of the ecosystem, could be met. The "Salmon Nation" concept, promoted by the Portland, Oregon based nonprofit Ecotrust, describes "a community of caretakers and citizens that stretches across arbitrary boundaries and bridges urban-rural divides," in their vision of a future that accommodates recovery of wild salmon stocks. The Endangered Species Act (ESA) itself, rather than promoting an incremental approach to the preservation of species, has as its stated purpose "to provide a means whereby the ecosystems upon which endangered species and threatened species depend may be conserved." The 2006 book "Salmon 2100: The Future of Wild Salmon" (Lackey et al. 2006) is a collection of twenty-three independent visions, each by separate authors, of alternative futures that would support the recovery of wild stocks of anadromous Columbia River fish.

One particularly thorough backcasting exercise is represented by a 15-year collaborative process among a diverse group of scientists led by Richard Williams, resulting in the book "Return to the River: Restoring Salmon to the Columbia

River" (Williams 2006). An example of "backcasting from principles," their process first determined principles, based on a dynamic ecosystem perspective, to guide the development of a desired "state of the world" that would accommodate the goal of the recovery of endangered salmon and steelhead runs in the Columbia River. Their vision, called the Normative River, consists of a "set of conditions for a fully functional river ecosystem." In contrast to planning inside the forecasting frame, their goal is not derived from incremental changes away from the current state of the river, but is guided by three conservation principles that reflect best practices in restoration ecology.

While the backcasting exercise led by Williams does consider social factors, it regards the problem of restoring salmon as an essentially biological problem. In contrast, the Bishops' "Listening Sessions" already described included a different and very broad "extended peer community" in order to develop a vision that incorporates values as well. Similarly, the Columbia Inter-Tribal Fish Commission recovery plan, Wy-kan-ush-mi Wa-kish-wit (Columbia River Inter-Tribal Fish Commission 1995) includes a vision based in cultural values and states that "respect and reverence for this perfect creation are the foundation for this plan."

All these processes began with aspirational goals rooted in a particular cultural context (conservation goals returning us to a Normative River; collected visions of a world allowing wild salmon survival in 2100; the vision of broad habitat restoration in the ESA, the Catholic vision of the Common Good, and the Native American vision of nature as relationship) and derive plans by backcasting from these normative goals.

It is worth examining the salmon recovery efforts headed by the National Oceanographic and Atmospheric Administration (NOAA) in the Pacific Northwest, to discern commonalities and contrasts with the three quite different values-based backcasting approaches described above. The stated goal of NOAA, the federal agency charged with fulfilling the mandate of the ESA in this case, is to reduce the risk of extinction for the evolutionary significant units (genetically distinct and geographically limited subpopulations) of the listed species to below 5% over the next 100 years (Kolmes and Butkus 2006). This goal contrasts sharply with the goals from the previous examples that posit a vision of a future state that supports healthy runs of wild salmon. The NOAA approach is well-suited to the incremental change methodologies inherent in the forecasting framework. Indeed, the main DSS tools used by NOAA are built around traditional forecasting models. Our research examines the degree to which such goal statements are influenced by the collection of tools in use.

Several of our ethnographic informants have demonstrated a backcasting point of view. For example, one subject while discussing his experiences with modeling complained that the modelers "get lost in the weeds." The situation in

Figure 4:
Perceived level of
uncertainty and
social distance
from production
of knowledge
(after MacKenzie
1990)

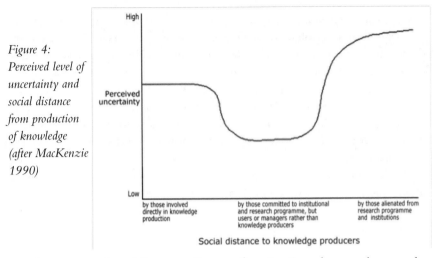

question concerned modeling mortality rates for migrating salmon as they pass the hydroelectric dams both upstream and downstream. In the view of this subject, the differences between the various methods of moving salmon past the dams are irrelevant. In his words "none of them get us where we need to go." The future state he had in mind was one resembling a "river of the past" in which the four hydroelectric dams on the lower Snake River had been breached.

In general, our interviews with stakeholders in the Columbia River Basin fit well with the perception of uncertainty pattern described by MacKenzie (MacKenzie 1990). In this study of individual's evaluations of uncertainty concerning the accuracy of nuclear missile guidance systems, MacKenzie describes a pattern in which perceived uncertainty varies with the subject's social distance from the production of knowledge according to the diagram in Figure 4.

Our ethnographic data revealed a similar pattern. We found that stakeholders far from knowledge production (in this case, knowledge in the form of model output) were generally more skeptical of the use of modeling and more likely to be alienated from planning processes driven by such tools. As well, they were generally more likely to have alternative perceptions about sources of uncertainty. Two such informants, employed as field biologists by a tribal agency, related that "some elders tell us that there used to be summer chinook" in a local river, "hundreds of thousands of them and the models will say, no, the temperature wasn't right and the habitat couldn't support that." A lack of transparency regarding uncertainty can privilege numerical model output, even when it is derived from qualitative input data, over ethnographic or oral history information. The importance of these issues for governance was clearly expressed by our interview subjects as they related their concerns that recovery goals for salmon could be impacted by using "model output to revise that escapement goal, saying that the sum of the parts is only 10 percent" of goals derived from traditional knowledge (escapement goals are numbers of fish allowed to survive fishery efforts to spawn in the wild). Holders of local knowledge, such as these informants, could play an important role in knowledge quality assessment as part of an extended peer review system. Berkes in a study of the role of indigenous knowledge in complex environmental problems (Berkes and Berkes 2009) writes that "Northern indigenous observations of climate change and abnormalities in animals (which may be related to Arctic ecosystem contamination), once dismissed by scientists, are being taken seriously. Indigenous observers do not carry out chemical tests for pollutants or gather quantitative data. But their ways of observing and assessing environmental changes provides insights regarding indigenous holism." It may be that in the post-normal context of high systems complexity, such perspectives are particularly important. Referring to an earlier study of traditional knowledge and climate change, Berkes (2009) writes that the data from Inuit informants "further illustrates the ability of indigenous knowledge to deal with multiple variables and complexity, and shows that local observations can provide information at the appropriate spatial scale to complement science."

The Precautionary Principle
As adopted in 1992 by the U.N. Conference on the Environment and Development, the Precautionary Principle holds that in situations of high risk for serious and irreversible damage, lack of certainty should not be used to justify delayed action. According to von Krauss et al. (2005), the following list of conditions is applicable in order to invoke the principle: (1) There are major uncertainties; (2) There is some evidence and a science-based scenario of possible harm; (3) The potential harm is significant, difficult to contain, and possibly irreversible; (4) The potential harm relates to an important value, e.g., human or environmental health; and (5) Uncertainties cannot be significantly reduced in the near future without thereby increasing the chances for the harm to occur and/or without making the control of the harm more difficult.

The circumstances of salmon recovery in the Columbia Basin constitute a remarkably good fit to these criteria, but awareness of its applicability is dependent on an awareness of the uncertainties involved. One obstacle to that awareness, in the case of salmon recovery, is that the scenario to which the Precautionary Principle might be applied is a "business as usual" scenario. Whereas the Precautionary Principle is typically applied in cases in which a novel, and risky, human action is proposed (for example, the introduction of a new technology or pharmaceutical), "business as usual" is a scenario in which no novel human action is proposed and therefore less consideration is applied to what the negative consequences of this "non-action" might

continued on page 356

The use of EDT to prioritize stream reaches and specific actions for salmon recovery, at the cost of other reaches and actions, is an implicit use of cost-benefit analysis. The cost is engaged through allocation of financial resources for recovery efforts; the benefit is predicted by the use of EDT. The precautionary principle stands in stark contrast to the use of cost-benefit analysis in public planning processes. Cost-benefit analysis, which has become in its implicit or explicit form the default decision method of choice, balances costs (represented here by allocation of salmon recovery funds) against an array of supposed benefits (as predicted here by EDT or another forecasting methodology), and chooses the relatively most "profitable" or least "costly" alternative(s). In a decision support system this may be articulated as looking for investments in salmon recovery efforts that would produce the greatest investment for the available resources. But what does cost-benefit analysis look like in the real world? Do the sorts of questions that are asked privilege certain answers over others? As Butkus and Kolmes (2011) have pointed out:

> It is clear that lacking infinite funds, a society cannot embrace all the activities that it would aspire to simultaneously, and that it is important to understand the relative social values of different ways of expending limited financial resources. Cost-benefit analysis is used for this end, which in fact it does not achieve because of inherent biases in the system that in the end favor shortsightedness and a laissez faire approach to environmental issues.

The three steps of cost-benefit analysis are (a) calculating the monetary value of the benefits that would be produced by a possible policy decision, (b) calculating the monetary costs of that policy being implemented, (c) adopting those policies for which benefits (expressed in monetary terms) exceed costs. The problems at each step of this process have been described in detail. The benefits associated with cost-benefit analysis of environmental policies are very hard to quantify (what is clean water worth? clean air? forests to hike in?). It is not clear that agreeing to monetize these sorts of benefits is appropriate at the outset. The costs of environmental policies are often hard to calculate, and exaggeration of costs can provide "cover" for unwilling business interests or for partisan political agendas. Given the difficulty in both determining the value of environmental benefits and projected costs that depend on assumptions about potential impacts of environmental policies on business, the third (decision) step of cost-benefit analysis is extraordinarily difficult to carry out objectively.

Philosophically, cost-benefit analysis turns out to be a conflict of worth and price, and in a region like the Columbia River Basin where different stakeholders have different values and cultures, this cannot help but become a clash of cultural values. Is relationship with an undisturbed section of the river for tribal people worth more or less than irrigation and hydroelectricity? And ultimately, is that a question that should be asked? Ackerman (2008) has answered in this way:

> There are no meaningful prices attached to protection of human life, health, nature, and the well-being of future generations, and no end of nonsense has resulted from the attempt to invent surrogate prices for them. The absence of prices is fatal to the cost-benefit project, but it is not the case that unpriced benefits are worthless: what is the cash value of your oldest friendship, your relationship with your children, or your right to vote and participate in a democratically governed country? As the German philosopher Immanuel Kant put it, some things have a price, or relative worth, while other things have a dignity, or inner worth. The failure of cost-benefit analysis, in Kantian terms, stems from the attempt to weigh costs, which usually have a price, against benefits, which often have a dignity.

Precluding serious engagement with the precautionary principle from current processes involving decision support and modeling comes, in part, from the privileged *laissez faire* position occupied by the alternative future states whose characteristics work well with decision support system boundaries and assumptions. Factors that did not enter into the initial formulation of a question cannot be protected by boundaries that recognize their importance. The privilege and power associated with asking the initial question cannot be overstated. In the Columbia River Basin there are many questions that different constituencies would choose to ask, but only the questions with which decision support and modeling are consistent have been given serious attention. As Butkus and Kolmes (2003) described:

> One approach in assessing salmon recovery efforts is to engage in an axiological analysis, that is, an analysis of the values that are driving the process. Ethically this is significant because from the standpoint of a de-ontological ethic, values hold the capacity of generating moral obligation and duty. In other words, humans are compelled to act ethically in relationship to that which they value. Salmon and steelhead carry a spectrum of value for the people in the Columbia River Basin ranging from scientific to economic, culinary-nutritional, recreational, aesthetic, cultural-religious, and sacred value. The latter is particularly important to the indigenous peoples of the Columbia River Basin. What these values have in common is that they are essentially instrumental in nature, that is, they reflect the use-value salmon hold for humans. There is nothing necessarily disturbing

about this, as all species enjoy an instrumental relationship with their biophysical environment. Simply put survivability requires it. Nonetheless what concerns this analysis of salmon recovery—as outlined above—is that the social context and process of recovery appears to accentuate and perhaps even accelerates the instrumentalization of salmon. If the institutional apparatus of salmon recovery is examined, it becomes clear that the social context is largely scientific-bureaucratic wherein the technological rationality of the empirical-analytical sciences predominates.

The Precautionary Principle for salmon recovery in the Columbia River Basin has something to do with protecting the remaining undammed reaches of the system, something to do with protecting old growth and riparian forests, and something to do with limiting development than encroaches closely on the river. All of these things impact salmon population numbers and distributions, and a goal of protecting what remains needs to be articulated. Exactly what this would look like would need to come from a dialogue with tribal members in particular, who were promised in the Treaty of 1855, where they ceded the great majority of their land to the government of the settlers, that they would have "the exclusive right of taking fish in the streams running through and bordering said reservation.... and at all other usual and accustomed stations in common with citizens of the United States." None of the salmon recovery planning being undertaken emphasizes this promise, and it seems likely that if it did the remaining wild salmon would be protected and at least given a chance to recover in numbers. For the Columbia River Basin, that would constitute "do no harm."

be. This is especially interesting to contemplate in circumstances such as the declining salmonid populations of the Columbia River, where "business as usual" activities are the ones that have led to progressive extirpation of the fish. Risk analysis of the known causes of ongoing salmon decline ought to be prioritized, but economic and political barriers to the application of the Precautionary Principle exist.

Our preliminary analysis suggests that the prevailing ways in which decision support and modeling are used works to exclude the framework of the Precautionary Principle from consideration in planning for salmon recovery. There are several possible reasons for this. In so far as modeling drives support of a forecasting frame for planning, it serves to privilege the "business as usual" scenario as described above. A likely result of this special status is that the status quo is not treated as a "proposed action" and thus, as noted, is not even eligible for evaluation by the Precautionary Principle. Typical forecasting approaches de-emphasize uncertainty associated with the status quo scenario. Questions of unintended consequences and uncertainty of outcomes are focused on scenarios that differ from status quo, while there is the implicit assumption that long-term consequences of continuing with current practices, or slight variations, can be projected with a high degree of certainty. Indeed the main modeling tool used in the sub-basin planning process does not report out, or even calculate, uncertainties. In a recent stakeholder meeting, questions arose about the role of global climate change as it might impact the modeling results. Modelers presenting the output of their work had not included global climate change predictions as a parameter in their calculations. Do uncertainties associated with global climate change make its inclusion in modeling suspect, but uncertainties associated with a more status quo scenario not need to be reported? If these uncertainties were included as a routine part of model-based decision support, the case for invoking the Precautionary Principle would be strengthened. As a proposed means of including the Precautionary Principle in planning, and to guide its potential implementation, we will briefly outline a re-framing of multi-stakeholder deliberation in such a way that it could more readily accommodate multiple analytical approaches, including, but not limited to modeling.

While there is a large literature on the circumstances under which the Precautionary Principle (PP) should be invoked, there is comparatively little written on the method of implementation of the principle. Sunstein proposes (2008) that the major shortcoming of the PP is "that the principle offers no guidance" and goes on to point out that the implementation of the principle is driven by underlying assumptions about goals. Here, the post-normal classification, amended with their associated management systems, may offer some guidance. Ravetz proposes (2004) that characteristics of the post-normal context, where "facts are uncertain, values in dispute, stakes high and decisions urgent" provide appropriate conditions for invoking the PP. Of the role of uncertainty, he writes that "the quality of results does not depend on the elimination of uncertainty," adding that "the skilled management of uncertainty, along with the recognition of decision stakes, is the key to quality, especially in the precautionary fields." However, guidance for the application of the PP goes beyond the management of uncertainty and it is here that resilience management could play an important role by informing the development of appropriate goals and methods of implementation. Sunstein argues (2008) that "a belief in the benevolence of nature plays a major role in the operation of the Precautionary Principle, especially among those who see nature as harmonious or in balance" and that "when people seek to protect nature against human intervention, it is often because the dangers of intervention are visible and familiar while the dangers of nonintervention are not." The theory underlying resilience management also holds that the assumption of balance in nature is indeed false, proposing instead a dynamic process, informed by an appreciation of fundamental uncertainties and an awareness of the potential for irreversible changes of state. Resilience management proposes instead (Walker et al. 2002) that managers "focus instead on maintaining the capacity of the system to cope with whatever the future brings, without the system changing in undesirable ways. This can be done by maintaining or increasing the system's resilience" where by "resilience" the authors mean "the potential of a system to remain in a particular configuration and to maintain its feedbacks and functions, and involves the ability of the system to reorganize following disturbance-driven change."

Also working in the context of complex adaptive systems, Cooney and Lang propose (2007) a framework of "adaptive governance" that has much in common with resilience management. "At its heart, adaptive governance accepts and responds to uncertainty by promoting learning in and through the policy-making process. It does so in a number of ways: by avoiding irreversible interventions and impacts, by encouraging constant monitoring of outcomes; by facilitating the participation of multiple voices in transparent policy-making processes; and by reflexively highlighting the limitations of the knowledge on which policy choices are based."

While the principle of resilience describes an important aspect of a desired state of affairs, and describes methods for its promotion, it does not provide a goal that incorporates necessary ethical and cultural elements. Walker et al. write (2002) that "(r)esilience is not necessarily desirable" citing "undesired ecological configurations" that "may indeed be both resistant and resilient," and going on to propose sustainability as "an overarching goal that generally includes assumptions or preferences about which system configurations are desirable." For such a goal to emerge, a deliberative process that incorporates cultural and ethical considerations in both framing the problem and setting goals is needed at the very inception of planning.[5] (Figure 5)

Figure 5: Components of Precautionary Planning.

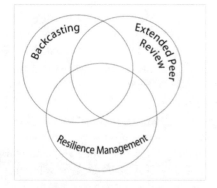

Conclusions

Uncertainty plays a central role in planning for complex social-ecological systems such as the case study of salmon recovery in the Columbia River Basin presented here. Unacknowledged uncertainties, or uncertainties masked by the reassuring appearance of precision provided by computer modeling and numerical benchmarks, can exclude valuable viewpoints, methods, and principles from planning. The forecasting frame can exclude methods of planning such as backcasting, scenario planning, and the precautionary principle that may be required to make difficult transitions to sustainable pathways.

Stakeholders bringing important ethical concerns and vital local knowledge may be excluded from, or drop out of, planning processes due to technical difficulties and inaccessible conceptual frameworks established by highly technological approaches to planning. But the points of view that members of these "extended peer communities" bring to the table can provide the very tools necessary to deal with the high levels of uncertainty inherent in today's complex social-environmental problem-solving contexts. A robust system of knowledge quality assessment, informed by such an extended peer community, and guided by the principles of adaptive management can improve both the quality and breadth of information upon which good decisions are based. Going beyond implications for planning, the involvement of an extended peer community can also lay the foundation for successful implementation.

The combination of high uncertainties and the risk of crossing thresholds argue against over-dependence on model forecasts and prediction as a foundation for management decisions, while arguing in favor of replacing optimization with resilience as a planning goal. A sustainable vision is one that gathers and attempts to integrate (or at least acknowledge) varied viewpoints through extended peer review, develops an inclusive aspirational vision through the process of backcasting, and addresses the challenges of an unpredictable future through the process of resilience management.

Acknowledgments

This chapter is based on research conducted by the Institute for Culture and Ecology and funded by the National Science Foundation [NSF 0722145].

Works Cited

Ackerman, F. *Poisoned for Pennies, the Economics of Toxics and Precaution* (Washington D.C.: Island Press, 2008).

Berkes, F, and M Berkes. "Ecological complexity, fuzzy logic, and holism in indigenous knowledge." *Futures* (Elsevier) 41, no. 1 (Feb 2009): 6–12.

Berkes, F, C Folke, and J Colding. *Linking social and ecological systems: management practices and social mechanisms for building resilience.* New York, USA: Cambridge University Press, 2000.

Bonneville Power Administration (BPA). Fish and Wildlife Implementation Plan Environmental Impact Statement, Appendix D. Bonneville Power Administration: Portland, OR, 2003.

Broman, G, J Holmberg, and KH Robèrt. "Simplicity without reduction: Thinking upstream towards the sustainable society." *Interfaces* (2000): 13–25.

Butkus, RA, and SA Kolmes. *Environmental Science and Theology in Dialogue* (Maryknoll New York: Orbis Press, 2011).

Butkus, RA, and SA Kolmes. 2003. Strategic interdisciplinarity: a scientific-theological analysis of salmon recovery in the Lower Columbia River Basin. *Colorado School of Mines Quarterly* 103:19–32.

Carpenter, S, and C Folke. "Ecology for transformation." *Trends in Ecology & Evolution* 21, no. 6 (2006): 309–15

Carpenter, S, B Walker, JM Anderies, and N Abel. "From metaphor to measurement: resilience of what to what?" *Ecosystems* (Springer) 4, no. 8 (2001): 765–81.

Carpenter, SR, and LH Gunderson. "Coping with Collapse: Ecological and Social Dynamics in Ecosystem Management." *BioScience* (BioOne) 51, no. 6 (2001): 451–57.

Catholic Bishops of the Pacific Northwest. "The Columbia River watershed: Caring for creation and the common good." 2001. Available at *www3.villanova.edu/mission/ CSTresource/…/columbiariver.pdf*

Coburn, Jason. Bringing local knowledge into environmental decision-making. *Journal of Planning Education and Research.* 22, no. 4 (2003): 420–33.

Columbia River Inter-Tribal Fish Commission. *Wy-kan-ush-mi Wa-kish-wit.* Columbia River Inter-Tribal Fish Commission, 1995. Available at http://www.critfc.org/text/trp.html

Cooney, R, and ATF Lang. "Taking Uncertainty Seriously: Adaptive Governance and International Trade." *European Journal of International Law* 18, no. 3 (Jun 2007): 523–51.

Dietz, T, and PC Stern. *Public Participation in Environmental Assessment and Decision Making.* The National Academy Press. 2009.

Environmental Protection Agency. *Guidance on the Development, Evaluation, and Application of Environmental Models.* 2009.

Fullerton, A H, A. Steel, Y Caras, M B Sheer, P Olson, and J Kaje. "Putting watershed restoration in context: alternative future scenarios influence management outcomes." *Ecological Applications* 19, no. 1 (2009): 218–35.

Functowicz, S, and J Ravetz. "Risk Management, Uncertainty and Post-Normal Science." In *Sustainable Fish Farming,* by Helge Reinertsen and Herborg Haaland, 261–271. Rotterdam: A.A. Balkema, 1995.

Funtowicz, SO, and JR Ravetz. "Uncertainty, complexity and post-normal science." *Environmental Toxicology and Chemistry* 13, no. 12 (1994): 1881–85.

Gallopin, GC, S Funtowicz, M O'Connor, and J Ravetz. "Science for the Twenty-First Century: From Social Contract to the Scientific Core." *International Social Science Journal (NWISSJ)* (Blackwell Synergy) 53, no. 168 (2001): 219–29.

Gordon, SN, KN Johnson, KM Reynolds, P Crist, and N Brown. *Decision-Support Systems for Forest Biodiversity: A Review.* United States Department of Agriculture Forest Service Technical Report, 2005.

Gunderson, L, and S Light. Adaptive management and adaptive governance in the Everglades. *Policy Sciences* 39 (2006):323–34.

Gunderson, L, G Peterson, and CS Holling. "Practicing Adaptive Management in Complex Social-Ecological Systems." In *Complexity Theory for a Sustainable Future*, by Jon Norberg and Graeme S. Cumming, 223–46. New York: Columbia University Press, 2008.

Hartley, Troy W, and Robert A Robertson. Stakeholder engagement, cooperative fisheries research and democratic science: The case of the Northeast Consortium. *Human Ecology* 13, no. 2 (2006): 161–71.

Holland, John H. *Hidden Order: How Adaptation Builds Complexity.* Reading, Mass: Addison-Wesley, 1995.

Holling, CS 1986. The resilience of terrestrial ecosystems; local surprise and global change. Pages 292–317 in WC Clark and RE Munn, editors. *Sustainable Development of the Biosphere.* Cambridge, U.K: Cambridge University Press.

Independent Scientific Advisory Board. *Model Synthesis Report: An analysis of decision support tools used in Columbia River Basin salmon management.* For the Northwest Power Planning Council and the National Marine Fisheries Service, 2001.

Jablonowski, M. "Avoiding Risk Dilemmas Using Backcasting." *Risk Management* 9 (2007), 118–27.

Janssen PHM, Petersen AC, JP Van der Sluijs, JS Risbey, JR Ravetz, and K von Krauss. "A guidance for assessing and communicating uncertainties." *Water Science and Technology* 52, no. 6 (2005): 125–34.

Kallis, G, et al. "Participatory methods for water resources planning." *Environment and Planning C: Government & Policy* 24, no. 2 (2006): 215–34.

Kates, RW, and WC Clark. "Environmental Surprise: Expecting the Unexpected?" *Environment* 38, no. 2 (1996): 6–11.

Kolmes, SA, and Butkus R. "Got wild salmon? A scientific and ethical analysis of salmon recovery in California and the Cascadia Bioregion." In *Salmon 2100 Project: The Future of Wild Pacific Salmon*, ed. RT Lackey, DH Lach, and SL Duncan. American Fisheries Society Press, 2006.

Kriebel, D, et al. "The precautionary principle in environmental science." *Environmental Health Perspectives* 109, no. 9 (2001): 871–876.

Kuhn, Thomas S. *The structure of scientific revolutions.* Chicago: University of Chicago Press, 1962.

Lackey, RT, DH Lach, and SL Duncan. *Salmon 2100: The Future of Wild Pacific Salmon.* American Fisheries Society, 2006.

Lakoff, GD. "A cognitive scientist looks at Daubert." *American Journal of Public Health* 95 (2005): 114–20.

Lee, KN, 1993. *Compass and Gyroscope: Integrating Science and Politics for the Environment.* Washington, D.C.: Island Press.

Lemons, J, K Shrader-Frechette, and C Cranor. "The precautionary principle: Scientific uncertainty and type I and type II errors." *Foundations of Science* 2, no. 2 (1997): 207–36.

Leopold, Aldo. *A Sand County Almanac, and Sketches Here and There.* New York: Oxford University Press, 1949.

Liu, Y, H Gupta, E Springer, and T Wagener. "Linking science with environmental decision making: Experiences from an integrated modeling approach to supporting sustainable water resources management." *Environmental Modelling & Software* 23, no. 7 (2008): 846–58.

MacKenzie, DA. *Inventing Accuracy: A Historical Sociology of Nuclear Missile Guidance.* Cambridge, Mass.: MIT Press, 1990.

Marcot, B G, WE McConnaha, PH Whitney, TA O'Neil, PJ Paquet, L. Mobrand, GR Blair, LCLestelle, KM Malone, KI Jenkins, 2002. A multi-species framework approach for the Columbia River Basin: integrating fish, wildlife, and ecological functions. Northwest Power Planning Council, Portland, Oregon. Available at http://www.nwcouncil.org/edt/framework/.

Marshall, BK, and JS Picou. "Postnormal Science, Precautionary Principle, and Worst Cases: The Challenge of Twenty-First Century Catastrophes." *Sociological Inquiry* 78, no. 2 (2008): 230–47.

Matthies, M, C Giupponi, and B Ostendorf. "Environmental decision support systems: Current issues, methods and tools." *Environmental modelling and software* 22, no. 2 (2007): 123–27.

McDaniel, RR, and DJ Driebe. *Uncertainty and Surprise in Complex Systems.* Berlin: Springer, 2005.

McDaniels, TL, and R Gregory. "Learning as an objective within a structured risk management decision process." *Environmental Science & Technology* 38, no. 7 (2004): 1921–26.

McElhany, P, C Busack, M Chilcote et al. *Revised viability criteria for salmon and steelhead in the Willamette and Lower Columbia Basins.* Seattle: Willamette/Lower Columbia Technical Recovery Team, Northwest Fisheries Science Center, 2006.

McElhany, P, M Chilcote, J Myers, and R Beamesderfer. "Viability Status of Oregon Salmon and Steelhead Populations in the Willamette and Lower Columbia Basins Part 1: Introduction and Methods."Report prepared for Oregon Department of Fish and Wildlife and National Marine Fisheries Service. 2007.

McElhany, P., T Backman, C Busack et al. *Interim Report on Viability Criteria for Willamette and Lower Columbia Basin Pacific Salmonids.* Seattle: Willamette/Lower Columbia Technical Recovery Team, Northwest Fisheries Science Center, 2003.

McLain R, Hill G, Kolmes S, Jones ET, "Adaptive Ecosystem Management in A Post-Normal Science Context," *Journal of Environmental Management* (publication pending).

Moller, Henrik, Fikret Berkes, Philip O'Brien Lyver, and Mina Kislalioglu. "Combining science and traditional ecological knowledge: Monitoring populations for co-management." *Ecology and Society* 9, no. 3 (2004).

National Research Council. *Models in Environmental Regulatory Decision Making.* Washington D.C.: National Academies Press, 2007.

Northwest Power Planning Council (NPPC), 1998a. Congressional update—November 3, 1998. Northwest Power Planning Council, Portland, OR. Available at http://www.nwcouncil.org/Library/cu/1998_1103.htm.

———. 1998b. An ecological framework for the multi-species planning process. Northwest Power Planning Council, Portland, OR. Available at http://www.nwcouncil.org/edt/framework/ecoframework.htm.

———. 2000b. The year of decision. Northwest Power Planning Council, Portland, OR. Available at http://www.nwcouncil.org/library/2000/yearofdecision/.

———. 2002c. Human effects analysis of the multi-species framework alternatives. Northwest Power Planning Council, Portland, OR. Available at http://www.nwcouncil.org/edt/framework/humaneffects/

Pahl-Wostl, C. "Participative and Stakeholder-based policy design, evaluation and modeling processes." *Integrated Assessment*, 3. no. 1 (2002): 3–14.

Pahl-Wostl, C, C Schlumpf, M Büssenschütt, A Schönborn, and J Burse. "Models at the interface between science and society: impacts and options." *Integrated Assessment* 1, no. 4 (2000): 267–80.

Pereira, AG, and SC Quintana. "3 pillars and 1 beam: Quality of river basin governance processes." *Ecological Economics* 68, no. 4 (2009): 940–54.

Peterson, GD, GS Cumming, and SR Carpenter. "Scenario planning: a tool for conservation in an uncertain world." *Conservation Biology*, 17, no. 2 (2003): 358–66.

Pinker, S, and G Lakoff. "Does language frame politics." *Public Policy Research* 14, no. 1 (2007): 59–71.

Ravetz, J. "The post-normal science of precaution." *Futures* (Elsevier) 36, no. 3 (Apr 2004): 347–57.

Refsgaard, J, J Van der Sluijs, A Hojberg, and P Vanrolleghem. "Uncertainty in the environmental modelling process—A framework and guidance." *Environmental Modelling & Software* 22, no. 11 (Nov 2007): 1543–56.

Refsgaard, J, JP Van der Sluijs, J Brown, and P Vanderkeur. "A framework for dealing with uncertainty due to model structure error." *Advances in Water Resources* 29, no. 11 (Nov 2006): 1586–97.

Risbey, JS, and M Kandlikar. "Expressions of likelihood and confidence in the IPCC uncertainty assessment process." *Climatic Change* 85, no. 1 (2007): 19–31.

Robert, K. "Tools and concepts for sustainable development, how do they relate to a general framework for sustainable development, and to each other?" *Journal of Cleaner Production* 8, no. 3 (2000): 243–54.

Robinson, J. "Energy backcasting: a proposed method of policy analysis," *Energy Policy* 10, no. 4 (1982): 337–44.

Rotmans, J, and MBA van Asselt. "Uncertainty in integrated assessment modelling: A labyrinthic path." *Integrated Assessment* 2, no. 2 (2001): 43–55.

Scheffe, M, S Carpenter, JA Foley, C Folke, and B Walker. "Catastrophic shifts in ecosystems." *Nature*, no. 413 (2001): 591–96.

Sunstein, CR. "Precautions and nature." *Daedalus* (MIT Press) 137, no. 2 (2008): 49–58.

van Asselt, MBA, and J Rotmans. "Uncertainty in Integrated Assessment Modelling." *Climatic Change* (Springer) 54, no. 1 (2002): 75–105.

van der Sluijs, J. "Uncertainty, assumptions and value commitments in the knowledge base of complex environmental problems." *Interfaces between Science and Society*, 2006: 64–81.

van der Sluijs, J, M Craye, S Funtowicz, and P Kloprogge. "Combining Quantitative and Qualitative Measures of Uncertainty in Model-Based Environmental Assessment: The NUSAP System." *Risk Analysis* 25, no. 2 (2005): 481–92.

van Eeten, MJG. "Bringing actors together around large-scale water systems: Participatory modeling and other innovations." *Knowledge, Technology, and Policy* (Springer) 14, no. 4 (2001): 94–108.

von Krauss, M, MBA van Asselt, M Henze, JR Ravetz, and MB Beck. "Uncertainty and precaution in environmental management." *Water Science & Technology* 52, no. 6 (2005): 1–9.

Vogel, Coleen, Susanne C. Moser, Roger E. Kasperson, and Geoffrey D. Dabelko. Linking vulnerability, adaptation, and resilience science to practice: Pathways, players, and partnerships. *Global Environmental Change* 17 (2007):349–64.

Walker, PA, R Greiner, D McDonald, and V Lyne. "The Tourism Futures Simulator: a systems thinking approach." *Environmental Modelling and Software* (Elsevier Ltd) 14, no. 1 (1998): 59–67.

Walker, B. and JA Meyers. "Thresholds in ecological and social–ecological systems: a developing database." *Ecology and Society* 9 no. 3 (2004).

Walker, W, P Harremoës, J Rotmans, J van der Sluijs, et al. "Defining Uncertainty: A Conceptual Basis for Uncertainty Management in Model-Based Decision Support." *Integrated Assessment* 4, no. 1, (2003):5–17.

Walters, CJ, and CS Holling. "Large-scale management experiments and learning by doing." *Ecology*, 1990: 2060–68.

Walz, A, et al. "Participatory scenario analysis for integrated regional modelling." *Landscape and Urban Planning* (Elsevier) 81, no. 1–2 (2007): 114–31.

Wardekker, J, van der Sluijs, J, P Janssen, P Kloprogge, and A Petersen. "Uncertainty communication in environmental assessments: views from the Dutch science-policy interface." *Environmental Science & Policy*, 11, no. 7 (2008): 627–41.

Westley, F. 2002. The devil in the dynamics: Adaptive management on the front lines. In: LH Gunderson and CS Holling, eds. *Panarchy: Understanding Transformations in Human and Natural Systems.* Washington, D.C.: Island Press.

Westley, F, SR Carpenter, WA Brock, CS Holling, and LH Gunderson. 2002. Why systems of people and nature are not just social and ecological systems. In LH Gunderson and CS Holling, eds. Panarchy: *Understanding Transformations in Human and Natural Systems.* Washington, D.C.: Island Press.

Wilkinson, A, and E Eidinow. 2008. Evolving practices in environmental scenarios: a new scenario typology. *Environmental Restoration Letters* 3 (4), 11.

Williams, Richard Nicholas. *Return to the River: Restoring Salmon to the Columbia River.* Amsterdam: Elsevier Academic Press, 2006.

Wilson, C, J Tansey, and S LeRoy. "Integrating Backcasting & Decision Analytic Approaches to Policy Formulation: A Conceptual Framework." *The Integrated Assessment Journal* 6, no. 4 (2006): 143–64.

The Columbia River Treaty in 2014 and Beyond: International Experiences and Lessons Learned

Stephen McCaffrey, Richard Paisley, Lynette de Silva, and Aaron Wolf

Introduction

This chapter identifies lessons learned from recent international experience with transboundary waters governance that may be relevant to the Columbia River Basin in 2014 and beyond, with particular reference to minimum stream flows; stream flow and other hydrological changes associated with climate change; and the role of third parties in negotiating new or adjusted governance mechanisms for international waters.

According to Kofi Annan, former U.N. Secretary General, "the water problems of our world need not be only a cause of tension: they can also be a catalyst for cooperation… If we work together, a secure and sustainable water future can be ours."[1] This is a message that has been echoed by academic scholars, including Geoffrey Dabelko, Ken Conca, and others (Carius et al. 2004; Bencala and Dabelko, 2008).[2] To help realize this cooperative potential in river basins, particularly as environmental, demographic, economic, and institutional changes challenge existing arrangements, it is helpful to share lessons from other riparian states that have successfully prevented conflict and mediated disputes. (Postel and Wolf, 2001).

Institutional Lessons from the International Community

A review of international water relations and institutional development over the past fifty years provides important insights into water conflict and the role of institutions. The historical record of water conflict and cooperation suggests that while international watercourses can cause tensions between co-riparian states, they are more likely to to focus the political will to craft institutions and create mechanisms to collaborate.

The centrality of institutions both in preventive hydro-diplomacy and in effective transboundary water management cannot be over-emphasized. Yet, while progress is indeed apparent, the past fifty years of treaty writing suggests that capacity-building opportunities still remain. Following are five characteristics of effective institutions:

1. An adaptable management structure. Effective institutional management structures incorporate a certain level of flexibility, which allows for public input, changing basin priorities, and new information and monitoring technologies. The adaptability of management structures must also extend to non-signatory riparians, by incorporating provisions addressing their needs, rights, and potential accession.

Clear and flexible criteria for water allocations and water quality management. Allocations, which are at the heart of most water disputes, are a function of water quantity and quality, as well as political fiat. Thus, effective institutions must identify clear allocation schedules and water quality standards that simultaneously provide for extreme hydrological events; new understanding of basin dynamics, including groundwater reserves; and changing societal values. Additionally, riparian states may consider prioritizing uses throughout the basin. Establishing catchment-wide water precedents may not only help to avert inter-riparian conflicts over water use, but will also protect the environmental health of the basin as a whole.

3. Equitable distribution of benefits. Distributing water benefits, a concept that is subtly yet powerfully different than pure water allocation, is at the root of some of the world's most successful institutions. Distributing benefits—whether from hydropower, agriculture, economic development, aesthetics, or the preservation of healthy aquatic ecosystems allows for positive-sum agreements, occasionally including even non-water-related gains in a "basket of benefits," whereas dividing the water itself only allows for winners and losers.

4. Concrete mechanisms to enforce treaty provisions. Once a treaty is signed, successful implementation is dependent not only on the actual terms of the agreement but also on the ability to enforce them. Appointing oversight bodies with decision-making and enforcement authority is one important step towards maintaining cooperative management institutions.

5. Detailed conflict resolution mechanisms. Many basins continue to experience disputes even after a treaty is negotiated and signed. Thus, incorporating clear mechanisms for resolving conflicts is a prerequisite for effective, long-term basin management.

The Columbia River and the Columbia River Treaty

In recognition of the general importance of cooperating with regard to their many shared water resources, Canada and the United States concluded an agreement in 1909 known as the Boundary Waters Treaty, which established an International Joint Commission (IJC) to govern their relations.[3] The regulation and management of the Columbia River began to first receive serious consideration in 1944 when the subject was referred to the IJC for study. One of the IJC recommendations was that the power production benefits in the United States from upstream storage in Canada be shared on a substantially equal basis, provided that an equal split of the benefits would result in an advantage to each country as compared with available alternatives. The IJC further recommended that when an equal split would not result in an advantage to each country, the countries would then have to negotiate and agree upon such other division of benefits as would be equitable to both countries and make cooperative development feasible.

The Columbia River Treaty (CRT) was subsequently developed from these recommendations, and explicitly recognized that the construction and operation of three treaty projects in Canada would increase the useable energy and dependable capacity of power plants in the United States, as well as provide potential irrigation and flood control benefits to both countries.[4] However, the benefits in the United States would clearly not be attainable at the same cost without the three treaty projects in Canada. In return for building the three projects in Canada, the CRT entitled Canada to, among other things, a lump sum payment for various downstream (flood control) benefits, as well as one half of the additional power generated by power plants in the United States that resulted from reservoir storage across the border in Canada.

The CRT has long served as a useful model for international cooperation. However, recent social, economic, political, and environmental circumstances present increasing challenges (Sanderson, 2009). Current issues include:

The legacy of the past: Local populations, including the Tribes and First Nations, feel that they were never adequately consulted and/or their interests accommodated with regard to the CRT, in particular about the flooding of fertile valley land and sites of cultural significance.

The appropriate role of hydroelectricity and other energy sources: The development of the region has led to energy demands that now outstrip supply.

Water quality and environment: The value that all the different stakeholders place on water quality is increasingly important—for both consumptive and non-consumptive uses.

Water quantity and water allocation: There are many different stakeholders, each with their own view on allocation and the reservoir levels.

Industry and livelihoods: The era of industrialization is starting to shift from heavy industry (e.g., mining) to a more diverse economic base that includes tourism.

Salmon and other anadromous fish habitat: There are numerous effects on salmon and their habitat by the physical barriers of the dams as well as the changes in the river ecosystems.

Public and aboriginal participation in environmental decision making and the role of civil society: Public expectations regarding participation have changed since the CRT was signed—the public and aboriginal groups on both sides of the border now expect to be consulted and accommodated .

Climate change: Climate change has the potential to seriously disrupt the sensitive ecosystem of the Columbia Basin through, for example, increasing temperatures and reduced snow pack.

All these issues are driving forces which will inevitably influence Columbia River governance in 2014 and beyond. Associated with each of these issues are also a multiplicity of stakeholder interests (Davidson and Paisley, 2009; McKinney, 2009).

International Experiences and Lessons Learned
Nearly half of the world's population is located within one or more of the 276 international drainage basins shared by two or more states (International Bureau of the Permanent Court of Arbitration 2002, at xix). Even more striking than the absolute number of international drainage basins, is a breakdown of each nation's land surface that falls within these watersheds. At least one hundred forty-five nations include territory within international basins (Wolf 2000). At least twenty-one nations lie in their entirety within international basins including thirty-three countries that have greater than 95 percent of their territory within these basins. All told, nineteen international drainage basins are shared by five or more riparian countries. The Danube has seventeen riparian nations. The Congo, Niger, Nile, Rhine, and Zambezi are shared by between nine and eleven countries. The remaining thirteen basins have between five and eight riparian countries.

With so many shared river basins it might be expected that there are some experiences worth learning from, particularly regarding negotiation, cooperation, and dispute resolution. While this is true, deriving lessons learned from international experiences is challenging because of differing objectives and criteria for measuring "success," as well as differing social, political, economic and cultural circumstances. Challenges also arise from issues of "scale," communicating across cultures, and the unpredictability of human nature. Nonetheless, with awareness of the contextual differences, some general lessons for the Columbia emerge (Figures 1, 2, and 3).

A review of treaties signed within the last twenty years also reveals some encouraging developments. At least 54 new bilateral and multilateral water agreements have been concluded since the 1992 United Nations Conference on Environment and Development in Rio de Janeiro, Brazil (the Rio Conference), representing basins in Asia, Africa, Europe, North America, and South America.[5] As in the past fifty years as a whole, European water accords continue to dominate. However, agreements from other regions, in particular Asia, have grown disproportionately. In addition to greater geographic representation, a number of improvements can be seen in this more recent set of treaties compared with the last half-century as a whole. First, a growing percentage of treaties address some aspect of water quality, a finding consistent with Rio's goal of both managing and protecting freshwater resources. Second, provisions concerning monitoring and evaluation, data exchange, and conflict resolution are included in many of the post-Rio treaties.

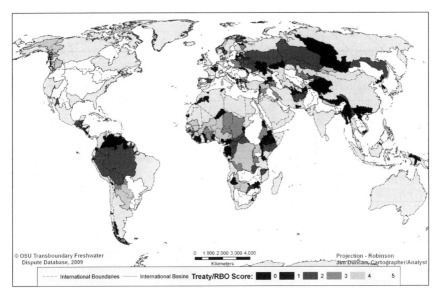

Figure 1. There are still many international river basins without agreements.

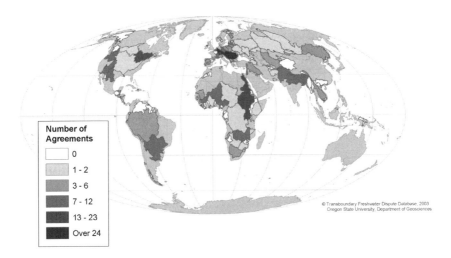

Figure 2 shows the areas where international river basin treaties have historically been concentrated.

Figure 3 illustrates a continuum of cooperation in international river basin agreements from unilateral action through coordination and collaboration to joint action.

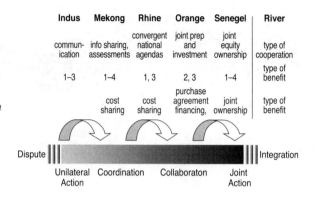

	Indus	Mekong	Rhine	Orange	Senegal	River
commun-ication		info sharing, assessments	convergent national agendas	joint prep and investment	joint equity ownership	type of cooperation
	1–3	1–4	1, 3	2, 3	1–4	type of benefit
		cost sharing	cost sharing	purchase agreement financing,	joint ownership	type of benefit

Dispute ||| ▭▭▭▭▭▭▭▭▭▭▭▭ ||| Integration

Unilateral Action   Coordination   Collaboraton   Joint Action

Third, a number of agreements establish joint water commissions with decision-making and/or enforcement powers, a significant departure from the traditional advisory standing of basin commissions. Fourth, country participation in basin-level accords appears to be expanding. Although few of the agreements incorporate all basin states, a greater proportion of treaties are multilateral and many incorporate all major hydraulic contributors. Finally, a 1998 agreement on the Syr Darya Basin, in which water management is exchanged for fossil fuels, provides a post-Rio example of basin states broadly capitalizing on their shared resource interests.

While a review of the past century's water agreements highlights a number of positive developments, institutional vulnerabilities remain. Notably, 158 of the world's 276 international basins lack any type of cooperative management framework. Furthermore, of the 106 basins with water institutions, approximately two-thirds have three or more riparian states, yet less than 20 percent of the accompanying agreements are multilateral. Also, despite the recent progress noted above, treaties with substantive references to water quality management, monitoring and evaluation, conflict resolution, public participation, and flexible allocation methods remain in the minority. As a result, most existing international water agreements continue to lack the tools necessary to promote long-term, holistic water management. Many treaties, for example, ignore issues of allocation, and of those that do, few possess the flexibility to handle changes in the hydrologic regime or in regional values. References to water quality, related groundwater systems, monitoring and evaluation, and conflict resolution mechanisms, while growing in numbers, are often weak in actual substance. Furthermore, enforcement measures and public participation, two elements that can greatly enhance the resiliency of institutions, are largely overlooked.

Finally, groundwater, with all its uncertainties and complexities, adds further challenges to these international regimes. A review of international water law specifically addressing groundwater reveals that many of the agreements were developed in the past fifty years, and have only recently adopted a definition of an aquifer. Matsumoto (2002) inventoried nearly 400 treaties listed in the Transboundary Freshwater Dispute Database (TFDD) and summarized the number of treaties recognizing groundwater by continent:[6] thirty-five treaties developed between European countries; thirteen treaties in Africa; ten treaties in the Middle East and Asia; four treaties in North America; and no treaties developed in South America.

Protection of groundwater quality has only been addressed in the past few years (Morris et al. 2003). The transboundary movement or "silent trade" of hazardous wastes into Lebanon described by Jurdi provides an example of the necessity to increase the need for "global harmonization" of international water and waste treaties (Jurdi 2002). Of the many international water treaties, few have monitoring provisions, and almost none have an enforcement mechanism (Chalecki et al. 2002).

There are three specific areas where recent experiences may be relevant to the Columbia River Basin: minimum stream flows; stream flow and other hydrological changes associated with climate change; and the role of third parties in negotiating new or adjusted governance mechanisms for international waters. These aspects of transboundary governance, which were not prominent when the original CRT was negotiated, are examined below.

Minimum Stream Flows[7]
The objective of minimum stream flow is to maintain water levels in rivers to protect the ecological, chemical, and physical integrity of ecosystems (Dyson et. al. 2008). Aesthetics may also be important in the context of establishing and maintaining minimum stream flows. Particularly in the western United States the increasing scarcity of water has led, in the last twenty years, to the development of a multiplicity of legal regimes designed to try to ensure adequate minimum stream flows. The same concerns that led to the development of the law in the western United States to protect minimum streams flows are now starting to appear at the international level, in a variety of treaty regimes and international practice (Dyson et. al. 2008).

According to Utton and Utton (1999):

[A]t the international level, numerous treaties provide for regulating the flow of transboundary rivers for a variety of purposes such as the production of hydroelectricity, protection of commercial fisheries, protection from flooding, and floatation of logs. Until recently, none of these treaties addressed the regulation of flows for the environmental goals of protecting the ecological, chemical, and physical integrity of international watercourses.

However, the Mekong Treaty, signed by Cambodia, Laos, Thailand, and Vietnam in 1995, specifically calls for the provision of "not less than the acceptable minimum monthly natural flow" of water in the Mekong River Basin. The minimum monthly natural flow requirement was included, in part, to "protect, preserve, enhance and manage the environmental and aquatic conditions and maintenance of the ecological balance exceptional to this river basin" (Dyson et al. 2008). A specific protocol regarding minimum stream flows was recently further negotiated for the Mekong in 2006 (Mekong River Commission 2006).[8]

In reaching consensus on the Mekong Agreement, the negotiators had to fashion a new framework for the Mekong regime that protected the interests of all parties, while faced with an era of growing water scarcity. As Browder and Orolano (2000) explain:

> *Among other concerns, the Vietnamese wanted to at least maintain the existing dry season flows into the Mekong Delta so they would be able to sustain rice harvests and combat salinity intrusion. The Lao were intent on preserving the dry season navigability of the Mekong River, which serves as Laos' main transportation artery. Finally, the Cambodians wanted to protect the hydrological and ecological integrity of the Tonle Sap (i.e., the Great Lake) by ensuring sufficient reverse wet season flows from the Mekong River into the Tonle Sap.*

This suggests that protection of in-stream flows may be conducive to the protection of other beneficial uses. Unfortunately, the outcome and application of this provision have not been closely explored.

The International Union for Conservation of Nature also notes that there are:

> *[S]everal agreements covering particular watercourses that contain general principles of international water law applicable to environmental flows. Yet others include similar principles but go a little further by establishing more specific provisions on the regulation of river flows. Examples of these agreements include: The Convention on the Protection and Use of Transboundary Watercourses and International Lakes (Helsinki Convention); The Mekong River Agreement; The Protocol on Shared Watercourses Systems in the Southern Africa Development Community; and The Convention on Co-operation for the Protection and Sustainable Use of the Waters of the Portuguese-Spanish River Basins (Utton and Utton 1999).*

In addition, Chapter V of the Berlin Rules deals with protection of the aquatic environment through the use of the precautionary approach (Salman 2007). "It [also] requires the states to take all measures to protect the ecological integrity necessary to sustain ecosystems dependent on particular waters, and to prevent, eliminate, reduce or control pollution and harm to the aquatic environment."

Further, it calls on states to take all appropriate measures to ensure flows adequate to protect the ecological integrity of the waters of a drainage basin, and to prevent the introduction of alien species, and hazardous substances (Salman 2007).

It is unfortunate that little has been written on the outcome of international agreements containing in-stream flow provisions, as these lessons from state experience would be valuable for the planning of future agreements between other riparian states. However, what can be drawn from these limited examples is a growing trend in international law that recognizes the importance of providing adequate protection for in-stream flows in transboundary water agreements.

Conspicuous by its absence in the CRT is any mention of minimum stream flows, which will potentially be an increasingly volatile issue. At issue is the fact that the Columbia River's flow in the United States portion of the river declined by about 14 percent during 1948–2004 largely because of a combination of reduced precipitation and higher water usage in the western United States.[9] Further increasing declines in flow in the United States portion of the Columbia River are anticipated due to the detrimental impact of climate change (Schreier 2006).

Stream Flow and other Hydrological Changes Associated with Climate Change
Many international rivers in the world are already suffering, or are increasingly likely to suffer, from the detrimental impact of climate change. In the volatile Middle East, for example, climate models predict hotter, drier, and less predictable weather, as a result of climate change (Brown and Crawford 2009). Higher temperatures will alter rainfall patterns, producing declines during the wet season, a northward shift in distribution, and more frequent unpredictable extreme rainfall events. Aquifers will replenish more slowly and river flows will be reduced with moderate scenarios suggesting a 30 percent reduction of the Euphrates and an 80 percent reduction of the Jordan by 2100. This means less water for agriculture and potential damages to infrastructure such as dams due to flooding or excess sedimentation. Timing for planting crops will alter as the first rains will occur at different times, and the frequency of rains will be less predictable during the growing season. A drier climate will also increase soil degradation due to higher evaporation rates, spreading desertification and generating a greater demand for irrigation water. A rise in sea level will contribute to saltwater intrusion into coastal aquifers, particularly the Gazan coastal aquifer that supplies 1.5 million Palestinians. Population growth will further contribute to water scarcity (a measure of the volume of water available per person). Without large-scale desalination, improved water efficiency or international water transfers, the region's renewable water resources will no longer provide for the needs of the region by 2020. Complicating these problems is the fact that a large proportion of available water is transboundary, and a long legacy of conflict

has led to a "'zero-sum approach" to resolving resource disputes. Access to water is very unevenly distributed across the region, causing bitter disputes such as those between Palestine and Israel. The culture of distrust causes the countries to ignore the benefits of co-management, while viewing resources as national rather than regional assets, and remaining unwilling to share data or engage in collaborative research. Rather than cooperate on regional projects such as a "water grid," countries rely on expensive national-level efforts. Efficient water management is discouraged by the fact that conservation gains have to be shared with rival neighbors.

Another example is the Orange-Senqu River Basin in southern Africa. Here, continuously changing patterns of water flow and utilization place strain on the conservation and management of shared water resources. Kistin and Ashton (2008) hypothesize that international agreements and institutions established for this basin must be designed for flexibility in the face of change. They identify four drivers of change: (1) Climate change; (2) Demographic change; (3) Socio-economic change; and (4) Management practices. These drivers are considered alongside various mechanisms that enhance treaty flexibility drawn from McCaffery (2003) and Fischhendler (2004). Evaluating a treaty against these mechanisms will, according to the authors, capture a treaty's ability to adapt to gradual and sudden changes. The mechanisms include: (1) allocation strategies; (2) drought response provisions; (3) amendment and review processes; (4) revocation clauses; and (5) institutional responsibilities. A review of international agreements among Botswana, Mozambique, Namibia, South Africa, and Zimbabwe reveals several flexibility mechanisms in existing treaties, indicating a high degree of adaptability.

Climate modeling for North America similarly predicts significant changes in the Canadian west and the American Pacific Northwest, particularly for water resources. Consequently, in seeking to deal appropriately with climate change, including issues relating to adaptability and resilience, the Columbia River Basin in 2014 and beyond would seem to have much to learn from these and other international examples.

Role of Third Parties[10]

It is not yet clear what role, if any, third parties might have in both Canada and the United States, leading to a possible decision as early as 2014 to terminate or adjust the current Columbia River Treaty. However, recent international experiences suggest that third parties can play an important role in promoting cooperation in international river basins as technical or legal advisors, financial backers, facilitators and/or political proponents (Paisley and Hearns 2006). The Mahakali River Treaty between Nepal and India is an example. While India rejected the involvement of foreign engineers or assistance during negotiations, Nepal felt it did not have the technical capacity to undertake the necessary hydrological research and feasibility studies on its own. Nepal was eventually able to conduct the necessary analysis in part with Canadian assistance (Marty 2001).

Third parties may even become party to a treaty, such as in the case of the World Bank in the Indus River Treaty between Pakistan and India (McCaffrey 2003).

Another example involving third parties as co-partners in project implementation is in the Chu-Talas system, where both the U.N. Economic Commission for Europe and the U.N. Economic and Social Commission for Asia and the Pacific have been involved in fund raising as well as project management, alongside the governments of Kazakhstan and Kyrgyzstan (Paisley and Hearns 2006).

Even in situations where there already appears to be a high level of trust and cooperation a third party can be instrumental in developing an agreement. In the case of the Iullemeden Aquifer System between Niger, Mali, and Nigeria, the role of third parties was highly developed. Funds for developing the data base and hydrogeological model and the associated meetings were made available through the Global Environment Facility (GEF), the activities were overseen by the United Nations Environmental Program, and technical assistance was provided by the United Nations Educational, Scientific and Cultural Organization and the Food and Agriculture Organization. The bulk of the managerial and administrative work. and its facilitation were undertaken by the Observatory of the Sahara and Sahal (OSS), based in Tunisia. Because its staff members were well respected for their technical abilities and were seen to be unbiased, the OSS was highly effective in bringing together all the relevant national ministries. Also, as a regional organization, it was viewed to have insight into the "'West African context," which assisted in negotiations and relation building.[11]

There is also a long and vibrant history at the international level in the area of "Track Two" diplomacy spearheaded by third parties. According to Diana Chigas:

Unofficial third-party intervention means different things to different people. At the inter-group or international level, the term encompasses a number of different terms: "track two diplomacy," citizen diplomacy, "multi-track diplomacy," supplemental diplomacy, pre-negotiation, consultation, interactive conflict resolution, back-channel diplomacy, facilitated joint brainstorming, coexistence work. While differing in emphasis, agenda, and theoretical approach, these initiatives share many common goals. They attempt to provide an environment that is low-key, non-judgmental, non-coercive, and safe, and to create a process in which participants feel free to share perceptions, fears and needs, and to explore ideas for resolution, free of the constraints of government positions. The process is designed to encourage the development of mutual understanding of differing perceptions and needs, the creation of new ideas, and strong problem-solving relationships.[12]

The potentially useful role of third parties in the successful negotiation and implementation of a major international water and energy agreement is further illustrated by the situation of the Nile River basin.[13] Eleven countries share the Nile River Basin: Burundi, D.R. Congo, Egypt, Eritrea, Ethiopia, Kenya, Rwanda, Sudan, Tanzania, Uganda, and, as of July 9, 2011, South Sudan. Historically heated disputes, sometimes involving the threat of armed force, have arisen between Nile Basin states, particularly Egypt and Sudan, and Egypt and Ethiopia. Even today, the various Lake Victoria Basin states complain that their use of the Nile is unreasonably restricted by a 1929 treaty between Egypt and Britain that requires Egypt's consent for any upstream development. In the mid-1990s, the Nile Basin states—under the auspices of an influential third party, the United Nations Development Programme—established the "Cooperative Framework" process, designed to produce an agreement on the use and management of the Nile. Subsequently in 1999 the Nile Council of Ministers (NILE COM), with the support of the World Bank, launched the Nile Basin Initiative (NBI) whose "shared vision" is: "[T]o achieve sustainable socio-economic development through equitable utilization of, and benefit from, the common Nile Basin water resources." The NBI is now an international organization with its headquarters in Entebbe, Uganda. Eight basin-wide or transboundary projects and programs have been planned under the NBI, and some are already underway. The eight projects are designed to produce "win-win" solutions through sharing benefits, thus avoiding "zero sum" situations that result from merely apportioning water. Third party involvement in the NBI process has also included the provision of professional international legal advice in connection with the preparation of the Cooperative Framework Agreement. The Agreement will enter into force when six Nile Basin states have ratified.

Similarly, in the case of the Senegal River, third-party involvement also appears to have enabled the parties to reach an agreement in challenging circumstances. The Senegal River basin is shared between Senegal, Mali, and Mauritania. In a series of treaties beginning in the early 1970s, these three nations formed the Senegal River Basin Organization and provided for the management of the river and the construction of two dams: one at Diama, a saltwater intrusion barrier at the mouth of the river; and the other at Manantali, in Mali, for hydropower production. One of the challenges in the Senegal River valley was that the new dams destroyed traditional flood recession agriculture and led to an explosion in water-born diseases with resulting health, economic, and cultural effects. Subsequently however, in 2002, the three countries agreed to a new Water Charter to manage the Manantali Dam in a way that allowed them to manipulate the flow regime in order to provide for, among other things, a crucial artificial flood to mimic pre-dam conditions (Vick 2006). In another positive development, Guinea, where the Senegal originates, has now joined the legal regime. These recent developments were facilitated by the World Bank, in whose headquarters the Water Charter was finalized.

Academic institutions may also have a key role to play as third parties through neutral fact finding, facilitation, and capacity building. An interesting example is the Universities Consortium on Columbia Basin Governance, which is comprised of one public university in each state and province in the Columbia River Basin—the University of British Columbia, the University of Idaho, the University of Montana, Oregon State University, and the University of Washington. This academic body came about to bring research informed by stakeholder input to bear on the issues of importance to both the U.S. and Canada in the Pacific Northwest (PNW) prior to the change in status of the Columbia River Treaty in 2024. Since either nation can terminate most of the provisions of the treaty as early as 2024 with a minimum ten years written notice (2014), the consortium recognized a timely opportunity for wide-ranging discussions on the challenges facing the basin.

Realizing that one of the goals of any comprehensive research effort by academic interests must be to assist in dialogue, these academic institutions sought to hold regional, cross-border dialogues involving tribes, First Nations, stakeholders, governments, and researchers throughout the PNW. To do so, the consortium initiated a series of symposia hosted by each of the universities. And, to encourage openness and the sharing of information, the world-famous Chatham House Rule was invoked at meetings to provide anonymity to speakers and participants when requested, which created a unique forum.

The symposia were designed to consider an interdisciplinary approach to Columbia River Basin water management in order to address a myriad of basin interests. Designed to help guide decision-making under uncertainties, where "uncertainties" refers to rapid changes, and social and economic instability such as, but not limited to, climate and environmental change; continued regional population growth; a threatened and deteriorating ecosystem; demand for non-fossil fuel energy; and deteriorating infrastructure.

The first symposium, hosted by the University of Idaho, was held April 2–4, 2009. It provided a unique opportunity to consider the effectiveness of the existing treaty under current and anticipated conditions, and to discuss whether it is desirable and appropriate for the two countries to open discussion on modification of the treaty and comprehensively address concerns of the basin and the region. Those invited to the symposium included representatives from universities, primarily professors and graduate students; nongovernment organizations; government agencies; and tribes from all parts of the basin.

Building on this forum of ideas, a second symposium sponsored by the Northwest Power and Conservation Council and the Columbia Basin Trust (CBT) was held at Oregon State University, on November 7–9, 2010. Its purpose was to address and provoke discussion around three key themes: needs and benefits, participatory processes, and transboundary governance mechanisms. The eighty-five attendees included the Columbia River Treaty Entities (the U.S. Army Corps of Engineers, the Bonneville Power Administration, and BC Hydro) and representatives from universities, nongovernment organizations, government agencies, companies, and tribes. The attendance of the tribes and various First Nations at this second symposium was particularly significant and welcomed.

Capitalizing on the expertise, working knowledge, and extensive experience in aspects of the Columbia River Basin, participants formed roundtables, with each table selecting a basin scenario option to explore. The possible scenarios included: (1) The treaty continues as currently written and implementation stays largely the same (except for built-in flood control changes); (2) The treaty continues as currently written, but implementation is modified solely through entity agreements under entity authority; (3) The treaty continues as currently written, but implementation is modified through both entity agreements under entity authority and other government approvals; (4) The treaty is amended; (5) The treaty is terminated and a new replacement treaty is negotiated. In terms of these scenarios, each roundtable considered possible actions, such as the required authority level, hydropower considerations, flood control considerations, and any other factors deemed critical.

The second symposium raised more questions than it answered, and highlighted some contrasting viewpoints. There were discussions about the actual process of the treaty decisions, and how that might be administered, and who could participate in the process. Other queries related to the symposium series, with participants questioning what such a forum can and cannot hope to accomplish.

Though most participants are experts in aspects of the Columbia River Basin, many were for the first time facing the very real issues posed in a scenario exercise and began to realize how an entity might or could react during the process of negotiating for advantage, and for a nation's real interests and concerns. For some being able to listen to different perspectives and converse openly was enlightening and appeared to be transforming, leaving more than one person to reconsider if this agreement can be more inclusive. For example, rather than thinking in terms of the basin's hydropower and irrigation needs, participants might try to address the more challenging solution of incorporating a larger suite of needs, which included hydropower, irrigation, and a healthy fish population. This shift in perspective was also evident when the U.S. Columbia River Tribes paid homage to their Canadian First Nations counterparts. This gesture of both acknowledgment and mutual respect prompted participants to reflect on collaborative approaches and ways of enhancing benefits for future generations.

In terms of the basin's future, participants also suggested that university graduate students become knowledgeable in the treaty, engage and interview stakeholders to learn the issues and interests, and develop scenarios and mock negotiation preparedness to inform the entities, as they move forward. Also emphasized was the continued value of this third-party approach being championed by the Universities Consortium on Columbia Basin Governance that encourages informed discussions of scenarios with a diverse group of participants.

Conclusion

Institutional lessons from the around the world that might be relevant to the future of the Columbia River Treaty include: adaptable management structures, clear and flexible criteria for water allocations and water quality management, and equitable distribution of benefits. More specifically, treaty research can benefit from in the areas of minimum stream flows; stream flow and other hydrological changes associated with climate change; and the role of third parties in negotiating new or adjusted governance mechanisms. These examples may inform future cooperation in the Columbia River Basin, so long as contextual factors, such as local populations, industry, and fish habitat are taken into consideration. Through careful examination of lessons learned from water management around the world, the Columbia Treaty experience can demonstrate best practices that will provide guidance for future transboundary water agreements.

Notes

1. See United Nations, *World's Water Problems Can Be 'Catalyst For Cooperation' Says Secretary-General In Message On World Water Day*, Press Release, SG/SM/8139 OBV/262 as found at http://www.un.org/News/Press/docs/2002/sgsm8139.doc.htm.

2. "There is considerable conflict over water, but it is not necessarily where politicians, journalists or advocates suggest we should expect it. Countries have historically been quick to rattle their sabers over water, but they have nevertheless been content to keep them sheathed. One hears of few—if any—actual cases of wars being fought over water. Instead, evidence from systematic assessments of bilateral and multilateral interactions over water suggests a cooperative narrative is more accurate than a violent one. Successful cooperation within many transboundary river basins has become a powerful counter-story to the ubiquitous water wars prediction." (Bencala and Dabelko 2008, p. 21) See also Conca and Dabelko 2002). Some of the case studies suggest that "a water peacemaking strategy can create shared regional identities and institutionalize cooperation on a broader range of issues. Examples of this dynamic include the institutionalized environmental cooperation around the Baltic Sea during the Cold War (Helsinki Commission) and the current cooperation in post-apartheid Southern Africa through the Southern African Development Community (SADC)."

3. See The International Joint Commission at http://www.ijc.org/rel/agree/water.html.

4. See the Columbia Basin Trust at http://www.cbt.org/crt/.

5. See United Nations Conference on Environment and Development, available at http://www.un.org/geninfo/bp/enviro.html.

6. See also Wolf 1999 and Morris et aL 2003. The Transboundary Freshwater Dispute Database is available at http://www.transboundarywaters.orst.edu/database/

7. Minimum stream flows are also known as minimum "in" stream flows. This section draws on the seminal article regarding the international law of minimum stream flows by Utton and Utton: Utton1999. See also Dyson et al. 2008.

8. See also: King et al. 1998 and Tharme and King 1999.

9. See University Cooperation for Atmospheric Research at http://www.ucar.edu/news/releases/2009/flow.jsp

10. This section is based in part on discussions with Dr. Glen Hearns.

11. Personal communication (Glen Hearns) with Issaka, I., General Secretary, Ministère de l'Hydrologique du Niger (14 November 2008); Maïga, S., Directrice Nationale de l'Hydraulique, Ministère des Mines, de l'Energie et de l'Eau du Mali (20–22 November 2008); and Chabo, J., Director of Nigeria Hydrological Service Agency, Ministry of Agriculture and Water, Nigeria (18 November 2008).

12. See Beyond Intractability at http://www.beyondintractability.org/essay/track2_diplomacy/.

13. While at the time of this writing significant uncertainty still surrounds the possibly successful conclusion of a comprehensive Nile River Basin Framework Agreement there are increasing signs, including the July 2009 approval by the Nile Council of Ministers (NILE COM) of a Nile Basin Data and Information Sharing and Exchange Interim Procedures Agreement, that suggest that there is a good chance that such a Framework Agreement will eventually be signed and implemented.

Works Cited

Bencala, Karin R., and Geoffrey D. Dabelko, 2008. "Water Wars: Obscuring Opportunities," *Journal of International Affairs* 61(2): 21–33.

Browder, Greg, and Leonard Orolano, The Evolution of an International Water Resources Management Regime in the Mekong River Basin, 40 *Natural Resources Journal* 499 (2000) at 519.

Brown, O., and Crawford, A. 2009. "Rising temperatures, rising tensions: Climate change and the risk of violent conflict in the Middle East". Winnipeg, MB: International Institute for Sustainable Development (1–42).

Carius, Alexander, Geoffrey Dabelko, and Aaron Wolf. Water, Conflict, and Cooperation, A Policy Brief—the United Nations and Environmental Security, ECSP Report , Issue 10 (2004).

Chalecki, Elizabeth L., Peter H. Gleick, Kelli L. Larson, Arian L. Pregenzer, and Aaron T. Wolf. Fire & Water: An Examination of the Technologies, Institutions, and Social Issues in Arms Control and Transboundary Water-Resources Agreements. Pacific Institute 2002.

Charte des Eaux du Fleuve Sénégal [Water Charter of the Senegal River], 18 May 2002. Official French text available at http://lafrique.free.fr/traites/omvs_200205.pdf Conca, Ken, and Geoffrey D. Dabelko (eds.). (2002). *Environmental Peacemaking*. Washington, D.C., and Baltimore: The Woodrow Wilson Center Press and Johns Hopkins University Press.

Davidson, Heather C., and Richard K. Paisley for the Canadian Columbia River Forum (discussion paper 2009). PDF available at *www.ccrf.ca/assets/.../issues-driving-forces-ccrf-final-march-2009.pdf*

Dyson, M., G. Bergkamp, and J. Scanlon (eds.). Flow—*The Essentials of Environmental Flows*, 2nd edition. Gland, Switzerland: IUCN (2008) found at http://www.iucn.org/dbtw-wpd/edocs/2003-021.pdf.

Fischhendler, I. Legal and institutional adaptation to climate uncertainty: a study of international rivers. *Water Policy* 6 (2004) 281–302.

International Bureau of the Permanent Court of Arbitration (ed.). The Resolution of International Water Disputes: Papers emanating from the Sixth PCA International Law Seminar 08 November 2002, Kluwer Law International, The Hague/London/New York.

Jurdi, M. Transboundary Movement of hazardous wastes into Lebanon: Part 2. Environmental impacts and the need for remedial actions. *Journal of Environmental Health* 64, no. 6: 15–19 (2002).

King, J. M., and D. Louw. 1998. Assessment of instream flow requirements for regulated rivers in South Africa using the Building Block Methodology. *Aquatic Ecosystem Health and Management* 1:109–24

Kistin, E. J., and Ashton, P. J. (2008). "Adapting to change in transboundary rivers: an analysis of treaty flexibility on the Orange-Senqu Basin." *Water Resources Development* 24(3): 385–400.

Marty, F., International River Management, Problems, Politics and Institutions. Peter Lang: Bern, Switzerland (2001).

Matsumoto, K., Transboundary groundwater and international law: Past practices and current implications. Unpublished research paper, Department of Geosciences, Oregon State University (2002).

McCaffrey, S., *The Law of International Watercourses: Non-Navigational Uses.* Oxford Monographs in International Law. Oxford University Press (2003).

McKinney, Matthew, 2009. Managing Transboundary Natural Resources: An Assessment of the Need to Revise and Update the Columbia River Treaty, prepared by author and the Natural Resources Conflict Resolution Program, University of Montana.

Mekong River Commission, Procedures for the Maintenance of Flows on the Mainstream (2006) as found at http://www.mrcmekong.org/download/agreement95/Procedures_ Guidlines/Procedures-Maintenance-Flows.pdf.

Morris, B. L., A. R. L. Lawrence, P. J. C. Chilton, B. Adams, R. C. Calow and B. A. Klinck, Groundwater and its susceptibility to degradation: A global assessment of the problem and options for management. Early Warning and Assessment Report Series, RS 03-3. United Nations Environment Programme, Nairobi, Kenya (2003).

Paisley, Richard K., and Glen Hearns, Some Observations from Recent Experiences with the Governance of International Drainage Basins in A. C. Corréa and Gabriel Eckstein (eds.): *Precious, Worthless or Immeasurable: the Value and Ethics of Water.* Center of Water Law and Policy and the International Center for Arid and Semi-arid Land Studies, Texas Tech University, 2006: 73–107.

Postel, Sandra L., and Aaron T. Wolf, "Dehydrating Conflict," *Foreign Policy*, No. 126. (Sept.–Oct., 2001), pp. 60–67.

Rutagwera, Patrick. About the NBI, October 2010, available at http://www.nilebasin.org/ newsite/index.php?option=com_content&view=section&id=5&layout=blog&Itemid= 68&lang=en

Sadoff, Claudia W., and Grey, D. "Beyond the river: the benefits of cooperation on international rivers." *Water Policy*, Vol. 4 (5) pp. 389–403. 2002.

Salman M. A. The Helsinki Rules, the UN Watercourses Convention and the Berlin Rules: Perspective on International Water Law, 23(4) *Water Resources Development* (2007): 625–37.

Sanderson, Chris. The Columbia River Treaty after 2014. A paper prepared for the Conference in Transboundary River Governance in the Face of Uncertainty, Coeur D'Alene, Idaho 2 to 4 April 2009.

Schreier, H. 2006. Climate change impact and adaptation options in the Columbia Basin. Climate Change in the Columbia Basin—Starting the dDialogue Conference. Columbia Basin Trust. Castlegar, B.C. Oct 19–20 (invited).

Tharme, R. E., and J. M. King. 1999. Development of the Building Block Methodology for Instream Flow Assessments, and supporting research on the effects of different magnitude flows on riverine ecosystems. *Water Research Commission Report* No. 576/1/98. 452 pp.

Utton, A. E. and J. Utton., International Law of Minimum Stream Flows, *Colorado Journal of International Environmental Law and Policy* (1999), volume 10, issue 1, p. 7–37.

Vick, Margaret J., The Senegal River Basin: A Retrospective and Prospective Look at the Legal Regime, 46 *Natural Resources Journal* 211 (2006).

Wolf, Aaron T. Development and Transboundary Waters: Obstacles and Opportunities: Report submitted to the World Commission on Dams, July, 2000, p. 30.

Wolf, A. T., 1999. The Transboundary Freshwater Dispute Database Project, *Water International* 24, no2.: 160–63. The Transboundary Freshwater Dispute Database is available at http://www.transboundarywaters.orst.edu/database/

Public Participation and Water Management in the European Union: Experiences and Lessons Learned

Lucia De Stefano and Guido Schmidt

Introduction

Public participation (PP) has been defined as "a process where individuals, groups and organizations choose to take an active role in making decisions that affect them" (Reed 2008) or as "allowing people to influence the outcome of plans and working processes" (CIS 2003) or, even more generically, as "the expectation that citizens have a voice in policy choices" (Bishop and Davis 2002). Regardless of the definition one uses, public participation increases the unavoidable uncertainty associated with any public decision-making process, but it also contributes to managing it. Indeed, contributions from the public and stakeholders, if truly taken into account, may lead to unexplored paths that could not be envisioned at the beginning of the participatory process (Arentsen et al. 2000; Newig et al. 2005). At the same time, public participation has been often been seen as a means to reduce uncertainty related to data availability and to the response of the society to particular public policies (Newig et al. 2005; Reed 2008; Kastens et al. 2007; Zorrilla et al. 2010). In particular, Newig et al. (2005) highlight that public participation can be an effective means to: a) profit from local knowledge; b) gain insight into the social system; c) profit from information about possible acceptance of alternatives; and d) mediate interest and goals. All this can contribute to formulate better informed decisions and more easily implemented solutions.

In the European Union (EU), its twenty-seven member states are currently in the midst of a historic restructuring of their water planning processes under the legal umbrella of the EU Water Framework Directive (WFD) (European Council 2000). The stated objective of this profound reform of the water management paradigm is the achievement of a significant improvement of the chemical and ecological/qualitative status of all surface, coastal, and groundwater bodies between 2015 and 2027 (Barreira 2006).

The EU Water Framework Directive is based on Integrated Water Resource Management (IWRM) principles and stresses the need for a holistic approach that takes into account economic, environmental, and ethical considerations in river basin water management (Rault and Jeffrey 2008; EC 2000; CIS 2003). In terms of ethical considerations, the WFD requires member states to base their actions on public information, consultation, and active involvement; as well as on the precautionary principle, the principle of subsidiary, and the principle of transparency (Rault and Jeffrey 2008). Among these principles, public participation stands out as key to ensure the success of the WFD: "The success of this Directive relies on close cooperation and coherent action at [European] Community, Member State and local level as well as on information, consultation and involvement of the public, including users" (EC 2000, Preamble).

A key milestone in the WFD planning process is the approval by the end of 2009 of new River Basin District Management Plans (RBMPs) for 215 River Basin Districts (RBDs) out of which 133 are international (Figure 1). "River Basin District" is the area of land and sea, made up of one or more neighboring river basins together with their associated groundwaters and coastal waters, as identified under article 3 (1) of the WFD as the main unit for management of river basins (EC 2000, Art. 1). The RBMPs should be the result of intense interactions among the different governmental and nongovernmental actors having a stake in the RBDs. In international basins the planning challenge is especially daunting. Indeed, if a river basin extends across international boundaries the directive specifically requires it to be assigned to an international RBD. When the international RBD covers EU territory only, member states should coordinate to produce a single RBMP. If the basin extends beyond the territories of the EU, the directive encourages member states to establish cooperation with nonmember states, "with the aim of

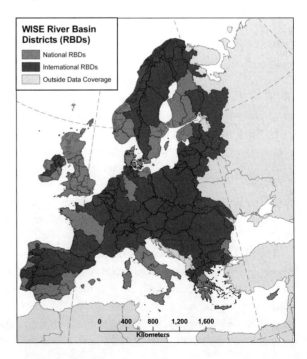

Figure 1. River Basin Districts for the EU Water Framework Directive management plans.

achieving the objectives of this Directive throughout the river basin district" (EC 2000, Art. 3). In international basins, the drafting of RBMPs should result not only from official interactions between countries but also from participatory processes occurring within the riparian basins and, ideally, across the borders (Keessen et al. 2008).

More than eight years have passed since the approval of the WFD in 2000 and EU member states are now almost at the end of the first of the three planning cycles foreseen by the directive (2003–2009, 2010–2015, 2016–2021). The experience gained so far permits us to identify challenges and successes in the on-going water-related participatory processes. This chapter presents the legal framework for public participation in EU water governance and offers an overview of the lessons learned so far.

The Legal Basis for Public Participation in EU Water Governance

The first international declaration that explicitly addresses the importance of public participation in water management dates back to January 1992, at the Dublin International Conference on Water and the Environment. Indeed, one of the principles of the Dublin Declaration states that "water development and management should be based on a participatory approach, involving users, planners and policy makers at all levels" (Dublin Declaration 1992). Shortly afterwards, the 1992 Rio Declaration reconfirmed the need for public involvement in environmental management by stating that "environmental issues are best handled with participation of all concerned citizens, at the relevant level" and that "States shall facilitate and encourage public awareness and participation by making information widely available." (U.N. General Assembly 1992, Principle 10). In the same year, the UNECE Convention on the Protection and Use of Transboundary Watercourses and International Lakes (Helsinki Convention 1992) addressed the need for comprehensive public information related to the transboundary waters management.

In 1998, the Aarhus Convention transposed the Rio principles related to access of information into a legally binding document. By then, the EU had already established minimum standards for public access to information and public participation in some environmental issues (EEC 1985), but the Aarhus Convention extended the EU requirements, giving broader definitions of environmental information and public authority, and recognizing the right of citizens and environmental nongovernmental organizations (NGOs) to turn to courts of justice when an environmental right has been infringed (Zaharchenko and Goldenman 2004). Since 2000, several EU directives have promoted important participatory elements, namely a directive on public access to environmental information (EC 2003), the partial transposition of the Aarhus Convention; a directive providing for public participation in respect to developing certain plans and programs relating to the environment (EC 2003a); and a directive on the assessment of certain plans and programs on the environment (EC 2001). All these directives complement the public participation provisions defined specifically for water resources planning and management in the Water Framework Directive.

The WFD text highlights the key role of public participation in the implementation process in Preamble 14 (see above) and in Preamble 46, which underscores that "to ensure the participation of the general public including users of water in the establishment and updating of river basin management plans, it is necessary to provide proper information of planned measures and to report on progress with their implementation with a view to the involvement of the general public before final decisions on the necessary measures are adopted."

Specific legal provisions are then described in Article 14, which distinguishes three forms of public participation with an increasing level of involvement: (1) information supply; (2) consultation; and (3) active involvement. According to the directive, the first two are to be ensured, while the latter should be encouraged.

Although these requirements refer to the whole WFD, explicit provisions were established at three specific steps in the planning process: Article 14 requires public consultation in writing on the following draft documents during a period of least six months: (1) The work program and time table for the production of the RBMPs, including the specific local/regional provisions regarding public consultation throughout the entire planning process (due by end of 2006 for the first planning process); (2) The definition of the Significant Water Management Issues (SWMI) identified in the River Basin District (RBD) and that will be addressed in the RBMP (due by the end of 2007); and (3) The River Basin Management Plan, including the program of measures designed to improve the chemical and ecological status of all waters in the RBD (due by the end of 2008). This consultation was to take place in the three planning cycles foreseen by the WFD (2003–2009, 2010–2015 and 2016–2021).

In order to facilitate a coherent and adequate interpretation of these and other legal settings of the directive, the EU member states, Norway and the European Commission established the Common Implementation Strategy (CIS), an initiative to develop, with the support of stakeholders and expert groups, "reference documents" for good practice in the implementation of the WFD. In 2003, the CIS completed the Guidance Document n. 8 on public participation (CIS 2003), to give specific help on how to implement public participation in the different steps of the management process. It is important to stress that, even though the Guidance Document has been endorsed by the EU water directors, it is not legally binding for member states, nor does it present itself as a "blueprint" for public participation. Rault et al. (2008) identify three reasons for this lack of blueprint. First, the inherent nature of public participation stresses the importance of the process and the need to adapt the employed tools to the evolution of that process, which precludes the use of pre-established "participatory recipes." Second, the WFD, as a directive, defines the objectives, but leaves the countries with the responsibility of achieving them. Third, specific methods should be defined based on "political culture decision making, culture of stakeholder involvement, organizational or institutional practices, budget and resources, history and previous attempts to engage stakeholders, environmental conditions and the scale of the project" (CIS 2003).

After creating a common understanding of the meaning of public participation in the context of the WFD, the CIS Document analyzes several issues that are considered to be crucial for designing and implementing a successful participatory planning process (Table 1).

In particular, the Guidance Document emphasizes the importance of a thorough preparation of the participatory process. This includes an accurate stakeholder analysis, the selection of activities and tools based on the questions 'why, what and who?', the availability of financial resources to support the process, and the inclusion of specific provisions to build the participants' capacity to effectively participate. In terms of the process itself, the Guidance Document stresses the value of managing participants' expectations on the process results, and the need to iteratively evaluate the process to allow for interim adjustments.

Public Participation under the WFD: The Implementation Process Since 2000 The implementation of the WFD requirements in terms of public participation is undoubtedly very challenging for EU member states: There is a wide range of water-related expertise and capacity within the EU, and each member state has a unique social, cultural, and economic history. Generally, northern and central European countries have a longer tradition of public participation in environmental issues, while southern and eastern European countries have historically been more state-driven (Timmerman and Langaas 2003; Maia 2003; World Wildlife Fund 2003).

In any case, a pan-European study assessing stakeholder participation in member states at the beginning of the WFD implementation showed common pitfalls of participatory practices in water-related decision-making processes throughout Europe(WWF 2003; De Stefano 2010). Table 2 highlights the problems identified in relation to selected issues of the CIS Guidance Document n.8.

Aware of the challenges posed by the directive, the European Commission has promoted a number of initiatives to support member states. In addition to the above mentioned Common Implementation Strategy, EC services set up a joint platform for online working and reporting procedures (WISE—Water Information System for Europe) and supported pilot exercises undertaken by member states to test different methods and tools for implementing the WFD (JRC/DGE 2005).

Table 1. Key considerations for designing and undertaking an effective participatory process for the WFD. Based on CIS (2003).

Issue	Key Considerations
Target	The public and the interested parties should be ensured access to background information and consultation of key draft documents
	Active involvement should be encouraged among stakeholders/ interested parties
Timing	The earlier participation starts the better, but there is a need to balance costs in terms of time and money and potential benefits (concept of proportionality)
Scale	Communication and coordination across scales and between units at the same scale is crucial
	Participation should take place at all the scales where activities will be implemented and also where their impact will be felt
	A comprehensive stakeholder analysis can help identifying the most appropriate scales for each step of the planning process
Information	Data transparency is essential to build trust and enable participation
Consultation	It should be clear who is being consulted, about what questions, in what time scale and for what purpose
	Reasoned feedback should be given to all the received comments
	The documents should be made widely available and effectively drawn to the attention of all interested groups and individuals
	The documents under consultation should be as simple and concise as possible
Active Involvement	Bilateral meetings, steering groups and advisory groups can be useful means to involves takeholders
	A comprehensive stakeholder analysis can help identifying the interested parties in each phase of the planning process
Others	Any forms of participation require capacity building
	The availability of financial resources is key to organize any participatory processes
	Need for iterative reporting and evaluation of the participatory process

The EC also funded several research projects on water and public participation and promoted a large number of conferences and other events on the WFD implementation processes.

Three EU-funded research projects deserve mention: HarmoniCop (HarmoniCop 2005; Tippett et al. 2005), which developed a practical handbook based on past approaches and experiences from nine European river basins; the NeWater project (Pahl-Wostl et al. 2005), which explored methods and tools for

Table 2. Main common difficulties to stakeholder participation at the beginning of the WFD implementation.

Issue	Problems identified
Target	The Authorities seek the participation of economic sectors (industry, water supply and agriculture) more proactively than that of environmental NGOs or research institutions
	Non-governmental stakeholders are actively involved in decision-making processes in less than 50% of the studied countries
Timing	Documents are issued for consultation only at the end of the decision-making process
Information	Bureaucracy or administrative obstacles make the access to background documents difficult and time-consuming
	Information is so technical, bulky or unorganized that it discourages involvement and consultation
Consultation	Schedule and objective of the consultation are often not clearly stated at the beginning of the participatory process
	Non-compulsory consultation on drafts is often irregular
	In 60% of the studied countries, responses to consultation and the outcomes of the consultation are rarely published before any progress is made in the decision-making process
Active Involvement	Financial support for non-governmental stakeholders attending participatory events
	Stakeholders often have insufficient human and financial capacity for continued active involvement on the part of stakeholders
Others	Stakeholders often have insufficient specialist knowledge to get truly involved in decision making process

Based on WWF (2003) and De Stefano (forthcoming). The studied countries were: Austria, Belgium, Bulgaria, Croatia, Estonia, Finland, France, Greece, Hungary, Ireland, Italy, Latvia, Poland, Portugal, Slovakia, Spain, Sweden, Switzerland, Turkey and the United Kingdom.

the participatory implementation of adaptive water management; and AquaStress (http://www.aquastress.net/), which looked for interdisciplinary solutions to situations where water resources are under stress (drought, flooding, and pollution), with special emphasis on public participation (Von Korff 2007).

As for activities focused on stakeholders and the public, the EC sponsored a large number of basin-wide conferences and events throughout Europe, and two pan-European conferences on water (2007 and 2009) gave stakeholders and authorities opportunities to share ideas and concerns on the implementation process (papers and results available at http://water.europe.es/participate).

The WFD process has undoubtedly strengthened the informal networks between water stakeholders both at national and European levels, creating "pools of expertise" on which they can rely on in critical times (Kranz and Vorwerk 2007). It has also spurred stakeholder activities to increase their influence on water policies. For instance, green NGOs have periodically produced pan-European assessments of directive progress (WWF, 2003; WWF/EEB, 2005; WWF/EEB, 2006; WWF/EEB, 2009) and have promoted events to raise awareness of the value of healthy rivers, which is the ultimate objective of the WFD (e.g., Big Jump, http://www.rivernet.org/bigjump/).

Despite numerous positive initiatives, the implementation of the WFD is still far from being complete. Progress is uneven within individual countries and inside international RBDs, both in terms of applied procedures and of expected results of public participation (Kampa et al. 2009). Interim assessments of the implementation process in 2008 (Kampa et al. 2009; WWF/EEB 2009) showed both advances and shortcomings in the WFD participatory process. On the positive side, approximately half of the EU basin authorities have actively involved stakeholders—mainly through working groups—in the preliminary formulation of documents submitted later for formal consultation, and around 50 percent of them have stressed the importance of this early involvement for detecting underlying conflicts and problems. In terms of consultation, the review by Kampa et al. (2009) highlights that there is no standard consultation "recipe," but that several elements influence the success of the process (Table 3).

On the negative side, that interim review revealed that in at least ten countries, consultations had been delayed by several months or had not yet started, and where consultations had started, their quality and effectiveness was often questioned (Kampa et al. 2009; WWF/EEB 2009).

The review also showed that some of the information provided—especially regarding economic data—was considered to be insufficient, unclear, or too technical. Also, data made available by authorities did not provide sufficient information to aid in assessing the environmental objectives stated in the RBMPs, the adequacy of the proposed measures, or the availability of funds to develop them (WWF/EEB 2009). This perceived lack of data or clarity should be carefully analyzed. In some cases it will be surely due to insufficient efforts or resources by authorities to make information available and transparent. In other cases, it is simply evidence of the huge challenge posed by the uncertainty inherent in any decision-making process on this scale. When making public data that are inevitably imperfect, authorities can choose to acknowledge that uncertainty or ignore it. The first position can expose them to strong criticism, as their proposed solutions may be perceived as poorly informed and lacking technical credibility. Conversely, the latter position can help keep authorities within their "comfort zones," but may seriously question the true

Table 3. Key conditions and good practices for successful consultation processes, modified from Kampa et al (2009).

Issues	Conditions for successful consultation
Timing	Involve important stakeholders before the official start of the public consultation
Financing	Provide sufficient financial resources, e.g. for TV spots and regional consultation, and for multinational and multilingual processes
Process organization	Organize the process well from the beginning, defining objectives and actors, and possible expectations
	Coordinate at national level with clear division of tasks and responsibilities
	Set thoroughly the objectives and content of public surveys
	Overcome historical distrust between some stakeholders
	Facilitate meetings by external experts to avoid any impression of a process controlled by administrations
	Working groups are a successful tool to provide suggestions by consensus. A combination of working groups and events at different (national, regional and local) scales is effective in covering a wide range of issues and collating feedbacks on both national and local issues and interests
	Discussions based on the development of scenarios, identifying needs, wishes, potential benefits and ranked priorities
	Establish links between WFD objectives and ecosystem services provided by the river to water users, stakeholders and the public
Public Information	Use a diversity of consultation methods
	Emphasize the local implications of the WFD
	Adapt national tools to local needs
	Account for specific difficulties to reach specific social groups (rural, elderly & young people)
	A good communication campaign using different types of media is essential for a successful consultation process. Radio can be an acceptable alternative to TV spots, in particular under tight budgetary conditions
	Use non-technical language to facilitate the access of the general public

openness of the participatory process. Examples of issues that are crucial for the RBMPs are: present and future water availability and variability, the definition of adequate in-stream flow regimes, or the sources of funding for implementing the planned management measures.

To summarize, it is still largely unclear whether the concerns of citizens and interested parties are taken into consideration as required under the RBMPs and there are doubts about whether the final plans will lead to significant improvements in water quality and sustainable use. The perception of lack of tangible benefits deriving from the participatory process may jeopardize the continued engagement of stakeholders and the public. Keeping the process focused on its final objectives and benefits, and not on its technical complexity, would undoubtedly contribute to maintaining the motivation of participants in the long run.

Transparency is also an important issue. According to Kampa et al. (2009), in 2009 the results of the six-month consultation phase of the Significant Water Management Issues documents (SWMI, previous to the draft RBMPs) had only been made public in 55 percent of the EU RBDs.

In international basins, the WFD implementation process has been particularly supported by the European Commission because of the added complexity of managing across borders and across administrations. Although there is no comprehensive overview of trans-border coordination and participation in the different RBDs available, there is no doubt that the WFD is beneficial for transboundary cooperation (EURO-INBO, 2008). Indeed, the implementation process has: (1) Strengthened the exchanges and sharing of data within European transboundary basins; (2) Led to the amendment or supplementation of existing international treaties or agreements, to make them comply with the new concepts or obligations; and (3) Widened the composition of existing international commissions to involve states that were not yet represented (EURO-INBO 2008).

In most cases member states have relied on existing international commissions (e.g., International Commission for the Protection of the Danube River, International Commission for the Protection of the Meuse) as a basis for cooperation and public participation. In some cases, this has been complemented with initiatives with strong bottom-up components. For example, in the Peipsi (Estonia, Russia) and Prespa lake basins (Greece, Republic of Macedonia, Albania) UNDP/GEF-funded projects have promoted the engagement of the local communities in the planning process. This comprehensive involvement—in particular NGOs—has led to improvements in participatory approaches, resulting in: a) the establishment of a joint vision between partners; b) the creation of informal transnational stakeholder networks with effective information-exchange structures; c) the early and tiered involvement of civil organizations in the planning process; and d) environmental education programs and public awareness-raising events (Mantziou and Gletsos 2008; Peipsi CTC 2009).

Despite the current shortcomings, WFD implementation should not be judged too early. Most of the EU environmental legislation implementation processes have faced significant delays and difficulties (Borzel 2000) and currently many of the water-related stakeholders have started focusing on the second and third planning cycles of the WFD. Keeping the next planning cycles in mind, the first decade of implementation is helping to identify which governance and participation issues should be promoted and supported as "good/best practices." The Common Implementation Strategy and the periodical reviews of the participatory processes should critically contribute to this learning process.

Although there is still a long way to go in order to establish fully functioning, active participation in all the river basin districts across Europe, the WFD already constitutes a powerful large-scale experiment of different ways to involve the public and stakeholders in river basin planning and water management processes. A Strengths-Weaknesses-Opportunities-Threats analysis (SWOT) of public participation in the WFD helps to identify advantages and inconveniences of the EU water management scheme in relations to participatory processes (Table 4).

Table 4. SWOT analysis of the WFD and its implementation process regarding public participation. Source: own elaboration.

Strengths	Weaknesses
Legal obligation to encourage/ensure participation	Poor participatory tradition in some countries
Increased financial resources associated to the planning process	Insufficient human and financial resources
3-step process, stimulating public and stakeholders and managing their expectations	Tight implementation schedule
Three planning cycles to gradually adjust approaches and tools	Technical complexity of the planning process
EU support to international coordination	Inadequate national transposition of EU legal requirements in some countries
Common Implementation Strategy guidance document	
Availability of updated information, on the RBD and EU levels	

Opportunities	Threats
Aarhus Convention requirements	Stakeholder fatigue and disillusionment
EU Directive on access to environmental information	Inertia of well-established planning processes to adapt to more participatory approaches
Support and input from research programs and projects	Resistance of traditionally powerful lobbying groups to the required increase in transparency and participation scope
	Difficulty of truly questioning current socio-economic models that are behind waters degradation
	Budget cuts due to the worldwide economic crisis

Table 4 shows that while the main strengths and opportunities are related to legal obligations to ensure participation and the long time frame of the planning process, human and political factors may seriously jeopardize the success of the public participation efforts. In particular, the inertia to change current practices and socio-economic development paradigms, as well as the complexity of the planning process, may create frustration and fatigue among stakeholders (Reed 2008). Indeed, too often the interest of stakeholders in participating is taken for granted, but the lack of impact of their engagement may eventually induce them to withdraw from the participatory processes. However, the legal obligation to provide participation opportunities should ensure consistency throughout the three planning processes, allowing for learning from past errors and building capacity within and among the authorities and the public.

Conclusion

The European Union is undertaking an ambitious river basin planning process towards sustainability. Uncertainty related to the baseline and projected data and to the effects of planning decisions is an inevitable component of this process. Public participation is at the same time a key element for success in managing uncertainty and a huge challenge for all the involved parties. The sincere acknowledgment of the existence of uncertainty is a necessary starting point for a truly open public participation process because it creates trust and allows for the design of creative solutions. At the same time, it requires additional efforts by all the parties to keep the process on track and, eventually, get to implementable solutions.

The analysis of EU member states' participatory practices at two different moments of the implementation process has highlighted elements that seem to be critical for success. The early engagement of stakeholders contributes to create trust in the process. Efforts to build participants' capacity and to keep the focus on the final benefits of the planning process are also crucial to ensure effective involvement over the long term. The participatory process should be designed thoroughly and should take into account funds and time constraints. A clear statement of the process objectives, target audience, responsibilities, and timetable contributes to managing participants' expectations.

These are among the recommendations drawn from the EU experience that should be kept in mind when designing and undertaking participatory processes. However, perhaps the most important premise for real participation in this case is the reestablishment of the too often lost connection of the European people with their water ecosystems. Only then will the EU society be ready to fully engage in the sustainable management of its waters.

Acknowledgments
The authors would like to thank Stefan Scheuer, Lynette de Silva and James Duncan for their valuable input and help in the elaboration of this chapter.

Works Cited

Arentsen, M. J., H. T. A. Bressers, and L. J. O"Toole. "Institutional and Policy Responses to Environmental Policy: A comparison of Dutch and U. S. Styles." *Policy Studies Journal* 28, no. 3 (2000): 597–611.

Barreira, A. "Water Governance at the European Union." *Journal of Contemporary Water Research & Education*, no. 135 (2006): 80–85.

Bishop, P, and G Davis. "Mapping Public Participation in Policy Choices." *Australian Journal of Public Administration* 61, no. 1 (2002): 14-29.

Borzel, T. A. "Why there is no 'Southern Problem': On Environmental Leaders and Laggards in The European Union." *Journal of European Public Policy* 7, no. 1 (2000): 141–62.

Common Implementation Strategy. *Guidance Document No 8: Public Participation in Relation to the Water Framework Directive.* Work Group 2.9—Public Participation, 2003.

De Stefano, L. "Facing the Water Framework Directive Challenge: A Baseline of Stakeholder Participation in the European Union." *Journal of Environmental Management.* (2010) Jun; 91(6):1332–40..

Dublin Declaration. 1992. *The Dublin Statement on Water and Sustainable Development.* U.N. International Conference on Water and the Environment. January 31, 1992. Available at: http://www.un-documents.net/h2o-dub.htm.

European Commission. 2000. *Directive 2000/60/EC of the European Parliament and of the Council Establishing a Framework for the Community Action in the Field of Water Policy.*
———. 2001. *Directive 2001/42/EC of the European Parliament and of the Council of 27 June 2001 on the Assessment of the Effects of Certain Plans and Programmes on the Environment.*
———. 2003. *Directive 2003/4/EC of the European Parliament and of the Council of 28 January. 2003 on Public Access to Environmental Information and Repealing Council Directive 90/313/EEC.*
———. 2003a. *Directive 2003/35/EC of the European Parliament and of the Council of 26 May 2003 Providing for Public Participation in Respect of the Drawing Up of Certain Plans and Programmes Relating to the Environment and Amending with Regard to Public Participation and Access to Justice Council Directives 85/337/EEC and 96/61/EC.*

European Economic Community. 1985. *Council Directive 85/337/EEC of 27 June 1985 on the Assessment of the Effects of Certain Public and Private Projects on the Environment.*

EURO-INBO. 2008. "WFD contributions to water management in transboundary river basins: interim evaluation and needs identified by the basin organisations. Note of the International Network of Basin Organisations for the Common Implementation Strategy of WFD." *IV International Symposium on Transboundary Waters Management. 1–4 October 2008, Sibiu, Romania.*

HarmoniCOP Team. *Learning Together to Manage Together: Improving Participation in Water Management.* Osnabrueck, Germany: University of Osnabrueck, 2009.
———. *Learning Together to Manage Together: Improving Participation in Water Management.* Edited by D. Ridder, E. Mostert and H. A. Wolters. 2006.
———. Handbook. Learning Together to Manage Together: Improving Participation in Water Management 2005. Available at http://www.harmonicop.uos.de/handbook.php

Helsinki Convention. 1992. *Convention on the Protection and Use of Transboundary Watercourses and International Lakes.* Helsinki, 17 March 1992.

JRC/DGE. Joint Research Centre and the Directorate General Environment of the European Commission. *Pilot River Basin Outcome Report Testing of the WFD Guidance Documents.* Luxembourg: Office for Official Publication of the European Communities, 2005.

Kampa, E., T. Dworak, B. Grandmougin, et al. "Active Involvement in River Basin Mangement: Plunge Into the Debate Conference Document." *Plunge into the Debate.* Brussels, Belgium, 2009.

Kastens, B., I. Borowski, and D. Ridder. "Transboundary River Basin Management in Europe: Legal Instruments to Comply with European Water Management Obligations in Case of Transboundary Water Pollution and Floods." *CAIWA 2007 International Conference on Adaptive & Integrated Water Management.* November 12–15, 2007.

Keessen, A. M., J. J. H. Jasper, and H. F. M. W. Van Rijswick. "Transboundary River Basin Management in Europe: Legal Instruments to Comply with European Water Management Obligations in a Case of Transboundary Water Pollution and Floods." *Urecht Law Review* 4, no. 3 (December 2008).

Kranz, N., and A. Vorwerk. "Public Participation in Transboundary Water Management." *Amsterdam Conference on the Human Dimensions of Global Environmental Change.* Berlin, 2007. 13.

Maia, R. "The Iberian Peninsula's Shared Rivers Harmonization of Use: A Portuguese Perspective." *Water International* 28, no. 3 (2003): 389–97.

Mantziou, D., and M. Gletso. "Transboundary Cooperation in the Prespa Basin: Experiences, Perspectives and Lessons Learned." *IV International Symposium on Transboundary Water Management.* Thessaloniki, Greece, October 15–18, 2008.

Newig, J., C. Pahl-Wostl, and K. Sigel. "The Role of Public Participation in Managing Uncertainty in the Implementation of the Water Directive." *European Environment* 15 (2005): 333–43.

Pahl-Wostl, C., T. Downing, P. Kabat, P. Magnuszewski, J. Meigh, M. Schüter, J. Sendzimir and S. Werners. *Transition to Adaptive Water Management: The NeWater Project. NeWater Report No. 1,* Osnabruck, Germany: University of of Osnabruck, 2005.

Peipsi Center for Transboundary Cooperation (CTC). "EU Water Framework Directive Implementation and Public Particpation in Transboundary Water Management in Estonia." 2009.

Rault, Pak, and P. J. Jeffrey. "Deconstructing Public Participation in the Water Framework Directive: Implementation and Compliance with the Letter or with the Spirit of the Law?" *Water and Environmental Journal* 22, no. 4 (2008): 241–49.

Reed, M. "Stakeholder Participation for Environmental Management: A Literature Review." *Biological Conservation* 141 (2008): 2417–31.

Timmerman, J., and S. Langaas. *Environmental Information in European Transboundary Water Management.* UK: IWA Publishing, 2003.

Tippett, J., B. Searle, C. Pahl-Wostl, and Y. Rees. "Social Learning in Public Participation in River Basin Management: Early Findings from HarmoniCOP European Case Studies." *Environmental Science & Policy* 8 (2005): 287–99.

U.N. Economic Commission For Europe. 1998. *Aarhus Convention. Convention on Access to Information, Public Participation in Decision-Making and Access to Justice in Environmental Matters. Aarhus, Denmark.* U..N General Assembly. 1992. Rio Declaration on Environment and Development, Agenda 21.

U.N. International Conference on Water and the Environment. 1992. Dublin Statement on Water and Sustainable Development.1992. Adopted January 31, 1992 in Dublin, Ireland. International Conference on Water and the Environment.

Von Korff, J. "Re-Focusing Research and Researchers in Public Participation." *CAIWA 2007 International Conference on Adaptive & Integrated Water Management.* November 12–15, 2007.

World Wildlife Fund. 2003. WWF's Water and Wetland Index: Summary of Water Framework Directive results. Available at:assets.panda.org/downloads/wwiwfdresultsdcm4.pdf.

WWF/EEB. 2005. EU Water Policy: Making the Water Framework Directive work. The Quality of National Transposition and Implementation of the Water Framework Directive at the end of 2004. A second "Snapshot" Report—Assessment of results from an environmental NGO questionnaire by the EEB and WWF. Available at www.eeb.org/activities/water/making-WFD-work-February05.pdf.

———. 2006. Complaint to the European Commission concerning failure of Austria, Belgium, Denmark, Estonia, Finland, Germany, Hungary, Ireland, Poland, Sweden and The Netherlands to comply with the provisions of the EU Water Framework Directive 2000/60/EC ("WFD") Article 5§1. Available at http://www.panda.org/wwf_news/news/?76520/WWF-and-EEB-submit-complaint-over-EU-water-infringements

———. 2009. What future for EU's water? Indicator based assessment of the Draft River Basin Management Plans under the EU Water Framework Directive. Brussels (Belgium).

Zaharchenko T., and G. Goldenman. "Accountability in Governance: The Challenge of Implementing the Aarhus Convention in Eastern Europe and Central Asia International Environmental Agreements." Politics, Law and Economics 4 (2004): 229–51.

Zorrilla, P., G. Carmona, A. De la Hera, C. Varela Ortega, P. Martínez-Santos, J. Bromley, and H. J. Henriksen. "Evaluation of Bayesian Networks in Participatory Water Resources Management, Upper Guadiana Basin, Spain." *Ecology and Society*, (2010). 15(3): 12. Available at http://www.ecologyandsociety.org/vol15/iss3/art12/.

The Impact of Institutional Design on the Adaptability of Governing Institutions: Implications for Transboundary River Governance

Craig W. Thomas

Introduction

"Transboundary" refers to political and ecological systems that cross international borders. "Transjurisdiction," by contrast, refers to political and ecological systems that cross jurisdictions within a country. Transboundary river governance necessarily includes transjurisdictional governance. Transjurisdictional problems must be resolved for transboundary governance to function effectively. This chapter accordingly sets aside the international component of treaties and diplomatic relations to focus on the transjurisdictional component of transboundary river governance.

The following sections examine three generic types of institutions for managing natural resources and ecosystems: (1) centralized bureaucracies, (2) politically appointed commissions, and (3) collaborative partnerships. They differ greatly in many respects, including the ease with which they can adapt to emergent transjurisdictional (and transboundary) problems. Bureaucracy is the classic type of institution for managing resource use *within* a jurisdiction. Bureaucratic institutions are well suited for addressing clearly defined problems through long-term planning by a single agency. In its pure form, bureaucracy is single-minded, because bureaucrats report through a chain of command following a highly formalized set of procedures that limit stakeholder participation and coordination with other agencies. By contrast, collaborative partnerships are open systems. In their pure form, collaborative partnerships lack a chain of command, membership is fluid, and informal norms largely govern decision-making processes. Collaborative partnerships are better suited for addressing transjurisdictional problems than bureaucracies. Politically appointed commissions, like bureaucracies, are highly formalized; but at the top of the hierarchy is a voting body appointed by elected officials. As political appointees, commission members represent specific interests. Commissions may also include seats reserved for public agencies. As voting bodies, politically appointed commissions are well suited for brokering the demands of competing interests through compromise and majority voting, particularly at larger scales. Collaborative partnerships, by contrast, seek consensus among interests, and are more effective at local scales where common interests are more likely to exist.

The Columbia River Basin is currently governed by a complex array of these three types of institutions. Bureaucratic institutions in the U.S. include the Bureau of Reclamation and Army Corps of Engineers, which operate thirty-one dams on the Columbia River and its tributaries; and the Bonneville Power Administration, which transmits electricity from these dams. These bureaucratic agencies were designed to address relatively straightforward problems—constructing and operating dams and transmission lines. By contrast, the Northwest Power and Conservation Council (NWPCC) is a politically appointed commission that balances environmental and energy needs in the four-state region (Washington, Oregon, Idaho and Montana).[1] The NWPCC has eight voting members, with the four governors each appointing two members. As a politically appointed body, the NWPCC is well suited for brokering interests across multiple jurisdictions. Many collaborative partnerships also exist in the Columbia River Basin, most operating at smaller scales. The Columbia Wetlands Stewardship Partners, for example, operates in the East Kootenay portion of British Columbia. It includes more than thirty-five public, private, and nonprofit organizations, with a mission to collaboratively manage the Columbia Wetlands ecosystem across jurisdictions.

Given the complex array of problems in the Columbia River Basin, it is appropriate to have multiple institutions operating at different scales under different missions. The purpose of this chapter is to provide a framework for analyzing how well each type of institution is suited for adapting to emergent transjurisdictional problems. The chapter closes with recommendations for modifying the existing set of institutions so the entire governance system can more readily adapt to emergent problems. Revisiting the Columbia River Treaty provides an excellent opportunity for examining and rethinking how the river basin can be governed in the twenty-first century.

Criteria for Assessing Institutional Adaptation

Institutions are sets of rules and norms that channel, enable, and constrain human behavior (Ostrom 2005). The key point is that institutions are *sets* of rules and norms. Hence, we need not change entire institutions, as is commonly claimed in political discourse, to achieve certain outcomes. We need only change those rules or norms that impede the ability of governing institutions to adapt as new problems arise. Some rules and norms enable adaptation, hence they can be nurtured within existing institutions. This section discusses three criteria for assessing the potential for institutional adaptation: (1) resiliency, (2) collaborative capacity, and (3) decision-making authority.

Resiliency is a crucial component of institutional adaptation. Resiliency refers to "the extent to which a system can absorb recurrent natural and human perturbations and continue to regenerate without slowly degrading or even unexpectedly flipping into less desirable states" (Folke et al. 2005).[2] While resiliency is a necessary condition for adaptation, it is not, however, a sufficient condition. If the problems are transjurisdictional, then institutional adaptation depends on collaborative capacity.

Collaborative capacity is a complex concept that refers to (1) the operational capacity of an institutional system to produce synergistic benefits for those involved; (2) the resource-raising capacity to leverage resources for operational purposes; and (3) the willingness and ability of individuals to make constructive use of the system's operational and resource-raising capacities (Thomas 2003; Bardach 1996). While resiliency exists at the institutional level of social-ecological systems, collaborative capacity exists at the level of individuals within institutions. Collaborative capacity is a necessary condition for managing transjurisdictional problems.

Decision-making authority refers to an institution's delegated authority to take specific actions. Bureaucracies and politically appointed commissions are generally granted broad decision-making authority by legislative bodies. High-level bureaucrats and political appointees may choose to delegate this authority to their staff. Collaborative partnerships, by contrast, often lack decision-making authority outside of the partnership. Moreover, agency officials working in a partnership may not have delegated authority to speak for their agency during collaborative planning processes, or to deliver agency resources for implementing projects. Lack of decision-making authority is a major limitation of collaborative partnerships.

The next three sections define and describe the three types of governing institutions in greater depth, and compare them in terms of the three criteria for institutional adaptation: resiliency, collaborative capacity, and decision-making authority. Underlying this chapter is the assumption that no single type of governing institution can perform well on all three adaptation criteria because there are fundamental trade-offs among them. Therefore, a mix of governing institutions is necessary for addressing different types of problems occurring at different scales.

Bureaucracy

In the U.S., rapid depletion of natural resources was the primary environmental problem during the late nineteenth and early twentieth centuries. Progressive reformers at that time sought new institutions to reduce corruption and enhance expertise in the public sector, and viewed bureaucracy as the best type of institution for managing public resources (Andrews 2006). Bureaucratic rules and norms are very good at reducing corruption, enhancing expert decision making, and distributing resources in an equitable manner (Knott and Miller 1987). They are highly resilient and authoritative in these regards, but they have much less collaborative capacity than other types of institutions. To understand why this is so, it is useful to review the rules and norms of bureaucratic institutions (Weber 1946).

When fully manifested, these institutional rules and norms constitute a pure bureaucracy. Most bureaucratic institutions, however, vary on each of these characteristics; hence, we need not replace a bureaucracy with another type of institution if we believe a particular public agency is insufficiently adaptable. We need only (de)bureaucratize by weakening or strengthening some of these characteristics. For example, divisions of labor need not be fixed, as in assembly lines. Loosening the division of labor (and the procedural rules governing these positions) enhances the ability of individuals to change administrative practices to adapt to new circumstances, including transjurisidictional environmental problems. The more we relax these bureaucratic rules and norms, the more we move towards new types of institutions. For example, politically appointed commissions relax several of these characteristics, because political appointees are appointed to represent political interests, which means they need not be experts (characteristics 6, 7, 8) or loyal to the institution (characteristics 10, 12).

It is now common to bash bureaucracy. Some have argued that bureaucracies serve few if any purposes—that they should be broken down (Barzelay 1992) to release entrepreneurial civil servants from procedural constraints (Moore 1995; Cohen and Eimicke 1995). Yet bureaucracies were designed with functional purposes in mind, and these purposes should always be considered before reform. Bureaucracies (such as the U.S. Bureau of Reclamation) are particularly effective for addressing straightforward problems that have narrowly defined problems (such as generating electricity) with narrowly defined solutions (such as building hydropower dams). Once institutionalized to address these narrowly defined problems and solutions, bureaucracies resist adaptation to new problems and solutions (Moe 1989).

Table 1. The Rules and Norms of Bureaucratic Institutions

1) Fixed division of labor (not shared responsibilities).
2) Hierarchical authority.
3) Management based on files (not personal thoughts).
4) Office-holding is a full-time job.
5) Office work follows stable rules.
6) Officials are experts (and thus competent for their offices).
7) Officials follow prescribed training and take exams.
8) Officials are appointed based upon merit (not politics).
9) Officials receive a fixed salary (and thus are not corrupt).
10) Officials are loyal to the functional purposes of the organization.
11) Officials have social status and esteem.
12) Officials have a career within the hierarchy.

Adapted from Weber 1946

The U.S. Forest Service, for example, was intentionally designed to resist internal innovation and stakeholder participation. It became the epitome of successful bureaucratic design during the first half of the twentieth century, extolled for its virtues in managing federal forests (Kaufman 1960). Yet, by the 1970s, the Forest Service confronted new, transjurisdictional problems for which it was poorly suited (Tipple and Wellman 1991). Hence, it has been slow to adapt, and remains so today. Internally, the Forest Service claims to pursue adaptive management; but it has been moderately successful, at best, in implementing adaptive management practices (Stankey et al. 2003). The bureaucratic form is compatible with adaptive management, but bureaucratic rules and norms must be designed accordingly, with a flattened hierarchy to allow more rapid changes on the ground and across jurisdictions (characteristic 2), and science-based expertise trumping technocratic expertise (characteristics 6 and 7) in Table 1.

Politically Appointed Commissions

Politically appointed commissions provide a means to incorporate multiple, competing interests into the highest level of the decision-making process. Unlike the bureaucratic model, commissioners are politically appointed rather than appointed solely based on merit. In practice, these distinctions are less clear. Bureaucratic rules can be relaxed at the top of the hierarchy to allow political appointments, while commission rules can require appointees to have relevant expertise. Regardless, the commission model allows for a wider range of interests to be represented at any one time at the top of the hierarchy. This means that commissions are more open to considering a wider range of problems and solutions, and thereby adapt to new problems.

Some have argued that the commission model is susceptible to capture by special interests through the political appointment process (Bernstein 1955). But that does not mean that the commission form of governance is inherently subject to capture, or that commissions are captured in the same way (Gormley 1983; Quirk 1981). Evidence indicates that commissions charged with regulating multiple industries, not a single industry, are less susceptible to capture (Gormley 1983). Legislation can also require proportional representation among interests on a commission, thereby reducing the likelihood of capture by single interests. Proportional representation of diverse interests also has the positive benefit of incorporating new thinking about problems and solutions.

Several politically appointed commissions now govern parts of the Columbia River in the U.S., including the Northwest Power and Conservation Council (discussed above), and the Pacific Fishery Management Council. Both operate under rules specifying proportional representation. The Pacific Fishery Management Council (one of eight such councils in the U.S.) has fourteen members, eight of which are political appointees nominated by the governors of Washington, Oregon, California, and Idaho. The other seats are reserved for federal and state agencies and one tribal representative. Given that the governors exert hierarchical authority over their respective state agency representatives, and that the governors nominate the political appointees, it is not surprising that interest-based voting occurs among members from the same states. On the Pacific Fishery Management Council, the Oregon bloc has been dominated by commercial interests; the Washington bloc, by recreational interests; and the Idaho bloc, by public interests (Thomas et al. 2010). Hence, a diverse range of interests in the Columbia River Basin are represented on the Pacific Fishery Management Council. Broad representation does not mean, however, that sustainable decisions are being made (Eagle et al. 2005; Okey 2003; Pew Oceans Commission 2003). But the commission model is better able than the bureaucratic model to address transjurisdictional problems due to the formal representation of multiple interests at the highest level of decision making.

As with bureaucratic institutions, we need only change some rules of a commission to enhance adaptability. For example, changing the nomination and apportionment rules can channel voting behavior to benefit public rather than special interests. This would enhance adaptability by reducing the power of special interests to protect the status quo. Apportionment rules can also require representatives from public agencies, as is the case with the Pacific Fishery Management Council. Hence, this type of institution is well-suited for collaborating across jurisdictions, or at least to broker interests across jurisdictions when consensus fails. The commission model provides much more collaborative capacity than the bureaucratic model, while providing the same level of decision-making authority. Thus, commissions are better able to adapt to transjurisdictional problems than bureaucracies.

Collaborative Partnerships

Collaborative partnerships are voluntary institutions that provide a venue for bringing public, private, and nonprofit stakeholders together to reach consensus on transjurisdictional problems. As collaborative partnerships mature, they increasingly develop collaborative capacity as partners develop norms of trust (Leach and Sabatier 2005a; 2005b) and discover synergistic benefits (Thomas 2003; Weber 2003). Yet, collaborative partnerships are less resilient than bureaucracies and commissions because rules, norms, and members are fluid. They may also lack sufficient decision-making authority to implement the policies, plans, and projects upon which the members agree. Thus, collaborative partnerships tend to be more effective when they are led or encouraged by public agencies that provide decision-making authority, as well as financial, technical, and human resources to enhance their resiliency (Koontz et al. 2004).

"Collaborative partnership" is a generic term that covers a wide variety of institutional rules, norms, and scales. The primary purpose of collaborative partnerships is to find joint solutions to transjurisdictional problems. Yet partnerships differ in terms of the decision-making rules used to reach joint solutions—such as consensus, voting, or third-party dispute resolution (O'Leary and Bingham 2003; Coglianese 1999; 2003). Partnerships also differ in scale, ranging from community-level partnerships (Weber 2003; Sabatier et al. 2005; Lubell 2004) to regional partnerships (Layzer 2008; Heikkila and Gerlak 2005; Thomas 2003).

In addition to decision-making rules, collaborative partnerships develop social capital, which allows them to work together on emergent problems (Leach and Sabatier 2005a; 2005b). Since partnerships are governed by fewer formal rules than bureaucracies and politically appointed commissions, informal norms play a much greater role. Maintaining these norms requires membership stability. Participation by public agency officials is often ephemeral, as officials change jobs, which decreases trust in public agencies as partners (Thomas 2003; 1998). Thus, while the fluidness of collaborative partnerships allows them to address new problems, it also reduces their resilience and decision-making authority as members and resources come and go.

Institutional Adaptation

Table 2 compares the three types of institutions in light of the criteria for institutional adaptation. The coding scheme (high, moderate, low) is a general assessment. Specific institutions may vary on some criteria, depending on the rules and norms they embody. No single institution can rank high on all of these criteria because fundamental trade-offs exist among them. Hence, a mix of institutions is better suited to ensure that a governance system—that is, a set of institutions—is adaptable.

Bureaucracies and politically appointed commissions rank high in resiliency because they have stable rules and funding. Collaborative partnerships, by contrast, rank low on resiliency because they are less formalized, their membership is fluid,

Table 2. The Adaptability of Governing Institutions

	Bureaucracies	Politically Appointed Commissions	Collaborative Partnerships
Resiliency	High	High	Low
Collaborative Capacity	Low	Moderate	High
Decision-Making Authority	High	High	Low

and resources can fluctuate greatly. Hence, partnerships decay much more readily than bureaucracies and commissions. An institution must be resilient to adapt to new problems. Resiliency, however, comes with a price, in that highly formalized, stable institutions, such as bureaucracies, constrain the ability of individuals to collaborate on transjurisdictional problems. The same features that enhance institutional resilience impede collaborative capacity. Politically appointed commissions lie somewhere between these poles. They are resilient, like bureaucracies; but they also provide a venue for developing collaborative capacity. The larger the commission (in terms of voting seats) and the wider the representation (in terms of apportionment) the more likely a commission can provide a resilient venue for developing transjurisdictional collaborative capacity. Commissions also rank high in terms of decision-making authority. Collaborative partnerships, by contrast, are subject to the whims of their voluntary members, who may or may not delegate decision-making authority from their agency, or provide sufficient resources, to the partnership. Hence, collaborative partnerships are much less resilient, and have much less decision-making authority, than bureaucracies and politically appointed commissions.

Governing the Columbia River Basin in the Twenty-first Century
This chapter reviewed three types of governing institutions: bureaucracies, politically appointed commissions, and collaborative partnerships. Each type of institution has strengths and weaknesses for adapting to emergent problems. None of them perform well on all criteria for institutional adaption (i.e., resiliency, collaborative capacity, and decision-making authority) so a mix of institutional forms is appropriate for addressing the wide variety of problems in the Columbia River Basin. Such a mix already exists. One way to improve institutional adaptation would be to modify existing institutions by changing rules and norms in ways that enhance resiliency, collaborative capacity, and/or decision-making authority. For bureaucracies, this would involve such things as decentralizing decision-making authority to enhance the ability of staff at local and regional levels to adapt to local problems. It would also mean delegating decision-making authority to collaborative partnerships, and providing partnerships with stable resources to enhance their resiliency. Reforming bureaucracies in this way would enhance adaptability, but would largely retain the status quo within these agencies.

Another alternative would be to create new institutions to fill gaps in the current governance system. One major gap is linkage among collaborative partnerships. Unlike bureaucracies and politically appointed commissions, collaborative partnerships operate at local scales, within tributary watersheds, or at the landscape level. Partnership activities may overlap in a geographic area, but there is no formal communication mechanism among the myriad partnerships in the Columbia River Basin to coordinate their actions or to voice their concerns at a larger regional level.

406 Craig W. Thomas

At least two options exist to increase communication and planning efforts among collaborative partnerships. One is to create a relatively informal learning network; another is to create a highly formalized commission.

Notes

1. Learning networks operate among collaborative partnerships (Butler and Goldstein 2010). They are relatively informal and self-generated. As such, they do not challenge the political status quo, and are thus more feasible than creating a formalized commission. A good example of a learning network is the International Model Forest Network, which shares information among members about model forestry practices. Each model forest exists within a local region, where public and private stakeholders come together to determine how to manage local forests for economic and environmental benefits. Public agencies may participate in these local partnerships, as they would in any collaborative partnerships; but the International Model Forest Network is not itself formally constituted by a public entity. Hence, the IMFN is a multi-scale learning network that bridges local partnerships.

A second option to achieve coordination among local collaborative partnerships in the Columbia River Basin would be to create a politically appointed commission, with seats reserved for collaborative partnerships representing sub-basins within the larger Columbia River Basin. This would be a unique type of commission, because it would empower collaborative partnerships with federal-level decision-making authority in the U.S. and/or Canada. It would also enhance the resilience of local partnerships by giving them a formal venue, with a particular set of rules and incentives, within which to operate.

From a resource management perspective, a new commission such as this would provide a means to integrate local efforts into regional—and perhaps transboundary—planning and management efforts. Collaborative partnerships would have a new formal venue within which to be heard, and would be given formal voting authority on the commission to represent local interests. Partnerships would also have an incentive to work together at the sub-basin level if seats were reserved for sub-basins, rather than partnerships per se. The incentive for partnerships to collaborate would be enhanced if they collectively nominated one of their members to represent their sub-basin interests.

Seats could also be reserved for the major bureaucracies in the Columbia River Basin, such as the U.S. Forest Service, Bureau of Reclamation, and Army Corps of Engineers. Each governor could also appoint one representative from one of their state agencies. Governors would have no political appointees outside the agencies, however, because those interests would be represented by the seats reserved for collaborative partnerships in the sub-basins. A seat could also be reserved for the International Joint Commission, or other transboundary organization, to represent transboundary concerns.

A new commission could be proposed as part of the negotiations for revisiting the Columbia River Treaty. Creating new institutions, of course, requires significant political effort; but there is a unique opportunity to rethink the governance system in the Columbia River Basin. The 1964 treaty focuses only on hydropower and flood control problems. The revised treaty could add environmental concerns, with a new commission serving as the formal institution to integrate planning and management efforts throughout the Columbia River Basin, or at least that part of the basin lying within the U.S.—or similarly in Canada. Unlike a traditional commission, with political appointees nominated from the top-down, some of the appointees would be nominated from the bottom-up, as representatives of collaborative partnerships at the sub-basin level. Doing so would not only better represent local ideas, but would also enhance the resilience and decision-making authority of these partnerships—and the collaborative capacity for these partnerships to work together at the sub-basin level.

"Commission" is used in this chapter as a generic term. Specific commissions may be called "councils," as are some collaborative partnerships.

2. For additional work on the resilience of social and ecological systems, see the special issue of *Ecology and Society* on this topic (http://www.ecologyandsociety.org/problems/view.php?sf=22.) For an alternative approach that draws on the idea of "robustness" from engineering, see Anderies et al. (2004). Both literatures differ from the idea of individual resilience within organizations (Vickers and Kouzmin 2001).

Works Cited

Anderies, John M., Marco A. Janssen, and Elinor Ostrom. "A Framework to Analyze the Robustness of Social-Ecological Systems from the Institutional Prospective." *Ecology and Society* 9, no. 1 (2004): art 18.

Andrews, Richard N. L. *Managing the Environment, Managing Ourselves: A History of American Environmental Policy.* New Haven, CT: Yale University Press, 2006.

Bardach, Eugene. *Getting Agencies to Work Together: The Practice and Theory of Managerial Craftsmanship.* Washington, D.C.: Brookings, 1998.

Bardach, Eugene. *Turf Barriers to Interagency Collaboration: In the State of Public Management.* Edited by Donald F. Kettl and H. Brinton Milward. Baltimore, MD: Johns Hopkins University Press, 1996.

Barzelay, Michael. *Breaking Through Bureaucracy: A New Version for Managing in Government.* Berkeley: University of California Press, 1992.

Bernstein, Marver. *Regulating Business by Independent Commission.* Princeton, NJ: Princeton University Press, 1955.

Butler, William Hale, and Bruce Evan Goldstein. "The US Fire Learning Network: Springing a Rigidity Trap through Multiscalar Collaborative Networks." *Ecology and Society* 15, no. 3, article 21 (2010) Available at http://www.ecologyandsociety.org/vol15/iss3/art21/.

Coglianese, Cary. "Is Satisfaction Success? Evaluating Public Participation in Regulatory Policymaking." In *The Promise and Performance of Environmental Conflict Resolution*, edited by Rosemary O'Leary and Lisa B. Bingham, 69–86. Washington, DC: Resources for the Future Press, 2003.

Coglianese, Cary. "The Limits of Consensus." *Environment* 41, no. 3 (1999): 28–33.

Cohen, Steven, and William Eimicke. *The New Effective Public Manager: Achieving Success in a Changing Government.* San Francisco, CA: Jossey-Bass, 1995.

Eagle, Josh, Sarah Newkirk, and Barton H. Thompson Jr. *Taking Stock of the Regional Fishery Management Councils.* Washington, D.C.: Island Press, 2003.

Folke, Carl, Thomas Hahn, Per Olsson, and Jon Norberg. "Adaptive Governance of Social-Ecological Systems." *Annual Review of Environmental Resources* 30 (2005): 441–73.

Gormley, William T. *The Politics of Public Utility Regulation.* Pittsburgh, PA: University of Pittsburgh Press, 1983.

Heikkila, Tanya, and Andrea K. Gerlak.. "The Formation of Large-Scale Collaborative Resource Management Institutions: Clarifying the Roles of Stakeholders." *Policy Studies Journal* 33, no. 4 (2005): 583–612.

Kaufman, Herbert. *The Forest Ranger: A Study in Administrative Behavior.* Washington, D.C.: Resources for the Future, 1960.

Knott, Jack H., and Gary J. Miller. *Reforming Bureaucracy: The Politics of Institutional Choice.* Englewood Cliffs, NJ: Prentice-Hall, 1987.

Koontz, Tomas M, Toddi A. Steelman, JoAnn Carmin, et al. 2004. *Collaborative Environmental Management: What Roles for Government?* Washington, DC: Resources for the Future Press, 2004.

Koontz, Tomas M., and Craig W. Thomas. "What Do We Know and Need to Know about the Environmental Outcomes of Collaborative Management." *Public Administration Review* 66, no. 6 (2006): 111–21.

Layzer, Judith A. *Natural Experiments: Ecosystem Management and the Environment.* Cambridge, MA: MIT Press, 2008.

Leach, William D., and Paul A. Sabatier. "Swimming Upstream: Collaborative Approaches to Watershed Management." In *Are Trust and Social Capital the Keys to Success? Watershed Partnership in California and Washington,* edited by Paul A. Sabatier, Will Focht, Mark Lubell and Zev Trachtenberg, 233–58. Cambridge, MA: MIT Press, 2005a.

Leach, William D., and Paul A. Sabatier. "To Trust an Adversary: Integrating Rational and Psychological Models in Collaborative Policymaking." *American Political Science Review* 99, no. 4 (2005b): 491–503.

Leach, William D., Neil W. Pelkey, and Paul A. Sabatier. "Stakehold Partnerships as Collaborative Policymaking: Evaluation Criteria Applied to Watershed Management in California and Washington." *Journal of Policy Analysis and Management* 21, no. 4 (2002): 645–70.

Lubell, Mark. "Collaborative Watershed Management: A View from the Grassroots." *Policy Studies Journal* 32, no. 3 (2004): 341–61.

Moe, Terry M. "The Politics of Bureaucratic Structure." In *Can the Government Govern?,* edited by John E. Chubb and Paul E. Peterson. Washington, D.C.: The Brookings Institution, 1989.

Moore, Mark. *Creating Public Value: Strategic Management in Government.* Cambridge, MA: Harvard University Press, 1995.

Okey, Thomas. "A Membership of the Eight Regional Fishery Management Councils in the United States: Are Special Interests Over-Represented?" *Marine Policy* 27 (2003): 193–206.

O'Leary, Rosemary, and Lisa B. Bingham. *The Promise and Performance of Environmental Conflict Resolution.* Washington, D.C.: Resources for the Future Press, 2003.

Ostrom, Elinor. *Understanding Institutional Diversity.* Princeton, N.J.: Princeton University Press, 2005.

Pew Oceans Commission. *America's Living Oceans: Charting a Course for Sea Change.* 2003. Available at http://www.pewtrusts.org/our_work_detail.aspx?id=130.

Quirk, Paul. *Industry Influence in Federal Regulatory Agencies.* Princeton, NJ: Princeton University Press, 1981.

Sabatier, Paul A., Will Focht, Mark Lubell, Zev Trachtenberg, Arnold Vedlitz, and Marty Matlock. *Swimming Upstream: Collaborative Approaches to Watershed Management.* Cambridge, MA: MIT Press, 2005.

Stankey, George H., Bernard T. Bormann, Clare. Shindler, Bruce Ryan, Victoria Sturtevant, Roger N. Clark, and Charles Philpot. "Adaptive Management and the Northwest Forest Plan." *Journal of Forestry,* 2003: 40–46.

Thomas, Craig W. *Bureaucratic Landscapes: Interagency Cooperation and the Preservation of Biodiversity.* Cambridge, MA: MIT Press, 2003.

Thomas, Craig W. "Maintaining and Restoring Public Trust in Government Agencies and Their Employees." *Administration and Society* 30 (1998): 166–93.

Thomas, Craig W., Arthur B. Soule, and Tyler B. Davis. "Special Interest Capture of Regulatory Agencies: A Ten-Year Analysis of Voting Behavior on Regional Fisheries Management Councils." *Policy Studies Journal* 38 (2010): 447–64.

Tipple, Terence, and J. Douglas Wellman. "Herbert Kaufman's Forest Ranger Thirty Years Later." *Public Administration Review* 51, no. 5 (1991): 421–27.

Vickers, Margaret, and Alexander Kouzmin. "Resilience in Organizational Actors and Rearticulating Voice." *Public Management Review* 3, no. 1 (2001): 95–119.

Weber, Edward P. *Bringing Society Back in: Grassroots Ecosystem Management, Accountability and Sustainable Communities.* Cambridge, MA: MIT Press, 2003.

Weber, Max. "Bureaucracy." In *From Max Weber: Essays in Sociology*, edited by H. H. Gerth and C. Wright Mills. New York: Oxford University Press, 1946.

APPENDIX

The Columbia River Treaty

Treaty between Canada and the United States of America relating to Cooperative Development of the Water Resources of The Columbia River Basin

The Governments of Canada and the United States of America

Recognizing that their peoples have, for many generations, lived together and cooperated with one another in many aspects of their national enterprises, for the greater wealth and happiness of their respective nations, and

Recognizing that the Columbia River Basin, as a part of the territory of both countries, contains water resources that are capable of contributing greatly to the economic growth and strength and to the general welfare of the two nations, and

Being desirous of achieving the development of those resources in a manner that will make the largest contribution to the economic progress of both countries and to the welfare of their peoples of which those resources are capable, and

Recognizing that the greatest benefit to each country can be secured by cooperative measures for hydroelectric power generation and flood control, which will make possible other benefits as well.

Have agreed as follows:

ARTICLE I

Interpretation

1. In the Treaty, the expression

(a) "average critical period load factor" means the average of the monthly load factors during the critical stream flow period;

(b) "base system" means the plants, works and facilities listed in the table in Annex B as enlarged from time to time by the installation of additional generating facilities, together with any plants, works or facilities which may be constructed on the main stem of the Columbia River in the United States of America;

(c) "Canadian storage" means the storage provided by Canada under Article II;

(d) "critical stream flow period" means the period, beginning with the initial release of stored water from full reservoir conditions and ending with the reservoirs empty, when the water available from reservoir releases plus the natural stream flow is capable of producing the least amount of hydroelectric power in meeting system load requirements;

(e) "consumptive use" means use of water for domestic, municipal, stock-water, irrigation, mining or industrial purposes but does not include use for the generation of hydroelectric power;

(f) "dam" means a structure to impound water, including facilities for controlling the release of the impounded water;

(g) "entity" means an entity designated by either Canada or the United States of America under Article XIV and includes its lawful successor;

(h) "International Joint Commission" means the Commission established under Article VII of the Boundary Waters Treaty, 1909, or any body designated by the United States of America and Canada to succeed to the functions of the Commission under this Treaty;

(i) "maintenance curtailment" means an interruption or curtailment which the entity responsible therefor considers necessary for purposes of repairs, replacements, installations of equipment, performance of other maintenance work, investigations and inspections;

(j) "monthly load factor" means the ratio of the average load for a month to the integrated maximum load over one hour during that month;

(k) "normal full pool elevation" means the elevation to which water is stored in a reservoir by deliberate impoundment every year, subject to the availability of sufficient flow;

(l) "ratification date" means the day on which the instruments of ratification of the Treaty are exchanged;

(m) "storage" means the space in a reservoir which is usable for impounding water for flood control or for regulating stream flows for hydroelectric power generation;

(n) "Treaty" means this Treaty and its Annexes A and B;

(o) "useful life" means the time between the date of commencement of operation of a dam or facility and the date of its permanent retirement from service by reason of obsolescence or wear and tear which occurs notwithstanding good maintenance practices.

2. The exercise of any power, or the performance of any duty, under the Treaty does not preclude a subsequent exercise of performance of the power or duty.

ARTICLE II

Development by Canada

1. Canada shall provide in the Columbia River basin in Canada 15,500,000 acre-feet of storage usable for improving the flow of the Columbia River.

2. In order to provide this storage, which in the Treaty is referred to as the Canadian storage, Canada shall construct dams:

(a) on the Columbia River near Mica Creek, British Columbia, with approximately 7,000,000 acre-feet of storage;

(b) near the outlet of Arrow Lakes, British Columbia, with approximately 7,100,000 acre-feet of storage; and

(c) on one or more tributaries of the Kootenay River in British Columbia downstream from the Canada-United States of America boundary with storage equivalent in effect to approximately 1,400,000 acre-feet of storage near Duncan Lake, British Columbia.

3. Canada shall commence construction of the dams as soon as possible after the ratification date.

ARTICLE III

Development by the United States of America Respecting Power

1. The United States of America shall maintain and operate the hydro electric facilities included in the base system and any additional hydroelectric facilities constructed on the main stem of the Columbia River in the United States of America in a manner that makes the most effective use of the improvement in stream flow resulting from operation of the Canadian storage for hydro-electric power generation in the United States of America power system.

2. The obligation in paragraph (1) is discharged by reflecting in the determination of down-stream power benefits to which Canada is entitled the assumption that the facilities referred to in paragraph (1) were maintained and operated in accordance therewith.

ARTICLE IV

Operation by Canada

1. For the purpose of increasing hydroelectric power generation in Canada and in the United States of America, Canada shall operate the Canadian storage in accordance with Annex A and pursuant to hydroelectric operating plans made thereunder. For the purpose of this obligation an operating plan if it is either the first operating plan or if in the view of either Canada or the United States of America it departs substantially from the immediately preceding operating plan must, in order to be effective, be confirmed by an exchange of notes between Canada and the United States of America.

2. For the purpose of flood control until the expiration of sixty years from the ratification date, Canada shall

(a) operate in accordance with Annex A and pursuant to flood control operating plans made thereunder

(i) 80,000 acre-feet of the Canadian storage described in Article II(2)(a),

(ii) 7,100,000 acre-feet of the Canadian storage described in Article II(2)(b),

(iii) 1,270,000 acre-feet of the Canadian storage described in Article II(2)(c),

provided that the Canadian entity may exchange flood control storage under subparagraph (ii) for flood control storage additional to that under subparagraph (I), at the location described in Article II(2)(a), if the entities agree that the exchange would provide the same effectiveness for control of floods on the Columbia River at The Dalles, Oregon;

(b) operate any additional storage in the Columbia River basin in Canada, when called upon by an entity designated by the United States of America for that purpose, within the limits of existing facilities and as the entity requires to meet flood control needs for the duration of the flood period for which the call is made.

3. For the purpose of flood control after the expiration of sixty years from the ratification date, and for so long as the flows in the Columbia River in Canada continue to contribute to potential flood hazard in the United States of America, Canada shall, when called upon by an entity designated by the United States of America for that purpose, operate within the limits of existing facilities any storage in the Columbia River basin in Canada as the entity requires to meet flood control needs for the duration of the flood control period for which the call is made.

4. The return to Canada for hydroelectric operation and the compensation to Canada for flood control operation shall be as set out in Articles V and VI.

5. Any water resource development, in addition to the Canadian storage, constructed in Canada after the ratification date shall not be operated in a way that adversely affect the stream flow control in the Columbia River within Canada so as to reduce the flood control and hydroelectric power benefits which the operation of the Canadian storage in accordance with the operating plans in force from time to time would otherwise produce.

6. As soon as any Canadian storage becomes operable Canada shall commence operation thereof in accordance with this Article and in any event shall commence full operation of the Canadian storage described in Article II(2)(b) and Article II(2)(c) within five years of the ratification date and shall commence full operation of the balance of the Canadian storage within nine years of the ratification date.

ARTICLE V

Entitlement to Downstream Power Benefits

1. Canada is entitled to one half the downstream power benefits determined under Article VII.

2. The United States of America shall deliver to Canada at a point on the Canada-United States of America boundary near Oliver, British Columbia, or such other

place as the entities may agree upon, the downstream power benefits to which Canada is entitled, less

(a) transmission loss,

(b) the portion of the entitlement disposed of under Article VIII(1), and

(c) the energy component described in Article VIII(4).

3. The entitlement of Canada to downstream power benefits begins for any portion of Canadian storage upon commencement of its operation in accordance with Annex A and pursuant to a hydroelectric operating plan made thereunder.

ARTICLE VI

Payment for Flood Control

1. For the flood control provided by Canada under Article IV(2)(a) the United States of America shall pay Canada in United States funds:

(a) 1,200,000 dollars upon the commencement of operation of the storage referred to in subparagraph (a)(i) thereof,

(b) 52,100,000 dollars upon the commencement of operation of the storage referred to in subparagraph (a)(ii) thereof, and

(c) 11,100,000 dollars upon the commencement of operation of the storage referred to in subparagraph (a)(iii) thereof.

2. If full operation of any storage is not commenced within the time specified in Article IV, the amount set forth in paragraph (1) of this Article with respect to that storage shall be reduced as follows:

(a) under paragraph (1)(a), 4,500 dollars for each month beyond the required time,

(b) under paragraph (1)(b), 192, 100 dollars for each month beyond the required time, and

(c) under paragraph (1)(c), 40,800 dollars for each month beyond the required time.

3. For the flood control provided by Canada under Article IV(2)(b) the United States of America shall pay Canada in United States funds in respect only of each of the first four flood periods for which a call is made 1,875,000 dollars and shall deliver to Canada in respect of each and every call made, electric power equal to the hydroelectric power lost by Canada as a result of operating the storage to meet the flood control need for which the call was made, delivery to be made when the loss of hydroelectric power occurs.

4. For each flood period for which flood control is provided by Canada under Article IV(3), the United States of America shall pay Canada in United States funds:

(a) the operating cost incurred by Canada in providing the flood control, and

(b) compensation for the economic loss to Canada arising directly from Canada foregoing alternative uses of the storage used to provide the flood control.

5. Canada may elect to receive in electric power, the whole or any portion of the compensation under paragraph 4(b) representing loss of hydroelectric power to Canada.

ARTICLE VII

Determination of Downstream Power Benefits

1. The downstream power benefits shall be the difference in the hydroelectric power capable of being generated in the United States of America with and without the use of Canadian storage, determined in advance, and is referred to in the Treaty as the downstream power benefits.

2. For the purpose of determining the downstream power benefits:

(a) the principles and procedures set out in Annex B shall be used and followed;

(b) the Canadian storage shall be considered as next added to 13,000,000 acre-feet of the usable storage listed in Column 4 of the table in Annex B;

(c) the hydroelectric facilities included in the base system shall be considered as being operated to make the most effective use for hydroelectric power generation of the improvement in stream flow resulting from operation of the Canadian storage.

3. The downstream power benefits to which Canada is entitled shall be delivered as follows:

(a) dependable hydroelectric capacity as scheduled by the Canadian entity, and

(b) average annual usable hydroelectric energy in equal amounts each month, or in accordance with a modification agreed upon under paragraph (4).

4. Modification of the obligation in paragraph (3)(b) may be agreed upon by the entities.

ARTICLE VIII

Disposal of Entitlement to Downstream Power Benefits

1. With the authorization of Canada and the United States of America evidenced by exchange of notes, portions of the downstream power benefits to which Canada is entitled may be disposed of within the United States of America. The respective general conditions and limits within which the entities may arrange initial disposals shall be set out in an exchange of notes to be made as soon as possible after the ratification date.

2. The entities may arrange and carry out exchanges of dependable hydroelectric capacity and average annual usable hydroelectric energy to which Canada is entitled for average annual usable hydroelectric energy and dependable hydroelectric capacity respectively.

3. Energy to which Canada is entitled may not be used in the United States of America except in accordance with paragraphs (1) and (2).

4. The bypassing at dams on the main stem of the Columbia River in the United States of America of an amount of water which could produce usable energy equal to the energy component of the down-stream power benefits to which Canada is entitled but not delivered to Canada under Article V or disposed of in accordance with paragraphs (1) and (2) at the time the energy component was not so delivered or disposed of, is conclusive evidence that such energy component was not used in the United States of America and that the entitlement of Canada to such energy component is satisfied.

ARTICLE IX

Variation of Entitlement to Downstream Power Benefits

1. If the United States of America considers with respect to any hydroelectric power project planned on the main stem of the Columbia River between Priest Rapids Dam and McNary Dam that the increase in entitlement of Canada to downstream power benefits resulting from the operation of the project would produce a result which would not justify the United States of America in incurring the costs of construction and operation of the project, Canada and the United States of America at the request of the United States of America shall consider modification of the increase in entitlement.

2. An agreement reached for the purposes of this Article shall be evidenced by an exchange of notes.

ARTICLE X

East-West Standby Transmission

1. The United States of America shall provide in accordance with good engineering practice east-west standby transmission service adequate to safeguard the transmission from Oliver, British Columbia, to Vancouver, British Columbia, of the downstream power benefits to which Canada is entitled and to improve system stability of the east-west circuits in British Columbia.

2. In consideration of the standby transmission service, Canada shall pay the United States of America in Canadian funds the equivalent of 1.50 United States dollars a year for each kilowatt of depend-able hydroelectric capacity included in the downstream power benefits to which Canada is entitled.

3. When a mutually satisfactory electric coordination arrangement is entered into between the entities and confirmed by an exchange of notes between Canada and the United States of America the obligation of Canada in paragraph (2) ceases.

ARTICLE XI

Use of Improved Stream Flow

1. Improvement in stream flow in one country brought about by operation of storage constructed under the Treaty in the other country shall not be used directly or indirectly for hydroelectric power purposes except:

   (a) in the case of use within the United States of America with the prior approval of the United States entity, and

   (b) in the case of use within Canada with the prior approval of the authority in Canada having jurisdiction.

2. The approval required by this Article shall not be given except upon such conditions, consistent with the Treaty, as the entity or authority considers appropriate.

ARTICLE XII

Kootenai River Development

1. The United States of America for a period of five years from the ratification date, has the option to commence construction of a dam on the Kootenai River near Libby, Montana, to provide storage to meet flood control and other purposes in the United States of America. The storage reservoir of the dam shall not raise the level of the Kootenai River at the Canada-United States of America boundary above an elevation consistent with a normal full pool elevation at the dam of 2,459 feet, United States Coast and Geodetic Survey datum, 1929 General Adjustment, 1947 International Supplemental Adjustment.

2. All benefits which occur in either country from the construction and operation of the storage accrue to the country in which the benefits occur.

3. The United States of America shall exercise its option by written notice to Canada and shall submit with the notice a schedule of construction which shall include provision for commencement of construction, whether by way of railroad relocation work or otherwise, within five years of the ratification date.

4. If the United States of America exercises its option, Canada in consideration of the benefits accruing to it under paragraph (2) shall prepare and make available for flooding the land in Canada necessary for the storage reservoir of the dam within a period consistent with the construction schedule.

5. If a variation in the operation of the storage is considered by Canada to be of advantage to it the United States of America shall, upon request, consult with Canada. If the United States of America determines that the variation would not be to its disadvantage it shall vary the operation accordingly.

6. The operation of the storage by the United States of America shall be consistent with any order of approval which may be in force from time to time relating to

the levels of Kootenay Lake made by the International Joint Commission under the Boundary Waters Treaty, 1909.

7. Any obligation of Canada under this Article ceases if the United States of America, having exercised the option, does not commence construction of the dam in accordance with the construction schedule.

8. If the United States of America exercises the option it shall commence full operation of the storage within seven years of the date fixed in the construction schedule for commencement of construction.

9. If Canada considers that any portion of the land referred to in paragraph (4) is no longer needed for the purpose of this Article Canada and the United States of America, at the request of Canada, shall consider modification of the obligation of Canada in paragraph (4).

10. If the Treaty is terminated before the end of the useful life of the dam Canada shall for the remainder of the useful life of the dam continue to make available for the storage reservoir of the dam any portion of the land made available under paragraph (4) that is not required by Canada for purposes of diversion of the Kootenay River under Article XIII.

ARTICLE XIII

Diversions

1. Except as provided in this Article neither Canada nor the United States of America shall, without the consent of the other evidenced by an exchange of notes, divert for any use, other than consumptive use, any water from its natural channel in a way that alters the flow of any water as it crosses the Canada-United States of America boundary within the Columbia River Basin.

2. Canada has the right, after the expiration of twenty years from the ratification date, to divert not more than 1,500,000 acre-feet of water a year from the Kootenay River in the vicinity of Canal Flats, British Columbia, to the headwaters of the Columbia River, provided that the diversion does not reduce the flow of the Kootenay River immediately downstream from the point of diversion below the lesser of 200 cubic feet per second or the natural flow.

3. Canada has the right, exercisable at any time during the period commencing sixty years after the ratification date and expiring one hundred years after the ratification date, to divert to the head-waters of the Columbia River any water which, in its natural channel, would flow in the Kootenay River across the Canada-United States of America boundary, provided that the diversion does not reduce the flow of the Kootenay River at the Canada-United States of America boundary near Newgate, British Columbia, below the lesser of 2500 cubic feet per second or the natural flow.

4. During the last twenty years of the period within which Canada may exercise the right to divert described in paragraph (3) the limitation on diversion is the lesser of 1000 cubic feet per second or the natural flow.

5. Canada has the right:

(a) if the United States of America does not exercise the option in Article XII(1), or

(b) if it is determined that the United States of America, having exercised the option, did not commence construction of the dam referred to in Article XII in accordance therewith or that the United States of America is in breach of the obligation in that Article to commence full operation of the storage, to divert to the headwaters of the Columbia River any water which, in its natural channel, would flow in the Kootenay River across the Canada-United States of America boundary, provided that the diversion does not reduce the flow of the Kootenay River at the Canada-United States of America boundary near Newgate, British Columbia, below the lesser of 1000 cubic feet per second or the natural flow.

6. If a variation in the use of the water diverted under paragraph (2) is considered by the United States of America to be of advantage to it Canada shall, upon request, consult with the United States of America. If Canada determines that the variation would not be to its disadvantage it shall vary the use accordingly.

ARTICLE XIV

Arrangements for Implementation

1. Canada and the United States of America shall each, as soon as possible after the ratification date, designate entities and when so designated the entities are empowered and charged with the duty to formulate and carry out the operating arrangements necessary to implement the Treaty. Either Canada or the United States of America may designate one or more entities. If more than one is designated the powers and duties conferred upon the entities by the Treaty shall be allocated among them in the designation.

2. In addition to the powers and duties dealt with specifically elsewhere in the Treaty the powers and duties of the entities include:

(a) coordination of plans and exchange of information relating to facilities to be used in producing and obtaining the benefits contemplated by the Treaty,

(b) calculation of and arrangements for delivery of hydroelectric power to which Canada is entitled for providing flood control,

(c) calculation of the amounts payable to the United States of America for standby transmission services,

(d) consultation on requests for variations made pursuant to Articles XII(5) and XIII(6),

(e) the establishment and operation of a hydrometeorological system as required by Annex A,

(f) assisting and cooperating with the Permanent Engineering Board in the discharge of its functions,

(g) periodic calculation of accounts,

(h) preparation of the hydroelectric operating plans and the flood control operating plans for the Canadian storage together with determination of the downstream power benefits to which Canada is entitled,

(i) preparation of proposals to implement Article VIII and carrying out any disposal authorized or exchange provided for therein,

(j) making appropriate arrangements for delivery to Canada of the downstream power benefits to which Canada is entitled including such matters as load factors for delivery, times and points of delivery, and calculation of transmission loss,

(k) preparation and implementation of detailed operating plans that may produce results more advantageous to both countries than those that would arise from operation under the plans referred to in Annexes A and B.

3. The entities are authorized to make maintenance curtailments. Except in case of emergency, the entity responsible for a maintenance curtailment shall give notice to the corresponding Canadian or United States entity of the curtailment, including the reason therefor and the probable duration thereof and shall both schedule the curtailment with a view to minimizing its impact and exercise due diligence to resume full operations.

4. Canada and the United States of America may by an exchange of notes empower or charge the entities with any other matter coming within the scope of the Treaty.

ARTICLE XV

Permanent Engineering Board

1. A permanent Engineering Board is established consisting of four members, two to be appointed by Canada and two by the United States of America. The initial appointments shall be made within three months of the ratification date.

2. The Permanent Engineering Board shall:

(a) assemble records of the flows of the Columbia River and the Kootenay River at the Canada–United States of America boundary;

(b) report to Canada and the United States of America whenever there is substantial deviation from the hydroelectric and flood control operating plans and if appropriate include in the report recommendations for remedial action and compensatory adjustments;

(c) assist in reconciling differences concerning technical or operational matters that may arise between the entities;

(d) make periodic inspections and require reports as necessary from the entities with a view to ensuring that the objectives of the Treaty are being met;

(e) make reports to Canada and the United States of America at least once a year of the results being achieved under the Treaty and make special reports concerning any matter which it considers should be brought to their attention;

(f) investigate and report with respect to any other matter coming within the scope of the Treaty at the request of either Canada or the United States of America.

3. Reports of the Permanent Engineering Board made in the course of the performance of its functions under this Article shall be prima facie evidence of the facts therein contained and shall be accepted unless rebutted by other evidence.

4. The Permanent Engineering Board shall comply with directions, relating to its administration and procedures, agreed upon by Canada and the United States of America as evidenced by an exchange of notes.

ARTICLE XVI

Settlement of Differences

1. Differences arising under the Treaty which Canada and the United States of America cannot resolve may be referred by either to the International Joint Commission for decision.

2. If the International Joint Commission does not render a decision within three months of the referral or within such other period as may be agreed upon by Canada and the United States of America, either may then submit the difference to arbitration by written notice to the other.

3. Arbitration shall be a tribunal composed of a member appointed by Canada, a member appointed by the United States of America and a member appointed jointly by Canada and the United States of America who shall be Chairman. If within six weeks of the delivery of a notice under paragraph (2) either Canada or the United States of America has failed to appoint its member, or they are unable to agree upon the member who is to be Chairman, either Canada or the United States of America may request the President of the International Court of Justice to appoint the member or members. The decision of a majority of the members of an arbitration tribunal shall be the decision of the tribunal.

4. Canada and the United States of America shall accept as definitive and binding and shall carry out any decision of the International Joint Commission or an arbitration tribunal.

5. Provision for the administrative support of a tribunal and for remuneration and expenses of its members shall be as agreed in an exchange of notes between Canada and the United States of America.

6. Canada and the United States of America may agree by an exchange of notes on alternative procedures for settling differences arising under the Treaty, including reference of any difference to the International Court of Justice for decision.

ARTICLE XVII

Restoration of Pre-Treaty Legal Status

1. Nothing in this Treaty and no action taken or foregone pursuant to its provisions shall be deemed, after its termination or expiration, to have abrogated or modified any of the rights or obligations of Canada or the United States of America under then existing international law, with respect to the uses of the water resources of the Columbia River basin.

2. Upon termination of this Treaty, the Boundary Waters Treaty, 1909, shall, if it has not been terminated, apply to the Columbia River basin, except insofar as the provisions of that Treaty may be inconsistent with any provision of this Treaty which continues in effect.

3. Upon termination of this Treaty, if the Boundary Waters Treaty, 1909, has been terminated in accordance with Article XIV of that Treaty, the provisions of Article II of that Treaty shall continue to apply to the waters of the Columbia River basin.

4. If upon the termination of this Treaty Article II of the Boundary Waters Treaty, 1909, continues in force by virtue of paragraph (2) of this Article the effect of Article II of that Treaty with respect to the Columbia River basin may be terminated by either Canada or the United States of America delivering to the other one year's written notice to that effect; provided however that the notice may be given only after the termination of this Treaty.

5. If, prior to the termination of this Treaty, Canada undertakes works usable for and relating to a diversion of water from the Columbia River basin, other than works authorized by or undertaken for the purpose of exercising a right under Article XIII or any other provision of this Treaty, paragraph (3) of this Article shall cease to apply one year after delivery by either Canada or the United States of America to the other of written notice to that effect.

ARTICLE XVIII

Liability for Damage

1. Canada and the United States of America shall be liable to the other and shall make appropriate compensation to the other in respect of any act, failure to act, omission or delay amounting to a breach of the Treaty or any of its provisions other that an act, failure to act, omission or delay occurring by reason of war, strike, major calamity, act of God, uncontrollable force or maintenance curtailment.

2. Except as provided in paragraph (1) neither Canada nor the United States of America shall be liable to the other or to any person in respect of any injury, damage or loss occurring in the territory of the other caused by any act, failure to act, omission or delay under the Treaty whether the injury, damage or loss results from negligence or otherwise.

3. Canada and the United States of America, each to the extent possible within its territory, shall exercise due diligence to remove the cause of and to mitigate the effect of any injury, damage or loss occur-ring in the territory of the other as a result of any act, failure to act, omission or delay under the Treaty.

4. Failure to commence operation as required by Articles IV and XII is not a breach of the Treaty and does not result in the loss of rights under the Treaty if the failure results from a delay that is not wilful or reasonably avoidable.

5. The compensation payable under paragraph (1):

(a) in respect of a breach by Canada of the obligation to commence full operation of a storage, hall be forfeiture of entitlement to downstream power benefits resulting from the operation of that storage, after operation commences, for a period equal to the period between the day of commencement of operation and the day when commencement should have occurred;

(b) in respect of any other breach by either Canada or the United States of America, causing loss of power benefits, shall not exceed the actual loss in revenue from the sale of hydroelectric power.

ARTICLE XIX

Period of Treaty

1. The Treaty shall come into force on the ratification date.

2. Either Canada or the United States of America may terminate the Treaty other than Article XIII (Except paragraph (1) thereof), Article XVII and this Article at any time after the Treaty has been in force for sixty years if it has delivered at least ten years written notice to the other of its intention to terminate the Treaty.

3. If the Treaty is terminated before the end of the useful life of a dam built under Article XII then, notwithstanding termination, Article XII remains in force until the end of the useful life of the dam.

4. If the Treaty is terminated before the end of the useful life of the facilities providing the storage described in Article IV(3) and if the conditions described therein exist then, notwithstanding termination, Articles IV(3) and VI(4) and (5) remain in force until either the end of the useful life of those facilities or until those conditions cease to exist, whichever is the first to occur.

ARTICLE XX

Ratification
The instruments of ratification of the Treaty shall be exchanged by Canada and the United States of America at Ottawa, Canada.

ARTICLE XXI

Registration with the United Nations
In conformity with Article 102 of the Charter of the United Nations, the Treaty shall be registered by Canada with the Secretariat of the United Nations.

This Treaty has been done in duplicate copies in the English language.

IN WITNESS WHEREOF the undersigned, duly authorized by their respective Governments, have signed this Treaty at Washington, District of Columbia, United States of America, this seventeenth day of January, 1961.

For Canada
John G. Diefenbaker (Prime Minister of Canada)
E. D. Fulton (Minister of Justice)
A. D. P. Heeney (Ambassador Extraordinary and Plenipotentiary of Canada to the United States of America)

For the United States of America
Dwight D. Eisenhower (President of the United States of America)
Christian A. Herter (Secretary of State)
Elmer F. Bennett (Under Secretary of the Interior)

ANNEX A

Principles of Operation

General:

1. The Canadian storage provided under Article II will be operated in accordance with the procedures described herein.

2. A hydrometeorological system, including snow courses, precipitation stations and stream flow gauges will be established and operated, as mutually agreed by the entities and in consultation with the Permanent Engineering Board, for use in establishing data for detailed programming of flood control and power operations. Hydrometeorological information will be made available to the entities in both countries for immediate and continuing use in flood control and power operations.

3. Sufficient discharge capacity at each dam to afford the desired regulation for power and flood control will be provided through outlet works and turbine installations as mutually agreed by the entities. The discharge capacity provided for flood control operations will be large enough to pass inflow plus sufficient storage releases during the evacuation period to provide the storage space required. The discharge capacity will be evaluated on the basis of full use of any conduits provide for that purpose plus one half the hydraulic capacity of the turbine installation at the time of commencement of the operation of storage under the Treaty.

4. The outflows will be in accordance with storage reservation diagrams and associated criteria established for flood control purposes and with reservoir-balance relationships established for power operations. Unless otherwise agreed by the entities the average weekly outflows shall not be less than 3000 cubic feet per second at the dam described in Article II(2)(a), not less than 5000 cubic feet per second at the dam described in Article II(2)(b), and not less than 1000 cubic feet per second at the dam described in Article II(2)(c). These minimum average weekly releases may be scheduled by the Canadian entity as required for power or other purposes.

Flood Control:

5. For flood control operation, the United States entity will submit flood control operating plans which may consist of or include flood control storage reservation diagrams and associated criteria for each of the dams. The Canadian entity will operate in accordance with these diagrams or any variation which the entities agree will not derogate from the desired aim of the flood control plan. The use of these diagrams will be based on data obtained in accordance with paragraph

2. The diagrams will consist of relationships specifying the flood control storage reservations required at indicated times of the year for volumes of forecast runoff. After consultation with the Canadian entity the United States entity may from time to time as conditions warrant adjust these storage reservation diagrams within the general limitations of flood control operation. Evacuation of the storages listed hereunder will be guided by the flood control storage reservation diagrams and refill will be as requested by the United States entity after consultation with the Canadian entity. The general limitations of flood control operation are as follows:

(a) The Dam described in Article II(2)(a) - The reservoir will be evacuated to provide up to 80,000 acre-feet of storage, if required, for flood control use by May 1 of each year.

(b) The Dam described in Article II(2)(b) - The reservoir will be evacuated to provide up to 7,100,000 acre-feet of storage, if required, for flood control use by May 1 of each year.

(c) The Dam described in Article II(2)(c) - The reservoir will be evacuated to provide up to 700,000 acre-feet of storage, if required, for flood control use by April 1 of each year and up to 1,270,000 acre-feet of storage, if required, for flood control use by May 1 of each year.

(d) The Canadian entity may exchange flood control storage provided in the reservoir referred to in subparagraph (b) for additional storage provided in the reservoir referred to in sub-paragraph (a) if the entities agree that the exchange would provide the same effectiveness for control of floods on the Columbia River at The Dalles, Oregon.

Power:

6. For power generating purposes the 15,500,000 acre-feet of Canadian storage will be operated in accordance with operating plans designed to achieve optimum power generation downstream in the United States of America until such time as power generating facilities are installed at the site referred to in paragraph 5(a) or at sites in Canada downstream therefrom.

7. After at-site power is developed at the site referred to in paragraph 5(a) or power generating facilities are placed in operation in Canada downstream from that site, the storage operation will be changed so as to be operated in accordance with operating plans designed to achieve optimum power generation at-site in Canada and downstream in Canada and the United States of America, including consideration of any agreed electrical coordination between the two countries. Any reduction in the downstream power benefits in the United States of America resulting from that change in operation of the Canadian storage shall not exceed

in any one year the reduction in downstream power benefits in the United States of America which would result from reducing by 500,000 acre-feet the Canadian storage operated to achieve optimum power generation in the United States of America and shall not exceed at any time during the period of the Treaty the reduction in downstream power benefits in the United States of America which would result from similarly reducing the Canadian storage by 3,000,000 acre-feet.

8. After at-site power is developed at the site referred to in paragraph 5(a) or power generating facilities are placed in operation in Canada downstream from that site, storage may be operated to achieve optimum generation of power in the United States of America alone if mutually agreed by the entities in which event the United States of America shall supply power to Canada to offset any reduction in Canadian generation which would be created as a result of such operation as compared to operation to achieve optimum power generation at-site in Canada and downstream in Canada and the United States of America. Similarly, the storage may be operated to achieve optimum generation of power in Canada alone if mutually agreed by the entities in which event Canada shall supply power to the United States of America to offset any reduction in United States generation which would be created as a result of such operation as compared to operation to achieve optimum power generation at-site in Canada and downstream in Canada and the United States of America.

9. Before the first storage becomes operative, the entities will agree on operating plans and the resulting downstream power benefits for each year until the total of 15,500,000 acre-feet of storage in Canada becomes operative. In addition, commencing five years before the total of 15,500,000 acre-feet of storage is expected to become operative, the entities will agree annually on operating plans and the resulting downstream power benefits for the sixth succeeding year of operation thereafter. This procedure will continue during the life of the Treaty, providing to both the entities, in advance, an assured plan of operation of the Canadian storage and a determination of the resulting downstream power benefits for the next succeeding five years.

ANNEX B

Determination of Downstream Power Benefits

1. The downstream power benefits in the United States of America attributable to operation in accordance with Annex A of the storage provided by Canada under Article II will be determined in advance and will be the estimated increase in dependable hydroelectric capacity in kilowatts for agreed critical stream flow periods and the increase in average annual usable hydroelectric energy output in kilowatt hours on the basis of an agreed period of stream flow record.

2. The dependable hydroelectric capacity to be credited to Canadian storage will be the difference between the average rates of generation in kilowatts during the appropriate critical stream flow periods for the United States of America base system, consisting of the projects listed in the table, with and without the addition of the Canadian storage, divided by the estimated average critical period load factor. The capacity credit shall not exceed the difference between the capability of the base system without Canadian storage and the maximum feasible capability of the base system with Canadian storage, to supply firm load during the critical stream flow periods.

3. The increase in the average annual usable hydroelectric energy will be determined by first computing the difference between the available hydroelectric energy at the United States base system with and without Canadian storage. The entities will then agree upon the part of available energy which is usable with and without Canadian storage, and the difference thus agreed will be the increase in average annual usable hydroelectric energy. Determination of the part of the energy which is usable will include consideration of existing and scheduled transmission facilities and the existence of markets capable of using the energy on a contractual basis similar to the then existing contracts. The part of the available energy which is considered usable shall be the sum of:

(a) the firm energy,

(b) the energy which can be used for thermal power displacement in the Pacific Northwest Area as defined in Paragraph 7, and

(c) the amount of the remaining portion of the available energy which is agreed by the entities to be usable and which shall not exceed in any event 40% of that remainder.

4. An initial determination of the estimated downstream power benefits in the United States of America from Canadian storage added to the United States base system will be made before any of the Canadian storage becomes operative. This determination will include estimates of the downstream power benefits for each year until the total of 15,500,000 acre-feet of Canadian storage becomes operative.

5. Commencing five years before the total of 15,500,000 acre-feet of storage is expected to become operative, estimates of downstream power benefits will be calculated annually for the sixth succeeding year on the basis of the assured plan of operation for that year.

6. The critical stream flow period and the details of the assured plan of operation will be agreed upon by the entities at each determination. Unless otherwise agreed upon by the entities, the determination of the downstream power benefits shall be based upon stream flows for the twenty year period beginning with July 1928 as contained in the report entitled Modified Flows at Selected Power

ANNEX B - TABLE - BASE SYSTEM

Project	Stream	Stream Mile Above Mouth	Usable Storage Acre-Feet	Normal Pool Feet	Elev. Tailwater Feet	Gross Head Feet	Initial Install. # of Units	Initial Install. Plant Kilowatts (Nameplate)	Estimated Ultimate Install. # of Units	Estimated Ultimate Install. Plant Kilowatts (Nameplate)
Hungry Horse	SFk Flathead	5	3,161,000[4]	3560	3083	477	4	285,000	4	285,000
Kerr	Flathead	73	1,219,000	2893	2706	187	3	168,000	3	168,000
Thompson Falls	Clark Fork	279	Pondage	2396	2336	60	6	30,000	8	65,000
Noxon Rapids	Clark Fork	170	Pondage	2331	2179	152	4	336,000	5	420,000
Cabinet Gorge	Clark Fork	150	Pondage	2175	2078	97	4	200,000	6	300,000
Albeni Falls	Pend Oreille	90	1,155,000	2062	2034	28	3	42,600	3	42,600
Box Canyon	Pend Oreille	34	Pondage	2031	1989	42	4	60,000	4	60,000
Grand Coulee	Columbia	597	5,232,000[4]	1290[3,4]	947	343	18	1,944,000	34	3,672,000
Chief Joseph	Columbia	546	Pondage	946	775	171	16	1,024,000	27	1,728,000
Wells (1)	Columbia	516	Pondage	775	707	68	6	400,000	10	666,700
Rocky Reach	Columbia	474	Pondage	707	614	93	7	711,550	11	1,118,150
Rock Island	Columbia	453	Pondage	608	570	38	10	212,100	10	212,100
Wanapum	Columbia	415	Pondage	570	486	84	10	831,250	16	1,330,000
Priest Rapids	Columbia	397	Pondage	486	406	80	10	788,500	16	1,261,600
Brownlee	Snake	285	974,000	2077	1805	272	4	360,400	6	540,600
Oxbow	Snake	273	Pondage	1805	1683	122	4	190,000	5	237,500
Ice Harbor	Snake	10	Pondage	440	343	97	3	270,000	6	540,000
McNary	Columbia	292	Pondage	340	265	75	14	980,000	20	1,400,000

John Day	Columbia	216	Pondage	265	161	104	8	1,080,000	20	2,700,000
The Dalles	Columbia	192	Pondage	160	74	86	16[2]	1,119,000	24[2]	1,743,000
Bonneville	Columbia	145	Pondage	74	15	59	10	518,400	16	890,400
Kootenay Lk	Kootenay	16	673,000	1745	--	--	--	--	--	--
Chelan	Chelan	0	676,000	1100	707	393	2	48,000	4	96,600
Couer d'Alene L.	Couer d'Alene	102	223,000	2128	--	--	--	--	--	--
TOTAL 24 PROJECTS			13,313,000[4]			3128		11,598,800	258	19,476,600

(1) The Wells project is not presently under construction; when this project or any other project on the main stem of the Columbia River is completed, they will be integral components of the base system.

(2) Includes two 13,500 kilowatt units for fish attraction water.

(3) With flashboards.

(4) In determining the base system capabilities with and without Canadian storage the Hungry Horse reservoir storage will be limited to 3,008,000 acre–feet (normal full pool elevation of 3560 feet) and the Grand Coulee project will not include the effect of adding flashboards, limiting the storage to 5,072 acre–feet (normal full pool elevation of 1288 feet). The total usable storage of the base system as so adjusted will be 13,000,000 acre–feet.

Sites - Columbia River Basin, dated June 1957. No retroactive adjustment in downstream power benefits will be made at any time during the period of the Treaty. No reduction in the downstream power benefits credited to Canadian storage will be made as a result of the load estimate in the United States of America, for the year for which the determination is made, being less than the load estimate for the preceding year.

7. In computing the increase in dependable hydroelectric capacity and the increase in average annual hydroelectric energy, the procedure shall be in accordance with the three steps described below and shall encompass the loads of the Pacific Northwest Area. The Pacific Northwest Area for purposes of these determinations shall be Oregon, Washington, Idaho, and Montana west of the Continental Divide but shall exclude areas served on the ratification date by the California Oregon Power Company and the Utah Power and Light Company.

Step I - The system for the period covered by the estimate will consist of the Canadian storage, the United States base system, any thermal installation operated in coordination with the base system, and additional hydroelectric projects which will provide storage releases usable by the base system or which will use storage releases that are usable by the base system. The installations included in this system will be those required, with allowance for adequate reserves, to meet the forecast power load to be served by this system in the United States of America, including the estimated flow of power at points of inter-connection with adjacent areas, subject to paragraph 3, plus the portion of the entitlement of Canada that is expected to be used in Canada. The capability of this system to supply this load will be determined on the basis that the system will be operated in accordance with the established operating procedures of each of the projects involved.

Step II - A determination of the energy capability will be made using the same thermal installation as in Step I, the United States base system with the same installed capacity as in Step I and Canadian storage.

Step III - A similar determination of the energy capability will be made using the same thermal installation as in Step I and the United States base system with the same installed capacity as in Step I.

8. The downstream power benefits to be credited to Canadian storage will be the differences between the determinations in Step II and Step III in dependable hydroelectric capacity and in average annual usable hydroelectric energy, made in accordance with paragraphs 2 and 3.

PROTOCOL

ANNEX TO EXCHANGE OF NOTES

*Dated January 22, 1964 Between the Governments of Canada
And The United States Regarding the Columbia River Treaty*

I. If the United States entity should call upon Canada to operate storage in the Columbia River Basin to meet flood control needs of the United States of America pursuant to Article IV(2)(b) or Article IV(3) of the Treaty, such call shall be made only to the extent necessary to meet forecast flood control needs in the territory of the United States of America that cannot adequately be met by flood control facilities in the United States of America in accordance with the following conditions:

(1) Unless otherwise agreed by the Permanent Engineering Board, the need to use Canadian flood control facilities under Article IV(2)(b) of the Treaty shall be considered to have arisen only in the case of potential floods which could result in a peak discharge in excess of 600,000 cubic feet per second at The Dalles, Oregon, assuming the use of all related storage in the United States of America existing and under construction in January 1961, storage provided by any dam constructed pursuant to Article XII of the Treaty and the Canadian storage described in Article IV(2)(a) of the Treaty.

(2) The United States entity will call upon Canada to operate storage under Article IV(3) of the Treaty only to control potential floods in the United States of America that could not be adequately controlled by all the related storage facilities in the United States of America existing at the expiration of 60 years from the ratification date but in no event shall Canada be required to provide any greater degree of flood control under Article IV(3) of the Treaty than that provided for under Article IV(2) of the Treaty.

A call shall be made only if the Canadian entity has been consulted whether the need for flood control is, or is likely to be, such that it cannot be met by the use of flood control facilities in the United States of America in accordance with subparagraphs (1) or (2) of this paragraph. Within ten days of receipt of a call, the Canadian entity will communicate its acceptance, or its rejection or proposals for modification of the call, together with supporting considerations. When the communication indicates rejection or modification of the call the United States entity will review the situation in the light of the communication and subsequent developments and will then withdraw or modify the call if practicable. In the absence of agreement on the call or its terms the United States entity will submit the matter to the Permanent Engineering Board provided for under Article XV of the Treaty for assistance as contemplated in Article XV(2)(c) of the Treaty. The entities will be guided by any instructions issued by the Permanent Engineering Board. If the Permanent Engineering Board does not issue instructions within ten days of receipt of a submission the United States entity may renew the call for any part or all of the storage covered in the original call and the Canadian entity shall forthwith honor the request.

II. In preparing the flood control operating plans in accordance with paragraph 5 of Annex A of the Treaty, and in making calls to operate for flood control pursuant to Articles IV(2)(b) and IV(3) of the Treaty, every effort will be made to minimize flood damage in both Canada and the United States of America.

III. The exchange of Notes provided for in Article VIII(1) of the Treaty shall take place contemporaneously with the exchange of the Instruments of Ratification of the Treaty provided for in Article XX of the Treaty.

IV. (1) During the period and to the extent that the sale of Canada's entitlement to downstream power benefits within the United States of America as a result of an exchange of Notes pursuant to Article VIII(1) of the Treaty relieves the United States of America of its obligation to provide east-west standby transmission service as called for by Article X(1) of the Treaty, Canada is not required to make payment for the east-west standby transmission service with regard to Canada's entitlement to downstream power benefits sold in the United States of America.

(2) The United States of America is not entitled to any payments of the character set out in subparagraph (1) of this paragraph in respect of that portion of Canada's entitlement to down-stream power benefits delivered by the United States of America to Canada at any point on the Canada-United States of America boundary other than at a point near Oliver, British Columbia, and the United States of America is not required to provide the east-west standby transmission service referred to in subparagraph (1) of this paragraph in respect of the portion of Canada's entitlement to downstream power benefits which is so delivered.

V. Inasmuch as control of historic streamflows of the Kootenay River by the dam provided for in Article XII(1) of the Treaty would result in more than 200,000 kilowatt years per annum of energy benefit downstream in Canada, as well as important flood control protection to Canada, and the operation of that dam is therefore of concern to Canada, the entities shall, pursuant to Article XIV(2)(a) of the Treaty, cooperate on a continuing basis to coordinate the operation of that dam with the operation of hydroelectric plants on the Kootenay River and elsewhere in Canada in accordance with the provisions of Article XII(5) and Article XII(6) of the Treaty.

VI. (1) Canada and the United States of America are in agreement that Article XIII(1) of the Treaty provides to each of them a right to divert water for a consumptive use.

Any diversion of water from the Kootenay River when once instituted under the provisions of Article XIII of the Treaty is not subject to any limitation as to time.

VII. As contemplated by Article IV(1) of the Treaty, Canada shall operate the Canadian storage in accordance with Annex A and hydroelectric operating plans made thereunder. Also, as contemplated by Annexes A and B of the Treaty and Article XIV(2)(k) of the Treaty, these operating plans before they are agreed to by the entities will be conditioned as follows:

(1) As the downstream power benefits credited to Canadian storage decrease with time, the storage required to be operated by Canada pursuant to paragraphs 6 and 9 of Annex A of the Treaty, will be that required to produce those benefits.

(2) The hydroelectric operating plans, which will be based on Step I of the studies referred to in paragraph 7 of Annex B of the Treaty, will provide a reservoir-balance relationship for each month of the whole of the Canadian storage committed rather than a separate relationship for each of the three Canadian storages. Subject to compliance with any detailed operating plan agreed to by the entities as permitted by Article XIV(2)(k) of the Treaty, the manner of operation which will achieve the specific storage or release of storage called for in a hydroelectric operating plan consistent with optimum storage use will be at the discretion of the Canadian entity.

(3) Optimum power generation at-site in Canada and downstream in Canada and the United States of America referred to in paragraph 7 of Annex A of the Treaty will include power generation at-site and downstream in Canada of the Canadian storages referred to in Article II(2) of the Treaty, power generation in Canada which is coordinated therewith, downstream power benefits from the Canadian storage which are produced in the United States of America and measured under the terms of Annex B of the Treaty, power generation in the Pacific Northwest Area of the United States of America and power generation coordinated therewith.

VIII. The determination of downstream power benefits pursuant to Annex B of the Treaty, in respect of each year until the expiration of thirty years from the commencement of full operation in accordance with Article IV of the Treaty of that portion of the Canadian storage described in Article II of the Treaty which is last placed in full operation, and thereafter until otherwise agreed upon by the entities, shall be based upon stream flows for the thirty-year period beginning July 1928 as contained in the report "Extension of Modified Flows Through 1958 – Columbia River Basin" and dated June 29, 1961, by the Water Management Subcommittee of the Columbia Basin Inter-Agency Committee.

IX. (1) Each load used in making the determinations required by Steps II and III of paragraph 7 of Annex B of the Treaty shall have the same shape as the load of the Pacific Northwest area as that area is defined in that paragraph.

(2) The capacity credit of Canadian storage shall not exceed the difference between the firm load carrying capabilities of the projects and installations included in Step II of paragraph 7 of Annex B of the Treaty and the projects and installations included in Step III of paragraph 7 of Annex B of the Treaty.

X. In making all determinations required by Annex B of the Treaty the loads used shall include the power required for pumping water for consumptive use into the Banks Equalizing Reservoir of the Columbia Basin Federal Reclamation Project but mention of this particular load is not intended in any way to exclude from those loads any use of power that would normally be part of such loads.

XI. In the event operation of any of the Canadian storages is commenced at a time which would result in the United States of America receiving flood protection for periods longer than those on which the amounts of flood control payments to Canada set forth in Article VI(1) of the Treaty are based, the United States of America and Canada shall consult as to the adjustments, if any, in the flood control payments that may be equitable in the light of all relevant factors. Any adjustment would be calculated over the longer period or periods on the same basis and in the same manner as the calculation of the amounts set forth in Article VI(1) of the Treaty. The consultations shall begin promptly upon the determination of definite dates for the commencement of operation of the Canadian storages.

XII. Canada and the United States of America are in agreement that the Treaty does not establish any general principle or precedent applicable to waters other than those of the Columbia River Basin and does not detract from the application of the Boundary Waters Treaty, 1909, to other waters.

CONTRIBUTORS

James D. Barton is the Chief of the Columbia Basin Water Management Division for the U.S. Army Corps of Engineers Northwestern Division in Portland, Oregon. He manages Corps of Engineers Water Management activities for water resource projects throughout the Columbia River Basin. He serves as the U.S. Co-Chair for the Columbia River Treaty Operating Committee. Mr. Barton previously served as the Chief of Water Management for the Corps of Engineers Southwestern Division in Dallas, Texas. Prior to that he worked in various other positions with the Department of Energy, Corps of Engineers, Bureau of Reclamation, and U.S. Geological Survey. Mr. Barton is a registered professional engineer in the state of Arkansas, and graduated from the University of Utah with a bachelor's degree in civil engineering. He has graduate degrees in business administration, project management, and strategic planning.

Aimee Brown earned a masters of science degree from Oregon State University's Department of Geosciences in 2009. Her thesis, *Understanding the Impact of Climate Change on Snowpack Extent and Measurement in the Columbia River Basin and Nested Sub Basins*, focused on creating a better understanding of current snowpack measurement systems, and the ability of these systems to accurately portray a variety of snow-covered areas under shifting precipitation regimes. Aimee's current work as a science journalist takes her around the globe reporting on the interactions between built and natural environments.

Barbara Cosens joined the law faculty at the University of Idaho in 2004, received tenure in 2009, and was promoted to full professor in 2010. She was the Shepard Professor for 2009–2010 at the College of Law. In 2009, she was the Stegner Young Scholar at the University of Utah. In 2008–2009 she received the William F. and Joan L. Boyd Teaching Award from the University of Idaho College of Law. She teaches water law, water policy, law and science, and property law and a leads a team-taught course in interdisciplinary methods in water resources. She is a principal investigator on development of the new Water Resources graduate degree program at UI, which includes options for concurrent J.D./M.S. and J.D./Ph.D. degrees. She represents the University of Idaho on the Universities Consortium on Columbia River Governance. Her research interests include the integration of law and science in water resource management and dispute resolution, water management and resilience, and the recognition and settlement of Native American water rights.

Lynette de Silva is Associate Director of the Program in Water Conflict Management and Transformation at Oregon State University. She also teaches courses in water resources management. She holds a master's degree in environmental geology (emphasizing hydrogeology) from Indiana University, Indianapolis; and a bachelor's degree in geology, from Brooklyn College, New York. Over the past fifteen years she has worked in areas emphasizing water resources and land management practices. This includes working on projects to assist in supporting water conflict prevention and resolution, both regionally and internationally; mitigating flooding in the Red River Basin; investigating and quantifying carbon in soils and various ecosystems in the U.S. northern Great Plains; and identifying sources of soil and groundwater contamination in Indiana. Her principal areas of interest include water conflict management, integrated watershed management, earth science education and outreach. Lynette volunteers as a mediation practitioner within the Benton County, Oregon court system. Among her primary goals these days are to inspire and enrich the lives of others.

Lucia De Stefano, Ph.D. in Geological Sciences, currently holds a teaching position at Complutense University of Madrid (Spain) and is a research fellow at the Water Observatory of the Botin Foundation. She has worked on water resources management for the private sector, nonprofit organizations and governmental bodies.

Andrea K. Gerlak is Director of Academic Development for the International Studies Association, Environmental Policy Faculty Associate with the Udall Center for Studies in Public Policy, and a researcher with the Institute of the Environment at the University of Arizona. She has more than ten years teaching experience as a faculty member at Columbia University and Guilford College. Her research interests are in environmental and natural resource policy, with particular attention to institutions and governance issues. She has published articles in *Policy Studies Journal, Natural Resources Journal, Journal of Public Administration Research and Theory, Water Policy, Society and Natural Resources, Global Environmental Politics,* the University of Denver *Water Law Review, Publius: The Journal of Federalism,* and *Environmental Management.*

AlaTanya Heikkila is an associate professor in the School of Public Affairs at the University of Colorado Denver and an associate research scientist at the Columbia Water Center. Heikkila's research has examined institutions for coordinating groundwater and surface water in the western United States, interstate water conflicts and cooperation, and collaborative watershed restoration programs. She has published articles in the *Journal of Policy Analysis and Management, Journal of Public Administration Research and Theory, Natural Resources Journal, Policy Studies Journal, Public Administration Review, Water Policy* and co-authored a book titled *Common Waters Diverging Streams: Linking Institutions and Water Management in Arizona, California, and Colorado* (Resources for the Future Press, 2004).

Gregory Hill is Chair of the Mathematics Department and Associate Professor of Mathematics and Environmental Science at the University of Portland. After careers as a researcher in pure mathematics and in mathematics education, he has focused his work on resilience science and issues of risk and uncertainty in environmental management. He serves as President of the Institute for Culture and Ecology (IFCAE), a nonprofit research organization based in Portland, Oregon that applies social science theory and methods to the study of ecosystem management and policy. Sponsored by the National Science Foundation, Dr. Hill and other scientists with the Institute are investigating the public planning process for salmon recovery in the Columbia River Basin. Their work focuses on the interface between science and policy, examining the effects of modeling and uncertainty on public participation, equity, and power.

Paul Hirt is a historian and senior sustainability scholar at Arizona State University. He specializes in the American West, global environmental history, and environmental policy. Hirt's publications include a monograph on the history of national forest management since WWII (*A Conspiracy of Optimism,* 1994), two edited collections of essays on Northwest history (*Terra Pacifica,* 1998 and *Northwest Lands, Northwest Peoples,* 1999), and more than two dozen articles and book chapters on environmental and western history. He is completing a monograph on the history of electric power in the U.S. Northwest and British Columbia scheduled for publication by the University Press of Kansas in 2013. Hirt also studies water, urban growth, and sustainability in the U.S. Southwest.

Eric T. Jones is a co-founder of the Institute for Culture and Ecology, a 501(c)3 nonprofit that applies social science theory and methods to the study of ecosystem management and policy. He is a native of Oregon and an environmental anthropologist involved in natural resource research, conservation, and policy analysis. He has researched and published on decision support systems, sustainable forestry, stewardship, land tenure, community-based environmental protection, interdisciplinary science, and political ecology.

Kelvin Ketchum manages the System Optimization section in BC Hydro's Generation group, responsible for operations planning for hydroelectric plants and reservoirs in B.C.'s portion of the Columbia and Peace river basins. He is also the Canadian Chair for the Columbia River Treaty Operating Committee and has been involved in Columbia River operations for over twenty years. Mr. Ketchum is a registered professional engineer and graduated from the University of British Columbia with a bachelor of applied science and a masters degree in engineering.

Steven A. Kolmes is Chair of the Environmental Science Department, Professor of Biology, and occupant of the Rev. John Molter, C.S.C., Chair in Science at the University of Portland. Dr. Kolmes has degrees in zoology from Ohio University and the University of Wisconsin–Madison. His interests are in the areas of salmon recovery planning, combining ethical and scientific analyses in environmental policy discussions, water and air quality issues, and the sublethal effects of pesticides. He has served on government scientific advisory panels (NOAA–Fisheries Technical Recovery Team for the Willamette and Lower Columbia

Rivers; Oregon Department of Environmental Quality Toxics Technical Advisory Committee) and on the steering committee for the Columbia River Pastoral Letter (The Columbia River Watershed: Caring for Creation and the Common Good). Dr. Kolmes teaches courses in environmental science, he team-teaches two courses (one entitled theology in ecological perspective and another entitled senior capstone seminar) with theologian Dr. Russell Butkus, and when he has time he occasionally teaches marine biology, invertebrate zoology, or animal behavior.

Stephen C. McCaffrey is Distinguished Professor and Scholar at the University of the Pacific, McGeorge School of Law. He was a member of the U.N. International Law Commission (ILC) from 1982–1991 and chaired the Commission's 1987 Session. He was the ILC's special rapporteur on international watercourses from 1985 until 1991, when the Commission adopted a complete first draft of articles on the topic. The final draft formed the basis for the negotiation of the 1997 U.N. Convention on the same subject. Professor McCaffrey served as Counselor on International Law in the U.S. Department of State from 1984–1985. He has served or is serving as counsel to Slovakia (*Gabcikovo-Nagymaros Project*), Nicaragua (*Navigational and Related Rights*, and *Certain Activities Carried Out by Nicaragua in the Border Area*) and Uruguay (*Pulp Mills on the River Uruguay*) in cases before the International Court of Justice. He also advised India in the *Baglihar HEP* case, brought under the Indus Waters Treaty and is counsel to India in the *Kishenganga* arbitration. He has been legal adviser to the Nile River Basin Cooperative Framework project and the Palestinian Authority/PLO and was a member of the National Research Council's Committee on the Management of the Colorado River Basin. Professor McCaffrey's books and articles concern subjects ranging from public international law and the law of international watercourses to transnational litigation and international environmental law.

Matthew McKinney is Director of the Center for Natural Resources and Environmental Policy at the University of Montana. His work focuses on collaborative approaches to natural resource and environmental conflict. As a practicing mediator, he has designed, facilitated, and mediated over fifty public processes over the past twenty years on a variety of local, state, regional, national, and international issues. He received a Ph.D. in Natural Resource Policy and Conflict Resolution from the University of Michigan; has published numerous articles and policy reports; co-authored *The Western Confluence: A Guide to Governing Natural Resources* (Island Press, 2004) and *Working Across Boundaries: People, Nature, and Regions* (Lincoln Institute of Land Policy, 2009)); and teaches workshops, seminars, and courses on natural resource policy and public dispute resolution. Matthew is an adjunct professor at the University of Montana's School of Law; Chair of the interdisciplinary Natural Resource Conflict Resolution Program; Member, U.S. Institute for Environmental Conflict Resolution Roster of Mediators; Senior Associate at the Lincoln Institute of Land Policy; and Senior Partner with the Consensus Building Institute. He serves on the Board of Advisors for the Rocky Mountain Land Use Institute, and is a Senior Advisor for the Sixth World Water Forum (2011–2010). When he is not working on natural resource and environmental issues, he can be found hiking, biking, fishing, floating, skiing, and otherwise enjoying the outdoors.

Rebecca McLain is a co-founder of the Institute for Culture and Ecology, a 501(c)3 nonprofit that applies social science theory and methods to the study of ecosystem management and policy. Over the past two decades, she has worked on land and resource tenure issues in the United States, West Africa, and Haiti. She has co-authored publications on adaptive management, natural resource governance, and large-scale socio-economic assessments. She is currently collaborating with the U.S. Forest Service on a project to create GIS data layers of socio-cultural values for use in national forest planning in western Washington. She is also part of an interdisciplinary team engaged in developing a research framework for understanding the role of non-timber forest products gathering in urban ecosystems.

Garry Merkel is Chair of the Columbia Basin Trust and a senior advisor/negotiator for the Ktunaxa Nation Council. Both of these organizations are key players in the Canadian portion of the Columbia basin affected by Columbia River Treaty operations.

Carmen Thomas Morse holds a B.S. (1992, University of Wisconsin-Madison) and an M.S. (1997, Oregon State University) in wildlife biology, and worked for the U.S. Fish and Wildlife Service for twelve years on environmental contaminants and endangered species issues before obtaining a J.D. with an emphasis in natural resources and environmental law (2010, University of Idaho College of Law). She is currently an attorney for Boise Inc. in Boise, Idaho, a position she has held since 2010.

Jeremy Mouat has been a professor of history and chair of the Department of Social Sciences at the Augustana Campus of the University of Alberta in Canada since 2005. He earned his B.A. and M.A. degrees in New Zealand, and completed his doctorate at the University of British Columbia. His research interests are chiefly in the history of technology and resource extraction in the nineteenth and early twentieth centuries, areas in which he has published widely.

Anne Nolin is an associate professor in the Department of Geosciences, Oregon State University. Her scientific interests focus on snow and ice in the climate system, particularly with regard to climate change and water resources. She received her Ph.D. in geography, University of California–Santa Barbara. Dr. Nolin spent ten years as a cryospheric researcher at the University of Colorado–Boulder. She has been involved in satellite remote sensing of snow and glaciers for over twenty years, working closely with NASA and also serving on the Space

Studies Board of the National Academy of Sciences. She serves as an Associate Editor for *Water Resources Research* and the *Journal of Hydrometeorology*. In 2009–2010 Dr. Nolin received an Erskine Fellowship at the University of Canterbury, New Zealand and held the Landolt Chair in Sustainable Futures at EPFL in Lausanne, Switzerland. Her recent publications focus on "at-risk" snow and glacier meltwater contributions to streamflow in a warming climate.

Richard Kyle Paisley is Director of the Global International Waters Governance Initiative at the University of British Columbia IAR in Vancouver, British Columbia, Canada. Richard's academic background includes undergraduate and graduate degrees in biochemistry, resource management, law and international law from U.B.C., the University of Washington, Pepperdine University School of Law and the London School of Economics. His research, teaching and consulting interests include governance of international waters, international business transactions, negotiations and environmental conflict resolution. Richard has directed a wide range of conferences, workshops and research projects, published extensively and been an advisor, trainer and special counsel on these subjects to governments, international agencies, universities, nongovernmental organizations and aboriginal groups. His outside interests include downhill and cross country skiing, cycling, backpacking, kayaking, and coaching his daughters' soccer teams.

Mary L. Pearson has returned to a small private practice in Spokane County, Washington after serving as Chief Judge for the Coeur d' Alene Tribes from October 2004 to August 2009. Prior, to that she served as Chief Judge for the Spokane Tribe from October 1997 to February 2002. Currently, she serves on the Court of Appeals for the Shoshone-Bannock Tribes of Idaho. This article of The River People was inspired by a larger work to be entitled *The Path of Life: an account of the loss of salmon caused by the dams on the Columbia River and the impact of those losses on the River People*. That manuscript will be used as an outline for a television production. The ultimate purpose is to change public opinion about Indian fishing rights and thereby change judicial review of Indian fishing cases. She is a mother of four children, one deceased, seventeen grandchildren, and three great grands. She is admitted to practice in Washington and Idaho and is on inactive status in Oregon. She attended Willamette University Law, Boise State University as an undergraduate, and has served as Vice President for the Northwest Tribal Judges as well as the Northwest Indian Bar of which she was a co-founder. She has written grants and is a grant reader, has taught domestic violence classes to judges and prosecutors, and edited several bench books for Tribal Judges. She was in private practice from 1978–1989 in Oregon and Idaho, served as a Tribal Judge for the Northwest Tribes from 1989–1994, served as a Tribal Attorney for the Suquamish Tribe from 1995–1997, and has retired twice.

Chris Peery is a fish biologist for the U.S. Fish and Wildlife Service at the Idaho Fisheries Resources Office. Prior to that he was a research faculty member in the Department of Fish and Wildlife Resources at the University of Idaho. He has graduate degrees in oceanography from the College of William and Mary and in fisheries from the University of Idaho. His research involves the study of the freshwater migration behavior of anadromous salmonids and Pacific lamprey and he is considered an expert on passage of fish at dams. He is a member of the Pacific Lamprey Technical Workgroup, which advises the Columbia Basin Fish and Wildlife Authority and U.S. Army Corps of Engineers on issues related to management of Pacific lamprey, and he is heavily involved in efforts to recover lamprey in the Columbia River basin.

Chris W. Sanderson, Q.C., Lawson Lundell LLP, focuses on government relations and regulation in the energy and resource sectors throughout western Canada. He advises utilities, independent power producers, marketers, mine and energy project developers, and governments. Chris appears frequently before regulatory boards in energy and environmental matters in British Columbia, Alberta, and the Northwest Territories. He represents clients in judicial proceedings arising in the regulatory context or, more generally, from the relationship between business and government heard by all levels of court in British Columbia and Alberta and in the federal court system including the Supreme Court of Canada. He has particular expertise in all aspects of public utility law. His recent experience includes a number of regulatory and judicial proceedings involving the extent of the Crown's obligation to consult First Nations. He was counsel to BC Hydro at all levels including the Supreme Court of Canada in *Rio Tinto Alcan and BC Hydro v. Carrier Sekani Tribal Council*, 2010 SCC 43.

Guido Schmidt is a Ph.D. environmental planning engineer with twenty years experience aimed at better understanding, preventing, and solving socio-environmental conflicts. His particular interest is in initiating and shaping policies that involve stakeholders and in management processes for mainstreaming ecosystem conservation.

John Shurts is General Counsel for the Northwest Power and Conservation Council. The Council is an interstate compact agency based in Portland, Oregon, authorized by the Northwest Power Act of 1980 and consisting of eight members appointed by the governors of Idaho, Montana, Oregon, and Washington. The Council develops and oversees a regional power plan for the Pacific Northwest and a fish and wildlife protection and mitigation program for the Columbia River Basin. The Council works extensively with federal, state, and tribal governments and agencies and nongovernmental organizations and the public in the development of these plans and programs and in the efforts of others to implement them. Part of Shurts' recent work for the Council has involved matters relating to the Columbia River Treaty between the United States and Canada, including assisting the Universities Consortium on Columbia Basin Governance in their series of public transboundary symposia on the Columbia River Treaty. Shurts has a Ph.D. degree in American history from the

University of Oregon with an emphasis on environmental and legal history, and is the author of *Indian Reserved Water Rights: The Winters Doctrine in its Social and Legal Context, 1880s–1930s.* As a consultant on Columbia River water, energy, and fish and wildlife issues, Shurts has briefed numerous delegations and study tours from around the world, including representatives from the Nile basin, the Mekong River, Vietnam, and the Ukraine. In 2008 Shurts served as a member of the Anadromous Fish Independent Review Panel organized by the Bureau of Reclamation and the U.S. Fish and Wildlife Service for a comprehensive review of the Central Valley Project's Anadromous Fish Restoration Program.

Adam M. Sowards is Associate Professor of History and Director of the Program in Pacific Northwest Studies at the University of Idaho. He also is a faculty member with the Water Resources and Environmental Science programs. Sowards has authored many articles on western environmental history and two books, including *The Environmental Justice: William O. Douglas and American Conservation* (Oregon State University Press, 2009). Along with interdisciplinary investigations on western water resources, his current book projects include an ecological history of North America (with Sterling Evans) and a study of the Canadian Arctic Expedition, 1913–18.

Eric Sproles is a doctoral candidate in Water Resources Science at Oregon State University. Eric's research focuses on the distribution of snow in mountain watersheds, and the impacts of projected climate on snowpack and stream runoff. Additionally, Eric is an instructor of geography and the GIS instructional designer at Lane Community College. His work at Lane helps middle and high school teachers integrate geographic thinking into the curriculum.

Craig W. Thomas is Associate Professor in the Daniel J. Evans School of Public Affairs at the University of Washington. He is the author of two books on collaborative environmental management: *Bureaucratic Landscapes: Interagency Cooperation and the Preservation of Biodiversity* (MIT Press, 2003) and *Collaborative Environmental Management: What Roles for Government?* (RFF Press, 2004). He has also published numerous articles and book chapters on institutional reform, performance management, habitat conservation planning, and research methods. He holds a doctorate in political science and a masters degree in Public Policy from the University of California, Berkeley. He also serves as editor of the *Journal of Public Administration Research and Theory.*

Eve Vogel moved to Portland, Oregon, in 1991 and was immediately fascinated by the Columbia River. After working for Portland Audubon (1991–1993), she earned an M.Ed. at Portland State, and taught high school biology, math, and Spanish in Portland and Vancouver. In 2007 she earned a Ph.D. in geography from the University of Oregon. While completing her degree, she interned for the Northwest Power Planning Council (now Northwest Power and Conservation Council), taught at Portland State, and had a son, Ari. She is currently an assistant professor of geography at the University of Massachusetts, but her research and writing—and her love of rivers—continue to focus on the Pacific Northwest.

Edward P. Weber (Ph.D., University of Wisconsin–Madison) is Professor in the School of Public Policy at Oregon State University, and former Professor and Director of the School of Environmental and Public Affairs at the University of Nevada–Las Vegas. Prior to his arrival at U.N.L.V. in December 2009, he was a professor of political science and the Edward Meyer Distinguished Professor of Public Administration and Policy at Washington State University, where he also served as the Director of the Thomas S. Foley Public Policy Institute (2001–2008). Weber has conducted and presented research in the U.S. and around the world in places as varied as France, Uzbekistan, New Zealand, Ukraine, Japan, Hungary, Slovenia, and the U.K. His research focuses on the reinvention of government, particularly the growing use of innovative regulatory programs, new ways to organize and control bureaucracy, and collaborative public management frameworks. These endeavors place a heavy emphasis on environmental/natural resource policy and sustainability.

Anthony G. White has been a public administrator in local, state, and federal governments for the past forty-two years. While he has worked with the Columbia River Treaty for the last sixteen, the work presented here grew out of his work as an adjunct associate professor at Portland State University, in particular a series of presentations to a history of Canada class over a period of several years. He wishes to thank Dr. Charles White, History Professor Emeritus, for the opportunities to make those presentations.

Aaron Wolf is a professor of geography in the Department of Geosciences at Oregon State University. He has an M.S. in water resources management (1988, emphasizing hydrogeology) and a Ph.D. in environmental policy analysis (1992, emphasizing dispute resolution) from the University of Wisconsin–Madison. His research focuses on issues relating transboundary water resources to political conflict and cooperation, where his training combining environmental science with dispute resolution theory and practice, have been particularly appropriate. Wolf has acted as consultant to the U.S. Department of State, the U.S. Agency for International Development, the World Bank, and several governments on various aspects of international water resources and dispute resolution. He has been involved in developing the strategies for resolving water aspects of the Arab–Israeli conflict, including co-authoring a State Department reference text, and participating in both official and "track II" meetings between co-riparians. Wolf, a trained mediator/facilitator, directs the Program in Water Conflict Management and Transformation, through which he has offered workshops, facilitations, and mediation in basins throughout the world.

Index

Page numbers written in italics denote illustrations.